AF581171

BIRKHÄUSER

ISNM
International Series of Numerical Mathematics

Volume 157

Inequalities and Applications

Conference on Inequalities and Applications, Noszvaj (Hungary), September 2007

Catherine Bandle
Attila Gilányi
László Losonczi
Zsolt Páles
Michael Plum
Editors

Birkhäuser
Basel · Boston · Berlin

Editors:

Catherine Bandle
Institut Mathematik
Universität Basel
Rheinsprung 21
4051 Basel
Switzerland
Email: c.bandle@gmx.ch

László Losonczi
Department Economic Analysis &
Information Technology for Business
University of Debrecen
Kassai út.26
4028 Debrecen
Hungary
Email: losi@math.klte.hu

Attila Gilányi
Zsolt Páles
Department of Analysis
Institute Mathematics & Informatics
University of Debrecen
Pf. 12
4010 Debrecen
Hungary
Email: gil@math.klte.hu
pales@math.klte.hu

Michael Plum
Institut für Analysis
Universität Karlsruhe
Kaiserstrasse 12
76128 Karlsruhe
Germany
Email: michael.plum@math.uni-karlsruhe.de

2000 Mathematics Subject Classification: 15-06, 15A39,15A42, 15A45, 26D05, 26D07, 26D10, 26D15, 26D20, 26E60, 30A10, 32A22, 34A40, 35J85, 35K85, 35L85, 35R45, 39B72, 41A17, 42A05, 47A30, 47A50, 47A63, 47J20, 49J40, 51M16, 52A40, 58E35, 60E15

Library of Congress Control Number: 2008935754

Bibliographic information published by Die Deutsche Bibliothek.
Die Deutsche Bibliothek lists this publication in the Deutsche Nationalbibliografie; detailed bibliographic data is available in the Internet at http://dnb.ddb.de

ISBN 978-3-7643-8772-3 Birkhäuser Verlag AG, Basel - Boston - Berlin

Basel · Boston · Berlin
P.O. Box 133, CH-4010 Basel, Switzerland
Part of Springer Science+Business Media
Printed on acid-free paper produced from chlorine-free pulp. TCF ∞
Printed in Germany

ISBN 978-3-7643-8772-3
9 8 7 6 5 4 3 2 1

e-ISBN 978-3-7643-8773-0
www.birkhauser.ch

Contents

Preface ix

Abstracts of Talks xi

Problems and Remarks xli

List of Participants xlvii

Part I: Inequalities Related to Ordinary and Partial Differential Equations

C. Bandle
A Rayleigh-Faber-Krahn Inequalitiy and Some Monotonicity Properties for Eigenvalue Problems with Mixed Boundary Conditions 3

H. Behnke
Lower and Upper Bounds for Sloshing Frequencies 13

B.M. Brown, V. Hoang, Michael Plum and I.G. Wood
On Spectral Bounds for Photonic Crystal Waveguides 23

C. Buşe
Real Integrability Conditions for the Nonuniform Exponential Stability of Evolution Families on Banach Spaces 31

K. Nagatou
Validated Computations for Fundamental Solutions of Linear Ordinary Differential Equations 43

Part II: Integral Inequalities

S. Barza and L.-E. Persson
Equivalence of Modular Inequalities of Hardy Type on Non-negative Respective Non-increasing Functions 53

R.C. Brown and D.B. Hinton
Some One Variable Weighted Norm Inequalities and Their Applications to Sturm-Liouville and Other Differential Operators 61

P. Cerone
Bounding the Gini Mean Difference 77

B. Gavrea
On Some Integral Inequalities 91

M. Johansson
A New Characterization of the Hardy and Its Limit Pólya-Knopp Inequality for Decreasing Functions 97

A. Čivljak, Lj. Dedić and M. Matić
Euler-Grüss Type Inequalities Involving Measures 109

W.D. Evans, A. Gogatishvili and B. Opic
The ρ-quasiconcave Functions and Weighted Inequalities 121

Part III: Inequalities for Operators

S.S. Dragomir
Inequalities for the Norm and Numerical Radius of Composite Operators in Hilbert Spaces 135

F. Kittaneh
Norm Inequalities for Commutators of Normal Operators 147

J. Matkowski
Uniformly Continuous Superposition Operators in the Spaces of Differentiable Functions and Absolutely Continuous Functions 155

T. Ogita and S. Oishi
Tight Enclosures of Solutions of Linear Systems 167

Part IV: Inequalities in Approximation Theory

I. Gavrea
Operators of Bernstein-Stancu Type and the Monotonicity of Some Sequences Involving Convex Functions 181

A.I. Mitrea and P. Mitrea
Inequalities Involving the Superdense Unbounded Divergence of Some Approximation Processes 193

C.P. Niculescu
An Overview of Absolute Continuity and Its Applications 201

Part V: Generalizations of Convexity and Inequalities for Means

S. Abramovich and S.S. Dragomir
Normalized Jensen Functional, Superquadracity and Related Inequalities ... 217

P. Burai
Comparability of Certain Homogeneous Means ... 229

M. Klaričić Bakula, M. Matić and J. Pečarić
On Some General Inequalities Related to Jensen's Inequality ... 233

A.W. Marshall and I. Olkin
Schur-Convexity, Gamma Functions, and Moments ... 245

Z. Daróczy and Zs. Páles
A Characterization of Nonconvexity and Its Applications in the Theory of Quasi-arithmetic Means ... 251

J. Mrowiec, Ja. Tabor and Jó. Tabor
Approximately Midconvex Functions ... 261

Part VI: Inequalities, Stability, and Functional Equations

W. Fechner and J. Sikorska
Sandwich Theorems for Orthogonally Additive Functions ... 269

R. Ger
On Vector Pexider Differences Controlled by Scalar Ones ... 283

Gy. Maksa and F. Mészáros
A Characterization of the Exponential Distribution through Functional Equations ... 291

D. Popa
Approximate Solutions of the Linear Equation ... 299

A. Varga and Cs. Vincze
On a Functional Equation Containing Weighted Arithmetic Means ... 305

PHB '07

Preface

Inequalities are found in almost all fields of pure and applied mathematics. Because of their various applications in areas such as the natural and engineering sciences as well as economics, new types of interesting inequalities are discovered every year. In the theory of differential equations, in the calculus of variations and in geometry, fields which are dominated by inequalities, efforts are made to extend and improve the classical ones.

The study of inequalities reflects the different aspects of modern mathematics. On one hand, there is the systematic search for the basic principles, such as the deeper understanding of monotonicity and convexity. On the other hand, finding the solutions to an inequality requires often new ideas. Some of them have become standard tools in mathematics. In view of the wide-ranging research related to inequalities, several recent mathematical periodicals have been devoted exclusively to this topic.

A possible way to speed up the communication between groups of specialists of the seemingly unconnected areas is to bring them together from many parts of the globe. Due to the efforts of János Aczél, Georg Aumann, Edwin F. Beckenbach, Richard Bellman and Wolfgang Walter, the first General Inequalities meeting was organized in Oberwolfach, Germany in 1976. Then six meetings were organized in Oberwolfach between 1978 and 1995 and one in Noszvaj, Hungary in 2002.

The *Conference on Inequalities and Applications* '07 also took place at the De La Motte Castle in Noszvaj, Hungary from September 9 to 15, 2007. It was organized by the Department of Analysis of the University of Debrecen.

The members of the Scientific Committee were Catherine Bandle (Basel), William Norrie Everitt (Birmingham, honorary member), László Losonczi (Debrecen), Zsolt Páles (Debrecen), Michael Plum (Karlsruhe) and Wolfgang Walter (Karlsruhe, honorary member).

The organizing Committee consisted of Zoltán Daróczy (honorary chairman), Attila Gilányi (chairman), Mihály Bessenyei (scientific secretary), Zoltán Boros, Gyula Maksa, Szabolcs Baják and Fruzsina Mészáros. There were 66 participants from 16 countries.

The talks at the symposium focused on the following topics: convexity and its generalizations; mean values and functional inequalities; matrix and operator inequalities; inequalities for ordinary and partial differential operators; integral and differential inequalities; variational inequalities; numerical methods. A number of sessions were, as usual, devoted to problems and remarks. The scientific program was complemented by several social events, such as a harpsichord recital of some masterpieces of Bach and Haydn, performed by Ágnes Várallyay.

This volume contains 33 research papers, about half of the works presented at the meeting. The material is arranged into six chapters ranging from *Inequalities related to ordinary and partial differential equations* to *Inequalities, stability, and functional equations.* The contributions given here reflect the ramification of inequalities into many areas of mathematics, and also present a synthesis of results in both theory and practice.

The editors of the volume are thankful to Mrs. Phyllis H. Brown for the artistic drawings made at the conference, which are illustrations to the six chapters. They thank Mihály Bessenyei for enthusiastically compiling the report of the meeting, Andrea Pákozdy for the preparation of the manuscripts and the publisher, Birkhäuser Verlag, for the careful typesetting and technical assistance.

The organization of the meeting and the publication of the proceedings were partially supported by the Hungarian Scientific Research Fund Grants NK–68040 and K–62316.

The Editors

International Series of Numerical Mathematics, Vol. 157, xi–xxxix

Abstracts of Talks

Abramovich, Shoshana: *Normalized Jensen functional and superquadracity.* (Joint work with Sever S. Dragomir.)

Normalized Jensen functional is the functional

$$J_n(f, \mathbf{x}, \mathbf{p}) = \sum_{i=1}^{n} p_i f(x_i) - f\left(\sum_{i=1}^{n} p_i x_i\right), \qquad \sum_{i=1}^{n} p_i = 1.$$

A superquadratic function is a function f defined on an interval $I = [0, a]$ or $[0, \infty)$ so that for each x in I there exists a real number $C(x)$ such that

$$f(y) - f(x) \geq f(|y - x|) + C(x)(y - x)$$

for all $y \in I$.

For example the functions x^p, $p \geq 2$ and the functions $-x^p$, $0 \leq p \leq 2$ are superquadratic functions as well as the function $f(x) = x^2 \log x$, $x > 0$, $f(0) = 0$.

Using these definitions we generalize for convex functions the inequality

$$MJ_n(f, \mathbf{x}, \mathbf{q}) \geq J_n(f, \mathbf{x}, \mathbf{p}) \geq mJ_n(f, \mathbf{x}, \mathbf{q})$$

dealt by S.S. Dragomir and we show that for superquadratic functions, we get nonzero lower bounds of $J_n(f, \mathbf{x}, \mathbf{p}) - mJ_n(f, \mathbf{x}, \mathbf{q})$ and nonzero upper bounds of $J_n(f, \mathbf{x}, \mathbf{p}) - MJ_n(f, \mathbf{x}, \mathbf{q})$.

Then we define an extended normalized Jensen functional

$$H_{n,k}(f, \mathbf{x}, \mathbf{q}, \mathbf{p}) = \sum_{i_1, \ldots, i_k = 1}^{n} p_{i_1} \cdots p_{i_k} f\left(\sum_{j=1}^{k} q_j x_{i_j}\right) \quad f\left(\sum_{i=1}^{n} p_i x_i\right),$$

$\sum_{i=1}^{n} p_i = 1$, $\sum_{j=1}^{k} q_j = 1$, and we establish bounds for

$$H_{n,k}(f, \mathbf{x}, \mathbf{q}, \mathbf{p}) - mH_{n,k}(f, \mathbf{x}, \mathbf{q}, \mathbf{r})$$

for different values of m when f is convex and when f is superquadratic.

At the end of the paper we get a reverse Jensen inequality for special cases of superquadratic functions.

Adamek, Mirosław: *On three-parameter families and associated convex functions.*

Using two-parameter families E.F. Beckenbach presented in [1] the concept of generalized convex functions. It is extended in [2] to two-dimensional case by the use of so-called three-parameter families. In particular, $\mathcal{F}$-midconvex functions are defined.

In this talk we define $(\mathcal{F},(t_1,t_2,t_3))$-convex functions and we show that any such a function must be $\mathcal{F}$-midconvex. We also consider some properties of three-parameter families.

References

[1] E.F. Beckenbach, *Generalized convex functions*, Bull. Amer. Math. Soc. **43** (1937), 363–371.

[2] M. Adamek, A. Gilányi, K. Nikodem, Zs. Páles, *A note on three-parameter families and generalized convex functions*, J. Math. Anal. Appl. **330** (2007), 829–835.

Baják, Szabolcs: *Invariance equation for generalized quasi-arithmetic means.* (Joint work with Zsolt Páles.)

We deal with the following equation, which is a generalization of the Matkowski-Sutô problem:

$$(\varphi_1+\varphi_2)^{-1}\big(\varphi_1(x)+\varphi_2(y)\big)+(\psi_1+\psi_2)^{-1}\big(\psi_1(x)+\psi_2(y)\big)=x+y\,,$$

where φ_1, φ_2, ψ_1, ψ_2 are monotonically increasing, continuous functions on the same interval and we assume that each function is four times continuously differentiable. First we establish the connection between φ_1 and φ_2, and ψ_1 and ψ_2 by comparing the derivatives up to the fourth order and taking $x=y$. Then we give the general solutions.

Bandle, Catherine: *An eigenvalue problem with mixed boundary conditions: open problems.*

In this talk we discuss the following non-standard eigenvalue problem:

$$\Delta u+(\lambda-q)u=0 \text{ in } D\subset\mathbb{R}^N,\ \frac{\partial u}{\partial n}=\lambda\sigma u \text{ on } \partial D.$$

Here $q>0$ and $\sigma<0$ are real numbers and D is a bounded domain. In a series of papers in collaboration with J. v. Below and W. Reichel the existence of positive and negative eigenvalues

$$\cdots<\lambda_{-2}<\lambda_{-1}<0<\lambda_1<\lambda_2<\cdots,\ \lambda_{\pm n}\to\pm\infty$$

was established. In addition it was shown that both λ_1 and λ_{-1} are simple which led to maximum and anti-maximum principles. The question arises to what extent do some of the classical inequalities for the membrane problem hold. Upper (lower) bounds for λ_1 (λ_{-1}) are easily obtained from the variational characterization. More difficult are lower (upper) bounds depending only on the geometry, q and σ. It turns out that this problem is much more involved and that many conjectures remain unsolved.

Barza, Sorina: *Sharp constants related to Lorentz spaces.*
(Joint work with Viktor Kolyada and Javier Soria.)

In the theory of Lorentz spaces $L^{p,s}(\mathbb{R},\mu)$ it is very natural to consider the following norm defined in terms of Köthe duality

$$\|f\|'_{p,s} = \sup\left\{\int_{\mathbb{R}} fg d\mu : \|g\|_{p',s'} = 1\right\}.$$

We call it the "dual norm". It was proved in [1] that the dual norm is equivalent with the usual norm defined in terms of the decreasing rearrangement but without giving the optimal constants. We will find the best constants in the inequalities relating these norms. This was done in a joint work with Viktor Kolyada and Javier Soria (see [2]).

References

[1] C. Bennett and R. Sharpley, *Interpolation of Operators*, Academic Press, Boston, 1988.

[2] S. Barza, V. Kolyada and J. Soria, *Sharp constants related to the triangle inequality in Lorentz spaces*, submitted.

Behnke, Henning: *Lower and upper bounds for sloshing frequencies.*

The calculation of the frequencies ω for small oscillations of an ideal liquid in a container results in the eigenvalue problem:

$$\begin{aligned} -\Delta\varphi &= 0 \text{ in } \Omega \quad \text{(liquid)},\\ \frac{\partial\varphi}{\partial\mathbf{n}} &= \lambda\varphi \text{ on } \partial_1\Omega \quad \text{(free surface)},\\ \frac{\partial\varphi}{\partial\mathbf{n}} &= 0 \text{ on } \partial_2\Omega \quad \text{(container wall)},\\ \lambda &> 0; \end{aligned} \tag{1}$$

$\mathbf{n}$ denotes the outward normal to the boundary $\partial\Omega$ of Ω, the relation between λ and ω is $\lambda = \omega^2/g$, g is the acceleration due to gravity. As an example for the diversity of possible domains with polygonal boundaries, let Ω be defined as

$$\begin{aligned} \Omega &:= \{(x,y) \in \mathbb{R}^2 : -1 < y < 0, -1 - \frac{y}{2} < x < 0\},\\ \partial_1\Omega &:= \{(x,y) \in \mathbb{R}^2 : -1 \le x \le 0, y = 0\}, \end{aligned}$$

and $\partial_2\Omega := \partial\Omega\backslash\partial_1\Omega$. The two-dimensional problem is a model for an infinitely long canal with cross section Ω.

Take $H_a := \{f \in H^1(\Omega) : \int_{\partial_1\Omega} f\, ds = 0\}$ and

$$\begin{aligned} a(f,g) &:= \int_\Omega (\text{grad} f)' \cdot \text{grad} g\, dx\, dy,\\ b(f,g) &:= \int_{\partial_1\Omega} ds \quad \text{for all } f,g \in H_a; \end{aligned}$$

now the weak form of (1) is:

$$\begin{aligned}&\text{Determine } \varphi \in H_a, \varphi \neq 0, \lambda \in \mathbb{R} \text{ such that}\\ &\qquad a(f,\varphi) = \lambda\, b(f,\varphi) \text{ for all } f \in H_a.\end{aligned} \tag{2}$$

A procedure for calculating lower and upper bounds to the eigenvalues of (2) is proposed. The calculation of upper bounds is done by means of the well-known Rayleigh-Ritz procedure. For the lower bound computation Goerich's generalization of Lehmann's method is applied, trial functions are constructed with finite elements. It is shown that Lehmann's method can not be applied in this context, whereas a specification of Goerisch's method is possible.

Rounding errors in the computation are controlled with interval arithmetic. Numerical results for different cross sections Ω are given.

Bessenyei, Mihály: *Hermite–Hadamard-type inequalities for Beckenbach-convex functions.*

Beckenbach structures, or as they are also termed, *Beckenbach families* are determined by the property that prescribing certain points on the plain (with pairwise distinct first coordinates) there exists precisely one member of the family that interpolates the points. Applying Beckenbach families, the classical convexity notion can be considerably generalized (see [1], [2] [3], [9]). The obtained convexity notion involves the notion of higher-order convexity due to Popoviciu (see [8]); in more general, the convexity notion induced by Chebyshev systems if the underlying Beckenbach family has a linear structure (see [7]).

The aim of the talk is to present some special support properties for generalized convex functions of Beckenbach sense, and, motivated by some earlier results (consult [4] [5], [6]), as direct applications of the support properties, to obtain Hermite–Hadamard-type inequalities for that kind of functions. The Markov–Krein representation problem of Beckenbach families is also investigated.

References

[1] E.F. Beckenbach, *Generalized convex functions*, Bull. Amer. Math. Soc. **43** (1937), 363–371.

[2] E.F. Beckenbach and R.H. Bing, *On generalized convex functions*, Trans. Amer. Math. Soc. **58** (1945), 220–230.

[3] E.F. Beckenbach, *Convex functions*, Bull. Amer. Math. Soc. **54** (1948), 439–460.

[4] M. Bessenyei, *Hermite–Hadamard-type inequalities for generalized convex functions*, Ph.D. dissertation (2005), RGMIA monographs (`http://rgmia.vu.edu.au/monographs`), Victoria University, 2006.

[5] M. Bessenyei and Zs. Páles, *On generalized higher-order convexity and Hermite–Hadamard-type inequalities*, Acta Sci. Math. (Szeged) **70** (2004), 13–24.

[6] M. Bessenyei and Zs. Páles, *Hermite–Hadamard inequalities for generalized convex functions*, Aequationes Math. **69** (2005), 32–40.

[7] S. Karlin and W.J. Studden, *Tchebycheff systems: With applications in analysis and statistics*, Pure and Applied Mathematics, Vol. XV, Interscience Publishers John Wiley & Sons, New York-London-Sydney, 1966.

[8] T. Popoviciu, *Les fonctions convexes*, Hermann et Cie, Paris, 1944.

[9] L. Tornheim, *On n-parameter families of functions and associated convex functions*, Trans. Amer. Math. Soc. **69** (1950), 457–467.

Boros, Zoltán: *Approximate convexity of van der Waerden type functions.*

Let $p \in]0, \infty[$. The Takagi–van der Waerden type function

$$T_p(x) = \sum_{n=0}^{\infty} \frac{(\operatorname{dist}(2^n x, \mathbb{Z}))^p}{2^n} \qquad (x \in \mathbb{R})$$

plays a specific role in the theory of approximately convex functions ([2], [3]). Motivated by this experience, we investigate whether T_p fulfils the inequality

$$T_p\left(\frac{x+y}{2}\right) \le \frac{T_p(x)+T_p(y)}{2} + c_p|x-y|^p \tag{1}$$

for every $x, y \in \mathbb{R}$ with some constant $c_p \ge 0$. If, for some fixed p, T_p satisfies the inequality (1), then, by [3, Theorem 6], we have $c_p \ge 2^{-p}$. It is established, for instance [1], that the Takagi–van der Waerden function T_1 satisfies the inequality (1) with $p = 1$ and $c_1 = 1/2$.

References

[1] Z. Boros, *An inequality for the Takagi function*, Math. Inequal. Appl., to appear.

[2] A. Házy and Zs. Páles, *On approximately midconvex functions*, Bull. London Math. Soc. **36** (2004), 339–350.

[3] A. Házy and Zs. Páles, *On approximately t-convex functions*, Publ. Math. Debrecen **66** (2005), 489–501.

[4] T. Takagi, *A simple example of the continuous function without derivative*, J. Phys. Math. Soc. Japan **1** (1903), 176–177.

[5] B.L. van der Waerden, *Ein einfaches Beispiel einer nichtdifferenzierbaren stetigen Funktion*, Math. Z. **32** (1930), 474–475.

Brown, Malcolm B.: *A Hardy Littlewood inequality for the p-Laplacian.*
(Joint work with Simon Aumann and Karl M. Schmidt.)

Hardy and Littlewood obtained the inequality

$$\left(\int_0^\infty |\, f' \,|^2 \, dx\right)^2 \le 4 \int_0^\infty |\, f \,|^2 \, dx \int_0^\infty |\, f'' \,| \, dx.$$

This has been generalized, by Everitt, to one involving the Sturm–Liouville operator $-y''+qy$: the HELP inequality. In this talk we show that the inequality may be further generalized to involve functions in L_p and in the spirit of the Everitt result may be considered as a generalization of the HELP inequality for the p-Laplacian

Brown, Richard C.: *An Opial-type Inequality with an integral boundary condition.* (Joint work with Michael Plum.)

Suppose that y is a real absolutely continuous function on the interval $[a,b]$, $-\infty < a < b < \infty$, $\int_a^b y\,dx = 0$, and $\int_a^b (y')^2\,dx < \infty$. We give a new proof that the best constant K of the inequality

$$\int_a^b |yy'|\,dx \le K(b-a)\int_a^b (y')^2\,dx$$

is $1/4$ and that equality holds if and only if $y = c(x-(a+b)/2$ for any constant c. The techniques employed are much more complicated than those required to prove the standard Opial inequality where the boundary conditions are $y(a) = 0 = y(b)$ and K is also $1/4$.

Bullen, Peter: *Equivalent inequalities.*
(Joint work with Li Yuan-Chuan and Yeh Cheh-Chi.)

This is a report on joint work in progress. It was pointed out in, [1, pp. 212–213] that many disparate looking inequalities are in fact equivalent. We are attempting to systematize this fact.

References

[1] P.S. Bullen *Handbook of Means and Their Inequalities*, Kluwer Academic Publishers, Dordrecht, 2003.

Burai, Pál: *Inequalities with Hölder and Daróczy means.*

We present some inequalities connected with Hölder means

$$\mathcal{H}_p(x,y) := \begin{cases} \left(\frac{x^p+y^p}{2}\right)^{1/p} & \text{if } p \neq 0 \\ \sqrt{xy} & \text{if } p = 0 \end{cases} \qquad x\,,y > 0\,,$$

and Daróczy means

$$\mathcal{D}_{w,p}(x,y) := \begin{cases} \left(\frac{x^p+w(\sqrt{xy})^p+y^p}{w+2}\right)^{1/p} & \text{if } p \neq 0 \text{ and } \infty > w \ge -1 \\ \sqrt{xy} & \text{if } p = 0 \text{ or } w = \infty \end{cases} \qquad x,\, y > 0\,.$$

Buşe, Constantin: *An ergodic version of the Rolewicz theorem on exponential stability in Banach spaces.*
(Joint work with Constantin P. Niculescu.)

We shall prove that a semigroup of operators acting on a Banach space X is uniformly exponentially stable, that is its exponential growth is negative, if and only if there exist two positive constants α and M such that for all positive t and each $x \in X$, one has:

$$\frac{1}{t}\int_0^t \phi(e^{\alpha s}||T(s)x||)ds \leq \phi(M||x||).$$

The result is proved under the general assumption that for each positive t and each $x \in X$, the map $s \mapsto ||T(s)x||$ is measurable on the interval $[0,t]$. Here ϕ is a proper function, i.e., a nonnegative and nondecreasing function on $[0,\infty]$ with $\phi(\infty) = \infty$. Similar result for evolution families are also proved.

Cerone, Pietro: *Bounding the Gini mean difference.*

Recent results on bounding and approximating the Gini mean difference for both general distributions and distributions supported on a finite interval are surveyed. It supplements the previous work utilizing the Steffensen and Karamata type approaches in approximating and bounding the *Gini mean difference* [1], given by

$$R_G(f) := \frac{1}{2}\int_{-\infty}^{\infty}\int_{-\infty}^{\infty} |x-y|\, dF(x)\, dF(y), \tag{1}$$

where $F(x)$ is the cumulative function associated with a density function $f(x)$. The mean difference has a certain theoretical attraction, being dependent on the spread of the variate-values among themselves rather than on the deviations from some central value. Further, its defining integral (1) may converge when the *variance* does not.
Some identities for the Gini Mean Difference, $R_G(f)$ will be stated here since they will form the basis for obtaining approximations and bounds.
Define the functions $e : \mathbb{R} \to \mathbb{R}$, $e(x) = x$, $F : \mathbb{R} \to \mathbb{R}_+$, $F(x) = \int_{-\infty}^{x} f(t)\, dt$, then the covariance of e and F is given by $\mathrm{Cov}(e, F) := E[(e - E(f))(F - E(F))]$.

Theorem. *With the above notation the following identities hold:*

$$R_G(f) = 2\,\mathrm{Cov}(e,F) = \int_{-\infty}^{\infty} (1 - F(y))\, F(y)\, dy = 2\int_{-\infty}^{\infty} xf(x)\, F(x)\, dx - E(f). \tag{2}$$

Steffensen type inequalities have attracted considerable attention in the literature given the variety of applications and its generality. The following result is obtained utilizing the Steffensen inequality [2] which provides some improvements over an earlier result of Gastwirth under less restrictive assumptions.

Theorem. *Let f be supported on $[a,b]$ and $E(f)$ exist. Then $R_G(f)$ satisfies $\int_a^{a+\lambda} (a+\lambda-x)\, f(x)\, dx \leq R_G(f) \leq \lambda - \int_{b-\lambda}^{b} [x-(b-\lambda)]\, f(x)\, dx$, where $\lambda = E(f) - a$.*

Below, we give bounds on $R_G(f)$ based on an inequality of Karamata type ([2, 3]).

Theorem. *Let $f(x)$ be a pdf on $[a,b]$ and $0 < m \leq f(x) \leq M$, then $R_G(f)$ satisfies* $\frac{E(f)+2M\left[a\left(\frac{a+b}{2}\right)-bE(f)\right]}{2bM-1} \leq R_G(f) \leq \frac{E(f)+2M\left[b\left(\frac{a+b}{2}\right)-aE(f)\right]}{2aM-1}$.

Theorem. *Let $f(x)$ be a pdf on $[a,b]$ with $a > 0$ and $0 < m \leq f(x) \leq M$, $x \in [a,b]$. Then $R_G(f)$ satisfies $\left(\frac{1-\rho\zeta}{1+\rho\zeta}\right) E(f) \leq R_G(f) \leq \left(\frac{\rho-\zeta}{\rho+\zeta}\right) E(f)$, where $\rho = \frac{M}{m}$, $\zeta = \frac{\mathcal{M}_2-a^2}{b^2-\mathcal{M}_2}$ and $\mathcal{M}_2 = \int_a^b x^2 f(x)dx$, the second moment about zero.*

References

[1] C. Gini, *Variabilità e Metabilità, contributo allo studia della distribuzioni e relationi statistiche*, Studi Economica-Gicenitrici dell' Univ. di Coglani **3** (1912), art 2, 1–158.

[2] J. Karamata, *O prvom stavu srednjih vrednosti odredenih integrala*, Glas Srpske Kraljevske Akademje **CLIV** (1933), 119–144.

[3] J. Karamata, *Sur certaines inégalités relatives aux quotients et à la différence de $\int fg$ et $\int f \cdot \int g$*, Acad. Serbe Sci. Publ. Inst. Math. **2** (1948), 131–145.

[4] J.F. Steffensen, *On certain inequalities between mean values and their application to actuarial problems*, Skandinavisk Aktuarietidskrift 1918, 82–97.

Chudziak, Jacek: *Stability of the equation originating from some kind of shift-invariance.*

A kind of invariance of n-attribute utility functions leads to the functional equation

$$U(x_1+z,\ldots,x_n+z) = k(z)U(x_1,\ldots,x_n) + l(z),$$

where k, l, U are unknown functions. Inspired by the stability problem for this equation, we consider the inequality

$$|U(x_1+z,\ldots,x_n+z) - k(z)U(x_1,\ldots,x_n) - l(z)| \leq \varepsilon,$$

where $\varepsilon \geq 0$ is fixed.

Daróczy, Zoltán: *A characterization of nonconvexity and its applications.*
(Joint work with Zsolt Páles.)

Given a nonconvex real function $f : I \to \mathbb{R}$, one can find elements $x, y \in I$ and $0 < t < 1$ such that

$$f(tx + (1-t)y) > tf(x) + (1-t)f(y) \tag{1}$$

holds. Assuming continuity, one can obtain the following sharper statement:

Theorem. *Let $f : I \to \mathbb{R}$ be continuous nonconvex function. Then there exist elements $x, y \in I$ such that* (1) *holds for all* $0 < t < 1$.

As an application, we show that this simple result has interesting and surprising consequences in the comparison theory of quasi-arithmetic means.

Dragomir, Sever Silvestru: *On the Grüss Type inequalities in Hilbert spaces with applications for operator inequalities.*

Let $(H, \langle \cdot, \cdot \rangle)$ be a Hilbert space over the real or complex number field $\mathbb{K}$, $B(H)$ the $\mathbb{C}^*$-algebra of all bounded linear operators defined on H and $A \in B(H)$. If A is invertible, then we can define the *Kantorovich functional* as $K(A;x) := \langle Ax, x\rangle \langle A^{-1}x, x\rangle$ for any $x \in H, \|x\| = 1$.

As pointed out by Greub and Rheinboldt in their seminal paper [3], if $M > m > 0$ and for the selfadjoint operator A we have $MI \geq A \geq mI$ in the partial operator order of $B(H)$, where I is the identity operator, then the *Kantorovich operator inequality* holds true

$$1 \leq K(A;x) \leq \frac{(M+m)^2}{4mM}, \qquad \text{for any } x \in H, \|x\| = 1. \tag{1}$$

On utilizing some recent Grüss' type inequalities in inner product spaces obtained by the author (see for instance [1]) we can obtain various Kantorovich operator type inequalities such as the following ones [2]:

Theorem. *Let $A \in B(H)$ and $\alpha, \beta \in \mathbb{K}$ be such that the transform*

$$C_{\alpha,\beta}(A) := (A^* - \overline{\alpha} I)(\beta I - A)$$

is accretive. If $\operatorname{Re}(\beta\overline{\alpha}) > 0$ *and the operator* $-i \operatorname{Im}(\beta\overline{\alpha}) C_{\alpha,\beta}(A)$ *is accretive, then*

$$|K(A;x) - 1| \leq \begin{cases} \frac{1}{4}\frac{|\beta-\alpha|^2}{|\beta\alpha|} - \left[\operatorname{Re}\langle C_{\alpha,\beta}(A)x, x\rangle \operatorname{Re}\left\langle C_{\frac{1}{\alpha},\frac{1}{\beta}}(A^{-1})x, x\right\rangle\right]^{\frac{1}{2}}, \\[2ex] \frac{1}{4}\frac{|\beta-\alpha|^2}{|\beta\alpha|} - \left|\left\langle\left(A - \frac{\alpha+\beta}{2} I\right)x, x\right\rangle\right| \left|\left\langle\left(A^{-1} - \frac{\alpha+\beta}{2\alpha\beta} I\right)x, x\right\rangle\right|, \\[2ex] \frac{1}{4}\frac{|\beta-\alpha|^2}{\operatorname{Re}(\beta\overline{\alpha})} |\langle Ax, x\rangle| \left|\langle A^{-1}x, x\rangle\right|, \\[2ex] \frac{|\beta+\alpha| - 2[\operatorname{Re}(\beta\overline{\alpha})]^{\frac{1}{2}}}{|\beta\alpha|^{\frac{1}{2}}} \left[|\langle Ax, x\rangle| \left|\langle A^{-1}x, x\rangle\right|\right]^{\frac{1}{2}}, \\[2ex] \frac{1}{4}\frac{|\beta-\alpha|^2}{|\beta\alpha|^{\frac{1}{2}}|\beta+\alpha|} \left[(\|Ax\| + |\langle Ax, x\rangle|)\left(\|A^{-1}x\| + \left|\langle A^{-1}x, x\rangle\right|\right)\right]^{\frac{1}{2}}, \end{cases} \tag{2}$$

for any $x \in H$, $\|x\| = 1$.

References

[1] S.S. Dragomir, *Advances in Inequalities of the Schwarz, Grüss and Bessel Type in Inner Product Spaces*, Nova Science Publishers Inc, New York, 2005, x+249 pp.

[2] S.S. Dragomir, *New Inequalities of the Kantorovich Type for Bounded Linear Operators in Hilbert Spaces*, Preprint, RGMIA Res. Rep. Coll. **9** (2006), Supplement, Article 4. `[Online http://rgmia.vu.edu.au/v9(E).html]`.

[3] W. Greub and W. Rheinboldt, *On a generalization of an inequality of L.V. Kantorovich*, Proc. Amer. Math. Soc. **10** (1959) 407–415.

Fazekas, Borbála: *Enclosure for the fourth order Gelfand-equation.*

We investigate the fourth order nonlinear biharmonic equation for $u \in H_0^2(\Omega)$ on the star-shaped domain $\Omega \subset \mathbb{R}^2$

$$\begin{aligned} \Delta^2 u = &\quad F(u) &\quad \text{on } \Omega, \\ u = &\quad 0 &\quad \text{on } \partial\Omega, \\ \frac{\partial u}{\partial \nu} = &\quad 0 &\quad \text{on } \partial\Omega. \end{aligned}$$

Our aim is to obtain an enclosure of a solution with a rigorous proof of existence. For that purpose, we use a computer-assisted approach based on a general method by M. Plum.

The main numerical tools are, e.g., C_0-finite element approximations for the solution, as well as for its gradient and its Laplacian, and homotopy methods (to get enclosures for certain eigenvalues).

We demonstrate our results on the example of the Gelfand-equation, i.e., in the case $F(u) = \lambda \exp(u)$.

Fechner, Włodzimierz: *A Sandwich theorem for orthogonally additive functions.* (Joint work with Justyna Sikorska.)

Let p be an orthogonally subadditive mapping, q an orthogonally superadditive mapping and assume that $q \leq p$. We prove that under some additional assumptions there exists a unique orthogonally additive mapping f such that $q \leq f \leq p$.

Gavrea, Bogdan: *On some integral inequalities.*

In this presentation we extend some results presented in [1] and we solve an open problem proposed here.

References

[1] Q.A. Ngo, D.D. Thang, T.T. Dat, D.A. Tuan, *Notes on an integral inequality,* J. Inequal. Pure Appl. Math. **7** (2006), Art. 120.

Gavrea, Ioan: *Operators of Bernstein-Stancu type and the monotonicity of some sequences involving convex functions.*

Using the properties of some sequences of positive linear operators of Bernstein-Stancu type, we establish some refinements of some inequalities obtained in [1].

References

[1] F. Qi, B.-N. Guo, *Monotonicity of sequences involving convex function and sequence,* Math. Inequal. Appl. **9** (2006), 247–254.

Ger, Roman: *On vector Pexider differences controlled by scalar ones.*

During several decades of the last age till now, stability type inequalities play a significant role in the general theory of inequalities. After many particular cases, Kil-Woung Jun, Dong-Soo Shin Byung-Do Kim presented the description of solutions of a general stability inequality of Pexider type (see [1]), while Yang-Hi Lee and Kil-Woung Jun investigated it assuming only a special form of the dominating function (see [2]). As it is well known, the standard approach (direct method), applied also in these two papers, is useless while dealing with the most delicate (singular) cases. Facing the lack of stability we then try to diminish the dominating function to get the desired result. The aim of our talk is to give the solutions of the inequality

$$\|F(x+y) - G(x) - H(y)\| \leq g(x) + h(y) - f(x+y)$$

where F, G, H map a given commutative semigroup $(S, +)$ into a Banach space and $f, g, h : S \to \mathbb{R}$ are given scalar functions. Reducing this inequality to the case where $G = H$ and $g = h$ we then apply sandwich type results obtained by Kazimierz Nikodem, Zsolt Páles and Szymon Wąsowicz in [3].

References

[1] K.-W. Jun, D.-S. Shin, B.-D. Kim, *On Hyers-Ulam-Rassias Stability of the Pexider Equation*, J. Math. Anal. Appl. **239** (1999), 20–29.

[2] Y.-H. Lee, K.-W. Jun, *A Generalization of the Hyers-Ulam-Rassias Stability of the Pexider Equation*, J. Math. Anal. Appl. **246** (2000), 627–638.

[3] K. Nikodem, Zs. Páles, Sz. Wąsowicz, *Abstract separation theorems of Rodé type and their applications*, Ann. Polon. Math. **72** (1999), 207–217.

Gilányi, Attila: *Regularity theorems for generalized convex functions.*
(Joint work with Zsolt Páles.)

One of the classical results of the theory of convex functions is the theorem of F. Bernstein and G. Doetsch [2] which states that if a real-valued Jensen-convex function defined on an open interval I is locally bounded above at one point in I then it is continuous. According to a related result by W. Sierpiński [4], the Lebesgue measurability of a Jensen-convex function implies its continuity, too.
In this talk we generalize the theorems above for (M, N)-convex functions, calling a function $f : I \to J$ (M, N)-convex if it satisfies the inequality

$$f(M(x, y)) \leq N_{x,y}(f(x), f(y))$$

for all $x, y \in I$, where I and J are open intervals, M is a mean on I and $N_{x,y}$ is a suitable mean on J for every $x, y \in I$ (cf., e.g., [3]) Our statements contain T. Zgraja's results on (M, M)-convex functions ([5]) and M. Adamek's theorems on λ-convex functions ([1]) as special cases.

REFERENCES

[1] M. Adamek, *On λ-quasiconvex and λ-convex functions*, Rad. Mat. **11** (2002/03), 171–181.

[2] F. Bernstein and G. Doetsch, *Zur Theorie der konvexen Funktionen*, Math. Ann. **76** (1915), 514–526.

[3] C. Niculescu and L.-E. Persson, *Convex Functions and Their Applications*, CMS Books in Mathematics, Springer, 2006.

[4] W. Sierpiński, *Sur les fonctions convexes mesurables*, Fund. Math. **1** (1920), 125–128.

[5] T. Zgraja, *Continuity of functions which are convex with respect to means*, Publ. Math. **63** (2003), 401–411.

Goldberg, Moshe: *Minimal polynomials and radii of elements in finite-dimensional power-associative algebras.*

We begin by revisiting the definition and some of the properties of the minimal polynomial of an element of a finite-dimensional power-associative algebra $\mathcal{A}$ over an arbitrary field $\mathbb{F}$. Our main observation is that p_a, the minimal polynomial of $a \in \mathcal{A}$, may depend not only on a, but also on the underlying algebra.

Restricting attention to the case where $\mathbb{F}$ is either $\mathbb{R}$ or $\mathbb{C}$, we proceed to define $r(a)$, the *radius* of an element a in $\mathcal{A}$, to be the largest root in absolute value of the minimal polynomial of a. As it is, r possesses some of the familiar properties of the classical spectral radius. In particular, r is a continuous function on $\mathcal{A}$.

In the third part of the talk we discuss stability of subnorms acting on subsets of finite-dimensional power-associative algebras. Our main result states that if $\mathcal{S}$, a subset of an algebra $\mathcal{A}$, satisfies certain assumptions, and f is a continuous subnorm on $\mathcal{S}$, then f is stable on $\mathcal{S}$ if and only if f majorizes the radius r.

Házy, Attila: *On a certain stability of the Hermite–Hadamard inequality.*
(Joint work with Zsolt Páles.)

In our talk, we investigate the connection between the stability forms of the functional inequalities related to Jensen-convexity, convexity, and Hermite–Hadamard inequality. In other words, we consider continuous functions $f : D \to \mathbb{R}$ satisfying

$$f\left(\frac{x+y}{2}\right) \leq \frac{f(x)+f(y)}{2} + \delta_J(\|x-y\|) \qquad (x, y \in D),$$

$$f\left(\frac{x+y}{2}\right) \leq \int_0^1 f\big(tx+(1-t)y\big)dt + \delta_H(\|x-y\|) \qquad (x, y \in D),$$

and

$$f\big(tx+(1-t)y\big) \leq tf(x) + (1-t)f(y) + \delta_C(t, \|x-y\|) \qquad (x, y \in D,\, t \in [0,1]),$$

where $\delta_J, \delta_H : [0, \infty[\to \mathbb{R}$, and $\delta_C : [0,1] \times [0, \infty[\to \mathbb{R}$ are given functions called the stability terms. The main results establish connections between these terms.

Hoang, Vu: *Enclosures for scalar nonlinear hyperbolic problems.*

Consider the Cauchy problem for the scalar conservation law in one space dimension:

$$\begin{aligned} u_t + f(u)_x &= 0 \\ u(x,0) &= u_0(x) \end{aligned}$$

where f is a given nonlinear function. Suppose an approximate solution w to this PDE has been computed by some numerical method. It is desirable to verify the existence of a weak solution to our problem lying close to w (in an appropriate norm). A challenge is presented by the well-known fact that weak solutions may develop discontinuities (shocks) and are not uniquely determined by their initial data; one is interested in enclosing "physically" correct solutions, characterized by additional (e.g., Entropy) criteria. We shall discuss an approach to this problem where one first computes an approximate solution to the perturbed equation $u_t - \varepsilon u_{xx} + f(u)_x = 0$ (with small $\varepsilon > 0$) and then obtains an solution of the conservation law by an iteration procedure.

Johansson, Maria: *A unified approach to Hardy type inequalities for non-increasing functions.*
(Joint work with Lars-Erik Persson and Anna Wedestig.)

Some Hardy type inequalities for non-increasing functions are characterized by one condition (Sinnamon), while others are described by two independent conditions (Sawyer). In this presentation we make a unified approach to such results and present a result which covers both the Sinnamon result and Sawyer's result for the case when one weight is non-decreasing. In all cases we point out that this condition is not unique and can even be chosen among some (infinite) scales of conditions.

Kittaneh, Fuad: *Norm inequalities for commutators of Hilbert space operators.*

Let A, B, and X be bounded linear operators on a complex separable Hilbert space. It is shown that if A and B are self-adjoint such that $a_1 \leq A \leq a_2$ and $b_1 \leq B \leq b_2$ for some real numbers a_1, a_2, b_1, and b_2, then for every unitarily invariant norm $|||\cdot|||$,

$$|||AX - XB||| \leq (\max(a_2, b_2) - \min(a_1, b_1))\, |||X|||$$

and

$$||AB - BA|| \leq \frac{1}{2}(a_2 - a_1)(b_2 - b_1),$$

where $||\cdot||$ is the usual operator norm. Consequently, if A and B are positive, then

$$|||AX - XB||| \leq \max(||A||, ||B||)\, |||X|||$$

and

$$||AB - BA|| \leq \frac{1}{2}\, ||A||\, ||B||.$$

Generalizations of these norm inequalities to commutators of normal operators are obtained, and applications of these inequalities are also given.

Klaričić Bakula, Milica: *Generalizations of Jensen–Steffensen's and related inequalities.*
(Joint work with Marko Matić and Josip Pečarić.)

Let $\varphi : (a,b) \to \mathbb{R}$ be a convex function and $p_i \in \mathbb{R}$, $i = 1, \ldots, n$ satisfying

$$0 \leq P_k \leq P_n, \quad k = 1, \ldots, n, \quad P_n > 0,$$

where

$$P_k = \sum_{i=1}^{k} p_i.$$

We prove that for any $x_i \in (a,b)$, $i = 1, \ldots, n$ such that

$$x_1 \leq x_2 \leq \cdots \leq x_n \quad \text{or} \quad x_1 \geq x_2 \geq \cdots \geq x_n$$

the following inequalities

$$\varphi(c) + \varphi'(c)\left(\frac{1}{P_n}\sum_{i=1}^{n} p_i x_i - c\right) \leq \frac{1}{P_n}\sum_{i=1}^{n} p_i \varphi(x_i)$$
$$\leq \varphi(d) + \frac{1}{P_n}\sum_{i=1}^{n} p_i \varphi'(x_i)(x_i - d),$$

hold for all $c, d \in (a,b)$.

We show that the discrete Jensen–Steffensen's inequality, as well as a discrete Slater type inequality, can be obtained from these general inequalities as special cases. We also prove that one of our general companion inequalities, under some additional assumptions on the function φ is tighter then the obtained Slater type inequality. The integral variants of the results are also established.

Kobayashi, Kenta: *A constructive a priori error estimation for finite element discretizations in a non-convex domain using mesh refinement.*

In solving elliptic boundary value problem by finite element method in a bounded domain which has a re-entrant corner, the convergent rate could be improved by using mesh refinement. In our research, we have obtained explicit a priori error estimation for finite element solution of the Poisson equation in a polygonal domain. Our result is important in a theoretical sense as well as practical calculations because the constructive a priori error estimation for linear problem are often used for computer-assisted proof for non-linear problems.

For $f \in L^2(\Omega)$, we consider the weak solution of the following partial differential equation.

$$\begin{cases} -\Delta u = f & \text{in} \quad \Omega, \\ u = 0 & \text{on} \quad \partial\Omega. \end{cases}$$

where Ω is the polygonal domain.

If Ω is a convex domain, we can obtain a priori error estimation as follows:

$$\|u - u_h\|_{H_0^1(\Omega)} \leq Ch\|f\|_{L^2(\Omega)},$$

where u_h is a finite element solution, h denotes maximum mesh size, and C is a constant which is calculated only by condition of mesh [3].

However, if Ω is a non-convex domain, we cannot obtain such $O(h)$ error estimation with uniform mesh because of the singularities at the re-entrant corner [1][2]. To deal with this difficulty, we use mesh refinement and furthermore, obtained a priori error estimation for finite element solutions.

References

[1] P. Grisvard, *Elliptic Problems in Nonsmooth Domains*, Pitman Publishing, Boston, 1985.

[2] P. Grisvard, *Singularities in Boundary Value Problems*, RMA **22** (1992), Masson, Paris.

[3] M.T. Nakao and N. Yamamoto, *A guaranteed bound of the optimal constant in the error estimates for linear triangular element*, Computing Supplementum, **15** (2001), 165–173.

Lemmert, Roland Heinrich: *Boundary value problems via an intermediate value theorem.*
(Joint work with Gerd Herzog.)

We use an intermediate value theorem([1], [2]) for quasimonotone increasing functions to prove the existence of a smallest and a greatest solution of the Dirichlet problem

$$u'' + f(t,u) = 0, \ u(0) = \alpha, \ u(1) = \beta$$

between lower and upper solutions, where $f : [0,1] \times E \to E$ is quasimonotone increasing in its second variable with respect to a regular cone.

References

[1] G. Herzog, R. Lemmert *Intermediate value theorems for quasimonotone increasing maps*, Numer. Funct. Anal. Optimization **20** (1999), 901–908.

[2] R. Uhl, *Smallest and greatest Fixed Points of quasimonotone increasing mappings*, Math. Nachr. **248-249** (2003), 204–210.

Losonczi, László: *Polynomials all of whose zeros are on the unit circle.*
(Joint work with Piroska Lakatos.)

According to a classical theorem of Cohn [1] *all zeros of a polynomial* $P \in \mathbb{C}[z]$ *lie on the unit circle if and only if*

- *P is self-inversive,*
- *all zeros of P' are in* **or** *on this circle.*

Here we discuss recent results (see, e.g., [2], [3]) which give simple sufficient conditions (inequalities in terms of the coefficients of P) for all zeros of P to be on the unit circle.

REFERENCES

[1] A. Cohn, *Über die Anzahl der Wurzeln einer algebraischen Gleichung in einem Kreise.* Math. Zeit. **14** (1922), 110–148.

[2] P. Lakatos, L. Losonczi, *Self-inversive polynomials whose zeros are on the unit circle.* Publ. Math. Debrecen **65** (2004), 409–420.

[3] L. Losonczi, A. Schinzel, *Self-inversive polynomials of odd degree*, Ramanujan J., to appear.

Makó, Zita: *On the equality of generalized quasi-arithmetic means.*
(Joint work with Zsolt Páles.)

Given a continuous strictly monotone function $\varphi : I \to \mathbb{R}$ and a probability measure μ on the Borel subsets of $[0,1]$, the two variable mean $M_{\varphi,\mu} : I^2 \to I$ is defined by

$$M_{\varphi,\mu}(x,y) := \varphi^{-1}\Big(\int_0^1 \varphi\big(tx+(1-t)y\big)d\mu(t)\Big) \qquad (x,y \in I).$$

This class of means includes quasi-arithmetic as well as Lagrangian means. The aim of my talk is to study their equality problem, i.e., to characterize those pairs (φ,μ) and (ψ,ν) such that

$$M_{\varphi,\mu}(x,y) = M_{\psi,\nu}(x,y) \qquad (x,y \in I)$$

holds. Under at most fourth-order differentiability assumptions for the unknown functions φ and ψ, a complete description of the solution set of the above functional equation is obtained.

Maksa, Gyula: *A decomposition of Wright convex functions of higher order.*
(Joint work with Zsolt Páles.)

In this talk we present the following generalization of a result of C.T. Ng [4]. Let $\emptyset \neq I \subset \mathbb{R}$ be an open interval, p be a fixed positive integer, and $f : I \to \mathbb{R}$ be a p-Wright convex function, that is, f satisfies the inequality

$$\Delta_{h_1} \dots \Delta_{h_{p+1}} f(x) \geq 0$$

for all $h_1 \dots h_{p+1} \in]0,+\infty[,\ x \in I,$ for which $x + h_1 + \dots + h_{p+1} \in I.$ Then

$$f(x) = C(x) + P(x) \qquad (x \in I)$$

where $C : I \to \mathbb{R}$ is (continuous) p-convex and $P : \mathbb{R} \to \mathbb{R}$ is a polynomial function of degree at most p, that is, $\Delta_h^{p+1} P(x) = 0$ for all $x, h \in \mathbb{R}$. In the proof we use some ideas, among others, from the works [1], [2], and [3].

REFERENCES

[1] N.G. de Bruijn, *Functions whose differences belong to a given class*, Nieuw Archief voor Wiskunde, **23**(1951), 194–218, 433–438.

[2] M. Kuczma, *An Introduction to the Theory of Functional Equations and Inequalities*, Państwowe Wydawnictwo Naukowe – Uniwersytet Śląski, Warszawa – Kraków – Katowice, 1985.

[3] Gy. Maksa, *Deviations and differences*, Thesis, L. Kossuth University, Debrecen, 1976.

[4] C.T. Ng, *Functions generating Schur-convex sums*, General Inequalities 5, **80** (1987), 433–438.

Matić, Marko: *Euler-Grüss type inequalities involving measures.*
(Joint work with Ambroz Čivljak and Ljuban Dedić.)

Let μ be a real Borel measure on Borel set $X \subset \mathbb{R}^m$ such that $\mu(X) \neq 0$ and let $f \in L_\infty(X, \mu)$ be such that

$$\gamma \leq f(t) \leq \Gamma, \qquad t \in X, \quad \mu\text{-a.e.},$$

for some $\gamma, \Gamma \in \mathbb{R}$. We prove that then

$$\left| \int_X f(t) \mathrm{d}\mu(t) \right| \leq \tfrac{1}{2}(\Gamma - \gamma) \left\| \mu \right\|, \tag{G}$$

where $\|\mu\|$ is the total variation of μ. Also, we discuss the equality case in (G). This inequality is a generalization of the key result from [2]. Using the inequality (G) and general Euler identities involving μ-harmonic sequences of functions that were proved in [1] we give various Euler-Grüss type inequalities.

References

[1] A. Čivljak, Lj. Dedić and M. Matić, *Euler harmonic identities for measures*, Nonlinear Funct. Anal. and Appl. In Press.

[2] M. Matić, *Improvement of some inequalities of Euler-Grüss type*, Computers and Mathematics with Applications **46** (2003) 1325–1336.

Matkowski, Janusz: *Generalization of some results on globally Lipschitzian superposition operators.*

Let $I \subset \mathbb{R}$ be an interval. A function $h : I \times \mathbb{R} \to \mathbb{R}$ generates the so-called superposition (or Nemytskij) operator $H : \mathbb{R}^I \to \mathbb{R}^I$ defined by

$$H(\varphi)(x) := h\left(x, \varphi(x)\right), \qquad \varphi \in J^I, \quad (x \in I).$$

It is known that there are some Banach function spaces $\mathcal{F}(I)$ of functions $\varphi : I \to \mathbb{R}$ with the norms $\|\cdot\|$ (stronger than the supremum norm) such that the global Lipschitz inequality

$$\|H(\varphi) - H(\psi)\| \leq c \left\|\varphi - \psi\right\|, \qquad \varphi, \psi \in \mathcal{F}(I),$$

implies that

$$h(x, y) = a(x)y + b(x), \qquad x \in I, \quad y \in \mathbb{R}$$

for some $a, b \in \mathcal{F}(I)$.

We shall present some results which show that the Lipschitz condition can replaced by the uniform continuity of H and weaker ones.

Mészáros, Fruzsina: *Functional equations connected with beta distributions.* (Joint work with Károly Lajkó.)

The functional equation

$$f_U(u)\, f_V(v) = f_X\left(\frac{1-v}{1-uv}\right) f_Y(1-uv)\,\frac{v}{1-uv} \qquad (u, v \in (0,1)) \tag{1}$$

for unknown density functions $f_X, f_Y, f_U, f_V : (0,1) \to \mathbb{R}_+$ was introduced by J. Wesołowski.

He determined the solution of (1) under the assumptions that the density functions are strictly positive and locally integrable on $(0,1)$ and he asked the measurable solution of (1) (see [5]).

The investigations of Wesołowski are based on the locally integrable real solutions $g_1, g_2, \alpha_1, \alpha_2 : (0,1) \to \mathbb{R}$ of the following general functional equation

$$g_1\left(\frac{1-x}{1-xy}\right) + g_2\left(\frac{1-y}{1-xy}\right) = \alpha_1(x) + \alpha_2(y) \quad (x, y \in (0,1)). \tag{2}$$

The main aim of this talk is to give the general solution of (1) for functions $f_X, f_Y, f_U, f_V : (0,1) \to \mathbb{R}_+$ and the general solution of (2) for functions $g_1, g_2, \alpha_1, \alpha_2 : (0,1) \to \mathbb{R}$. Furthermore we determine the solution of (1) under the following natural assumptions:

1. the density functions are measurable,
2. (1) is satisfied for $(u, v) \in (0,1)^2$ almost everywhere.

References

[1] N.C. Giri, *Introduction to probability and statistics*, Marcel Dekker, New York, 1974.

[2] A. Járai, *Measurable solutions of functional equations satisfied almost everywhere*, Mathematica Pannonica **10** (1999), 103–110.

[3] A. Járai, *Regularity Properties of Functional Equations in Several Variables*, Springer, Dordrecht, 2005.

[4] Gy. Maksa, *Solution on the open triangle of the generalized fundamental equation of information with four unknown functions*, Utilitas Math. **21** (1982), 267–282.

[5] J. Wesołowski, *On a functional equation related to an independence property for beta distributions*, Aequationes Math. **66** (2003), 156–163.

Mitrea, Alexandru: *Inequalities related to the error-estimation and superdense unbounded divergence of some approximation procedures.* (Joint work with Paulina Mitrea.)

Let consider the approximation procedures of interpolatory type, described by the relations:

(1) $$Af = D_n f + R_n f, \qquad f \in C^s[-1,1]; n \geq 1,$$

where A is a given continuous linear functional, $s \geq 0$ is an integer and D_n, $n \geq 1$, are approximating functionals associated to a triangular node matrix $\{x_n^k : n \geq 1; 1 \leq k \leq i_n\} \subseteq [-1, 1]$ of the form:

$$D_n f = \sum_{j=0}^{m} \sum_{k=1}^{i_n} a_n^{kj} f^{(j)}(x_n^k),$$

with a given integer m, $0 \leq m \leq s$.

Our aim is to obtain theorems concerning the convergence or the superdense unbounded divergence of the approximation procedures (1) and to give, in this framework, estimations of the approximation error $R_n f$.

To this end, we shall establish various inequalities regarding the norm of the functionals D_n for the following approximation procedures:

(i) Numerical differentiation formulas, i.e., $Af = f^{(s)}(0)$

(ii) Quadrature formulas, i.e., $Af = \int_{-1}^{1} f(x)dx$, with Jacobi or equidistant interpolatory nodes x_n^k.

References

[1] K. Balazs, *On the convergence of the derivatives of Lagrange interpolation polynomials*, Acta Math. Acad. Sci. Hungar. **55**(1990), 323–325.

[2] S. Cobzaş, I. Muntean, *Triple condensation of singularities for some interpolation processes*, Proc. of ICAOR, vol.1, 1997, 227–232.

[3] A.I. Mitrea, *On the convergence of a class of approximation procedures*, P.U.M.A. **15** (2005), 225–234.

Mohapatra, Ram Narayan: *Sharp inequalities for rational functions in the complex plane.*
(Joint work with John Boncek.)

Markov and Bernstein Inequalities for real and complex polynomials have been studied extensively. Their analogues for rational functions have been studied by Borwein and Erdelyi, and Li, Mohapatra and Rodriguez separately. Extremal rational functions have been determined to show that these inequalities are sharp. In the case of polynomials an extension of the concept of the derivative is achieved by considering polar derivatives. In that case Laguerre's theorem help establish many nice results. In this talk we shall discuss some of the known results and define an analogue of polar derivative for rational functions and obtain an inequality for self inversive rational functions. Some other related results will also be mentioned.

Moslehian, Mohammad Sal: *Asymptoticity aspect of the quadratic functional equation in multi-normed spaces.*

The notion of multi-normed space was introduced by H.G. Dales and M.E. Polyakov in [2]. This concept is somewhat similar to operator sequence space and has some connections with operator spaces and Banach latices. In this talk, we investigate the stability (see [3]) of the quadratic functional equation for mappings from linear spaces into multi-normed spaces (see [1]). We then study an asymptotic behavior of the quadratic equation in the framework of multi-normed spaces.

REFERENCES

[1] H.G. Dales, M.S. Moslehian, *Stability of mappings on multi-normed spaces*, Glasgow Math. J., to appear.

[2] H.G. Dales, M.E. Polyakov, *Multi-normed spaces and multi-Banach algebras*, preprint.

[3] Th.M. Rassias, *On the stability of functional equations and a problem of Ulam*, Acta Appl. Math. **62** (2000), 23–130.

Mrowiec, Jacek: *On generalized convex sets and generalized convex functions (in the sense of Beckenbach).*

A natural way to generalize the concept of a convex function was introduced by E.F. Beckenbach in [1]. In this talk this concept is extended to higher-dimensional cases. We will present and compare two definitions of generalized convex sets and their properties. Then two ways of introducing generalized convex functions are presented.

REFERENCES

[1] E.F. Beckenbach, *Generalized convex functions*, Bull. Amer. Math. Soc. **43** (1937), 363–371.

Nagatou, Kaori: *Eigenvalue problems on 1-D Schrödinger operators.*

In this talk we consider the following eigenvalue problem

$$-u'' + q(x)u + s(x)u = \lambda u, \qquad x \in \mathbb{R}, \tag{1}$$

where we assume that $q(x) \in L^\infty(\mathbb{R})$ is a periodic function and $s \in L^\infty(\mathbb{R})$ satisfies $s(x) \to 0$ $(|x| \to \infty)$. This kind of operator has essential spectrum with band-gap structure, and depending on the perturbation it may have, in addition, isolated eigenvalues in the spectral gaps.

This kind of problem is very important not only in practical physical problems but also in relation to a nonlinear problem

$$-u'' + q(x)u + f(u) = 0 \text{ on } \mathbb{R}. \tag{2}$$

In order to enclose a weak solution $u \in H^1(\mathbb{R})$ of (2) we need to estimate an eigenvalue (especially with the smallest absolute value) of the linearized operator in some spectral gap. Due to the lack of appropriate variational characterizations and to the "spectral pollution" problem, it is difficult to locate these eigenvalues analytically or numerically. We will show how a mathematically rigorous treatment of such a problem could be done by computer-assisted means, and especially we focus on excluding eigenvalues in spectral gaps.

REFERENCES

[1] E.B. Davies, M. Plum, *Spectral pollution*, IMA J. Num. Anal. **24** (2004), 417–438.

[2] M.S.P. Eastham, *The Spectral Theory of Periodic Differential Equations*, Scottish Academic Press, 1973.

[3] T. Kato, *Perturbation Theory for Linear Operators*, Springer Verlag, Berlin, Heidelberg 1966, 1976.

[4] K. Nagatou, *A numerical method to verify the elliptic eigenvalue problems including a uniqueness property*, Computing **63** (1999), 109–130.

[5] K. Yosida, *Functional Analysis*, Springer Verlag, Berlin 1995.

Nakao, Mitsuhiro T.: *The guaranteed a priori error estimates in the finite element method and the spectral method with applications to nonlinear PDEs.*

In this talk, we first consider the guaranteed a priori error estimates in the finite element method for Poisson's equation and for bi-harmonic problems. In these error estimates, actual values of constants in the various inequalities play important and essential roles. Next, as an application of the results, we show a numerical verification method of solutions for nonlinear elliptic problems and Navier-Stokes equations as well as other applications. We also show similar kinds of error estimates can also be possible for the spectral Galerkin method with applications to prove the bifurcating solutions of two and three dimensional heat convection problems. Several numerical examples which confirm the actual effectiveness of our method will be presented.

REFERENCES

[1] K. Hashimoto, K. Kobayashi, M.T. Nakao, *Verified numerical computation of solutions for the stationary Navier-Stokes equation in nonconvex polygonal domains*, MHF Preprint Series, Kyushu University, MHF2007-2 (2007), 15 pp.

[2] K. Nagatou, K. Hashimoto, M.T. Nakao, *Numerical verification of stationary solutions for Navier-Stokes problems*, J. Comput. Appl. Math. **199** (2007), 424–431.

[3] M.T. Nakao and K. Hashimoto, *On guaranteed error bounds of finite element approximations for non-coercive elliptic problems and its applications*, J. Comput. Appl. Math., to appear.

[4] Y. Watanabe, N. Yamamoto, M.T. Nakao T. Nishida, *A Numerical Verification of Nontrivial Solutions for the Heat Convection Problem*, J. Math. Fluid Mech. **6** (2004), 1–20.

Niculescu, Constantin P.: *An overview of absolute continuity and its applications.*

The basic idea of *absolute continuity* is to control the behavior of a function $f : X \to \mathbb{R}$ via an estimate of the form

$$|f| \leq \varepsilon q + \delta(\varepsilon)p, \quad \varepsilon > 0, \tag{1}$$

where $p, q : X \to \mathbb{R}$ are suitably chosen nonnegative functions. This appears as an useful relaxation of the condition of domination

$$|f| \le p,$$

The aim of our paper was to illustrate the usefulness of the notion of absolute continuity in a series of fields such as Functional Analysis, Approximation Theory and PDE.

References

[1] F. Altomare, M. Campiti, *Korovkin-Type Approximation Theory and Its Applications*, de Gruyter Studies in Mathematics, de Gruyter, Berlin, 1994.

[2] A.V. Arkhangelskii, L.S. Pontryagin, *General Topology I.*, Springer-Verlag, Berlin, 1990.

[3] H. Brezis and E.H. Lieb, *A relation between pointwise convergence of functions and convergence of functionals*, Proc. Amer. Math. Soc. **88** (1983), 486–490.

[4] B. Chabat, *Introduction à l'analyse complexe.* Tome 1, Ed. Mir, Moscou, 1990.

[5] K.R. Davidson and A.P. Donsig, *Real Analysis with Real Applications*, Prentice Hall, Inc., Upper Saddle River, N. J., 2002.

[6] J. Diestel, H. Jarchow and A. Tonge, *Absolutely Summing Operators*, Cambridge University Press, 1995.

[7] L.C. Evans, *Partial Differential Equations*, American Mathematical Society, Providence, R.I., 3rd Printing, 2002.

[8] H. Hueber, *On Uniform Continuity and Compactness in Metric Spaces*, Amer. Math. Monthly **88** (1981), 204–205.

[9] P.P. Korovkin, *On convergence of linear positive operators in the space of continuous functions*, Doklady Akad. Nauk. SSSR (NS) **90** (1953), 961–964. (Russian)

[10] H.E. Lomeli and C.L. Garcia, *Variations on a Theorem of Korovkin*, Amer. Math. Monthly **113** (2006), 744–750.

[11] C.P. Niculescu, *Absolute Continuity and Weak Compactness*, Bull. Amer. Math. Soc. **81** (1975), 1064–1066.

[12] C.P. Niculescu, *Absolute Continuity in Banach Space Theory*, Rev. Roum. Math. Pures Appl. **24** (1979), 413–422.

[13] C.P. Niculescu and C. Buşe, *The Hardy-Landau-Littlewood inequalities with less smoothness*, J. Inequal. Pure Appl. Math. **4** (2003), Art. 51, 8 pp.

[14] H.H. Schaefer, *Banach Lattices and Positive Operators*, Springer Verlag, New York, 1974.

Nikodem, Kazimierz: *Notes on t-quasiaffine functions.*
(Joint work with Zsolt Páles.)

Given a convex subset D of a vector space and a constant $0 < t < 1$, a function $f : D \to \mathbb{R}$ is called *t-quasiaffine* if, for all $x, y \in D$,

$$\min\{f(x), f(y)\} \le f\big(tx + (1-t)y\big) \le \max\{f(x), f(y)\}.$$

If, furthermore, both of these inequalities are strict for $f(x) \neq f(y)$, f is called *strictly t-quasiaffine*. We show that every t-quasiaffine function is also $\mathbb{Q}$-quasiaffine (i.e., t-quasiaffine for every rational number t in $[0,1]$). An analogous result is established for strict t-quasiaffinity. The basic role in the proof of these results is played by a theorem stating that if A and B are disjoint t-convex sets such that $D = A \cup B$, then A and B are also $\mathbb{Q}$-convex.

Ogita, Takeshi: *Lower and upper error bounds of approximate solutions of linear systems.*
(Joint work with Shin'ichi Oishi.)

This talk is concerned with the problem of verifying the accuracy of an approximate solution $\tilde{x}$ of a linear system

$$Ax = b, \tag{1}$$

where A is a real $n \times n$ matrix and b is a real n-vector. If A is nonsingular, there exists a unique solution $x^* := A^{-1}b$. We aim on verifying the nonsingularity of A and calculating some $\underline{\epsilon}, \overline{\epsilon} \in \mathbb{R}^n$ such that

$$\mathbf{o} \leq \underline{\epsilon} \leq |x^* - \tilde{x}| \leq \overline{\epsilon}, \tag{2}$$

where $\mathbf{o} := (0, \ldots, 0)^T \in \mathbb{R}^n$. A number of fast self-validating algorithms (cf., for example, [1, 3, 5, 6]) have been proposed to verify the nonsingularity of A and to compute $\overline{\epsilon}$ in (2). In this talk, computing the lower bound $\underline{\epsilon}$ in (2) is also considered. If $\underline{\epsilon}_i \approx \overline{\epsilon}_i$, then we can *verify* that the error bounds (and the verification) for $\tilde{x}_i$ are of high quality. The main point of this talk is to develop a fast method of calculating both $\underline{\epsilon}$ and $\overline{\epsilon}$ satisfying (2), which are as tight as we need. It is possible when using new accurate algorithms for summation and dot product [2, 4]. Numerical results are presented elucidating properties and efficiencies of the proposed verification method.

References

[1] T. Ogita, S. Oishi, Y. Ushiro, *Computation of sharp rigorous componentwise error bounds for the approximate solutions of systems of linear equations*, Reliable Computing **9** (2003), 229–239.

[2] T. Ogita, S.M. Rump, S. Oishi, *Accurate Sum and Dot Product*, SIAM J. Sci. Comput. **26** (2005), 1955–1988.

[3] S. Oishi, S.M. Rump, *Fast verification of solutions of matrix equations*, Numer. Math. **90** (2002), 755–773.

[4] S.M. Rump, T. Ogita, S. Oishi, *Accurate floating-point summation*, Part I and II, submitted.

[5] S.M. Rump, *Verification methods for dense and sparse systems of equations*, Topics in Validated Computations – Studies in Computational Mathematics (J. Herzberger ed.), Elsevier, Amsterdam, 1994, 63–136.

[6] T. Yamamoto, *Error bounds for approximate solutions of systems of equations*, Japan J. Appl. Math. **1** (1984), 157–171.

Olkin, Ingram: *Characterizations of some probability distributions.*

In the theory of life distributions in engineering reliability and in medical survival analysis semiparametric families are somewhat central. These are families that have both a real parameter and a parameter that is itself a distribution. Examples of such families are scale parameter families, power parameter families, moment parameter families, and many others. The coincidence of two families provides a characterization of the underlying distribution. In this talk we introduce these families and obtain some characterizations. Each characterization is obtained by solving a functional equation.

Opic, Bohumír: *The ρ-quasiconcave functions and weighted inequalities.*
(Joint work with William D. Evans and Amiran Gogatishvili.)

Let ρ be a positive, continuous and strictly increasing function on the interval $I := (a, b) \subseteq \mathbb{R}$. A non-negative function h is said to be ρ-quasiconcave on I – notation $h \in Q_\rho(I)$ – if h is non-decreasing on I and h/ρ is non-increasing on I. (Note that when $I = (0, +\infty)$ and the function ρ is the identity map on I, then the class $Q_\rho(I)$ coincides with the well-known class $Q((0, +\infty))$ of all quasiconcave functions on the interval $(0, +\infty)$.) We present a representation of ρ-quasiconcave functions on I by means of non-negative Borel measures on I.

Let $h \in Q_\rho(I)$. We decompose the interval I on a system $\{I_k\}$ of disjoint subintervals I_k with the property that, for all $x, y \in I_k$, either $h(x) \approx h(y)$ or $(h/\rho)(x) \approx (h/\rho)(y)$.

Given the weighted Lebesgue space $L^q(w, I, \mu)$ (w is a weight on I, μ is a non-negative Borel measure on I, $q \in (0, +\infty]$), we show that the ρ- fundamental function of this space is ρ-quasiconcave on I. This fact is used to discretize $L^q(w, I, \mu)$-quasinorms of ρ-quasiconcave functions on I.

The operator T, whose domain $\mathcal{D}(T)$ is a subset of all non-negative functions on I, is called ρ-quasiconcave provided that $Tf \in Q_\rho(I)$ for all $f \in \mathcal{D}(T)$.

We apply our results to characterize the validity of weighted inequalities involving ρ-quasiconcave operators. Our method consists in a discretization the of inequalities in question. We solve them locally (which represents an easier task) to obtained a discrete characterization of the original problem. Finally, we use the antidiscretization to convert the discrete characterization to a continuous one.

Páles, Zsolt: *Comparison in a general class of means.*
(Joint work with László Losonczi.)

Means of the form

$$M_{f,g;\mu}(x,y) := \left(\frac{f}{g}\right)^{-1}\left(\frac{\int_0^1 f\big(tx+(1-t)y\big)d\mu(t)}{\int_0^1 g\big(tx+(1-t)y\big)d\mu(t)}\right) \qquad (x, y \in I)$$

are considered, where I is an open real interval, $f, g : I \to \mathbb{R}$ are continuous functions such that g is positive and f/g is strictly monotonic, and μ is a Borel

probability measure on $[0,1]$. This class of means generalizes and includes quasi-arithmetic, Lagrangian, Cauchy, Gini, and Stolarsky means if the generating functions f, g and the measure μ are chosen properly.
The aim is to study the comparison problem of these means, i.e., to find necessary conditions and sufficient conditions for the functions (f, g) and (h, k) and for the measures μ, ν such that

$$M_{f,g;\mu}(x,y) \leq M_{h,k;\nu}(x,y) \qquad (x, y \in I)$$

holds.

Pearce, Charles E.M.: *The alternative lattice: oriented bond percolation, phase transitions inequalities.*

Consider a lattice with the sites (atoms) connected by bonds. Each bond is, independently of every other, open with probability p and closed with probability $1-p$. When there exists a connected path of open bonds of like orientation from the origin to infinity, *percolation* is said to occur. The probability $\theta(p)$ of percolation is a nondecreasing function of p.

A common phenomenon is for there to exist a critical value $p = p_{cb} \in (0,1)$ such that

$$\theta(p) \begin{cases} = 0 & \text{for } 0 \leq p < p_{cb} \\ > 0 & \text{for } p_{cb} < p \leq 1 \end{cases}$$

When this phenomenon occurs, the process is said to undergo a *phase transition* at $p = p_{cb}$. The exact determination of the critical probability p_{cb} is usually difficult and has been achieved for relatively few graph configurations.

Much effort has gone into finding good upper and lower bounds for critical probabilities. Work in this area is characterized by subtle and intricate probabilistic arguments and sometimes also heavy computation. The problem has proved more difficult in oriented graphs. Here each bond has an orientation and paths are required to proceed in the direction of that orientation on each link.

A particular case of some interest is the square lattice on the positive quadrant of the plane. The critical probability is still not known exactly for its oriented bond graph, although some upper and lower estimates have been obtained. Recently some theoretical interest has focussed on a more complicated related lattice, the so-called alternative lattice.

We address the question of finding a rigorous lower bound for the critical probability on the alternative lattice. Our estimate employs an analytical procedure which centres on the use of inequalities. The procedure utilizes a technique used recently on the corresponding site lattice, where it was used to obtain a considerable improvement on existing estimates. See Pearce and Fletcher [1].

References

[1] C.E.M. Pearce and F.K. Fletcher, *Oriented site percolation, phase transitions and probability bounds*, J. Inequal. Pure Appl. Math. **6** Art. 135 (2005), 1–15.

Perić, Ivan: *Frequency variant of Euler type identities and the problem of sign constancy of the kernel in associated quadrature formulas.*

Extended Euler identities generalize the well known formula for the expansion of an function in Bernoulli polynomials. Quadrature formulas are obtained using affine combinations of the extended Euler identities for symmetric nodes. The main step in obtaining the best possible error estimates is to prove that in this manner obtained kernel has some "nice" zeros. Generally, the problem of distribution of nodes such that this kernel has controlled zeros it seems to be difficult. Based on Multiplication Theorem the frequency variants of the extended Euler identities are given. Analogously obtained kernel appears to be more tractable for the investigation of zeros. The case of m-adic frequencies and the case of frequencies with no gaps are completely solved. The general case is considered using some interesting convexity arguments.

Plum, Michael: *A computer-assisted existence proof for photonic band gaps.* (Joint work with Vu Hoang and Christoph Wieners.)

The investigation of monochromatic waves in a periodic dielectric medium ("photonic crystal") leads to a spectral problem for a Maxwell operator. It is well known that the spectrum is characterized as a countable union of compact real intervals ("bands") which may or may not be separated by gaps, and the occurrence of such gaps is of great practical interest but difficult to prove analytically. In this talk, we will attack this problem, for the 2D case of polarized waves, by computer-assisted means. First we reduce the problem, using an analytical perturbation type argument, to the computation of enclosures for finitely many eigenvalues of finitely many periodic eigenvalue problems. This task is then carried out by computer-assisted variational methods.

Popa, Dorian: *Approximate solutions of linear equation.*

In this paper we give some results on the stability of some linear functional equations. A functional equation with the unknown function φ

$$E(\varphi) = F(\varphi) \tag{1}$$

is said to be Hyers–Ulam stable if for an approximate solution φ_a, i.e.,

$$|E(\varphi_a)(x) - F(\varphi_a)(x)| \leq \varepsilon \tag{2}$$

for some fixed constant $\varepsilon \geq 0$, there exists a solution of equation (1) such that

$$|\varphi(x) - \varphi_a(x)| \leq \delta.$$

We investigate the Hyers–Ulam stability of the linear equation of the higher order in single variable and the linear recurrence with nonconstant coefficients.

References

[1] D. Popa, *Hyers-Ulam stability of the linear recurrence with constant coefficients*, Adv. Diff. Equations **2** (2005), 101–107

[2] J. Brzdek, D. Popa, B. Xu, *Note on the stability of the linear recurrence*, Abh. Math. Sem. Univ. Hamburg **76** (2006), 183–189.

Reichel, Wolfgang: *A priori bounds for non-linear finite difference boundary value problems.*
(Joint work with Joe McKenna.)

On a bounded domain $\Omega \subset \mathbb{R}^N$ we consider positive solutions of the non-linear boundary value problem

$$-\Delta u = f(x, u) \text{ in } \Omega, \quad u = 0 \text{ on } \partial\Omega \tag{1}$$

and its finite-difference discretization on an equidistant mesh:

$$-\Delta_h u = f(x, u) \text{ in } \Omega_h, \quad u = 0 \text{ on } \partial\Omega_h. \tag{2}$$

A prototype nonlinearity is given by $f(x, s) = s^p$ for some exponent $p > 1$. Much research has been carried out on the question of a priori bounds for (1), and very little is known about a priori bounds for (2). We will show a method to obtain a priori bounds in the discrete setting based on discrete versions of the Hardy and the Sobolev inequality and the Moser iteration method. If Ω is an N-dimensional cube, then a priori bounds for (2) hold if $1 < p < \frac{N}{N-1}$, which is considerably smaller than the exponent $\frac{N+2}{N-2}$ related to the Sobolev-embedding.

Sadeghi, Ghadir: *Mazur–Ulam theorem in non-Archimedean normed spaces.*
(Joint work with Mohammad Sal Moslehian.)

The theory of isometric mappings was started by paper [2] by S. Mazur and S. Ulam, who proved that every isometry of a normed real vector space onto another normed real vector space is a linear mapping up to translation. The hypothesis surjectivity is essential. Without this assumption, J.A. Baker [1] proved that every isometry from a normed real space into a strictly convex normed real space is linear up to translation. A number of mathematicians have used the Mazur–Ulam theorem; see [3] and references therein.

The Mazur–Ulam Theorem is not valid in the contents of non-Archimedean normed spaces, in general. As a counterexample, take $\mathbb{R}$ with the trivial non-Archimedean valuation and define $f : \mathbb{R} \to \mathbb{R}$ by $f(x) = x^3$. Then f is clearly a surjective isometry and $f(0) = 0$, but f is not linear.

In this talk, we establish a Mazur–Ulam type theorem in the framework of non-Archimedean normed spaces over valuation fields (see [4]).

References

[1] J.A. Baker, *Isometries in normed spaces*, Amer. Math. Monthly **78** (1971), 655–658.

[2] S. Mazur and S. Ulam, *Sur les transformation isometriques d'espaces vectoriels normes*, C. R. Acad. Sci. Paris **194** (1932), 946–948.

[3] Th.M. Rassias and P. Semrl, *On the Mazur–Ulam theorem and the Aleksandrov problem for unit distance preserving mapping*, Proc. Amer. Math. Soc. **114** (1992), 989–993.

[4] A.C.M. van Rooij, *Non-Archimedean functional analysis*, Monographs and Textbooks in Pure and Applied Mathematics vol. 51, Marcel Dekker, New York, 1978.

Sperb, René: *Bounds for the solution in reaction-diffusion problems with variable diffusion coefficient.*

The type of problems describes a steady state of a reaction diffusion process modelled by the equation

$$\operatorname{div}(\sigma(x)\,\nabla u) + f(u) = 0 \qquad \text{in} \quad \Omega \in \mathbb{R}^N$$

with Dirichlet boundary conditions.

Bounds for the solution are derived by using a maximum principle for an appropriately chosen functional of the solution.

Székelyhidi, László: *Spectral synthesis problems on locally compact groups.*

Spectral analysis and spectral synthesis problems are formulated and solved on noncommutative locally compact groups.

Tabor, Jacek: *Characterization of convex functions.*
(Joint work with Józef Tabor.)

As is well known, there are many inequalities which in the class of continuous functions are equivalent to convexity (for example the Jensen inequality, Hermite-Hadamard inequalities and so on). We show that this is not a coincidence, namely, we prove that an arbitrary nontrivial linear inequality which is valid for all convex functions is valid only for convex functions. In other words we obtain the following

Theorem. *Let K be a compact subset of $\mathbb{R}^n$ and let ν, μ, $\nu \neq \mu$ be Borel measures in K. We assume that*

$$\int_K f d\nu \le \int_K f d\mu$$

for every continuous convex function f such that $K \subset \operatorname{dom}(f)$ (where dom *denotes the domain). Let W be a convex subset of a Banach space and let $h \in C(W, \mathbb{R})$ be such that*

$$\int_K (h \circ a) d\nu \le \int_K (h \circ a) d\mu$$

for every affine function a such that $a(K) \subset W$. Then h is convex.

Varga, Adrienn: *On a functional equation containing four weighted arithmetic means.*

In this talk we study the functional equation

$$f(\alpha x + (1-\alpha)y) + f(\beta x + (1-\beta)y) = f(\gamma x + (1-\gamma)y) + f(\delta x + (1-\delta)y)$$

which holds for all $x, y \in I$ where $I \subset \mathbb{R}$ is a non-void open interval, $f : I \to \mathbb{R}$ is an unknown function and $\alpha, \beta, \gamma, \delta \in (0, 1)$ are arbitrarily fixed.

REFERENCES

[1] Z. Daróczy, *Notwendige und hinreichende Bedingungen für die Existenz von nichtkonstanten Lösungen linearer Funktionalgleichungen*, Acta Sci. Math. Szeged **22** (1961), 31–41.

[2] A. Gilányi and Zs. Páles, *On Dinghas-type derivatives and convex functions of higher order*, Real Anal. Exchange **27** (2001/2002), 485–493.

[3] Zs. Páles, *Extension theorems for functional equations with bisymmetric operations*, Aequationes Math. **63** (2002), 266–291.

[4] K. Lajkó, *On a functional equation of Alsina and García-Roig*, Publ. Math. Debrecen **52** (1998), 507–515.

[5] L. Székelyhidi, *On a class of linear functional equations*, Publ. Math. Debrecen **29** (1982), 19–28.

[6] L. Székelyhidi, *Convolution type functional equations on topological Abelian groups*, World Scientific Publishing Co. Inc., Teaneck, NJ, 1991.

[7] Z. Daróczy, K. Lajkó, R.L. Lovas, Gy. Maksa and Zs. Páles, *Functional equations involving means*, Acta Math. Hungar., in press.

Wąsowicz, Szymon: *On error bounds of quadrature operators.*

In [1] using a theorem of support-type we obtained for convex functions of higher order (defined on $[-1, 1]$) some Hadamard-type inequalities of the form

$$\mathcal{L}(f) \leq \mathcal{I}(f) \leq \mathcal{U}(f), \tag{$*$}$$

where $\mathcal{I}(f) = \int_{-1}^{1} f(x)dx$ and $\mathcal{L}$, $\mathcal{U}$ stand for some operators connected with quadrature rules. In this talk we show that the operator in the middle of $(*)$ need not to be an integral and only two its properties are important. Namely, we obtain the inequalities of the form

$$\mathcal{L}(f) \leq \mathcal{T}(f) \leq \mathcal{U}(f),$$

where $\mathcal{T}$ is any operator such that

(i) $\mathcal{T}$ is nondecreasing and

(ii) $\mathcal{T} = \mathcal{I}$ for polynomials of degree not greater than some fixed positive integer.

Observe that many quadrature-type operators fulfil (i) and (ii).

We also show that such inequalities allow us to obtain the error bounds of quadrature operators for less regular functions than in the classical results of numerical analysis.

REFERENCES

[1] S. Wąsowicz, *Support-type properties of convex functions of higher order and Hadamard-type inequalities*, J. Math. Anal. Appl. **332** (2007), 1229–1241.

(Compiled by MIHÁLY BESSENYEI)

International Series of Numerical Mathematics, Vol. 157, xli–xlv

Problems and Remarks

1. Remark

(*Reply to the remarks on my talk on Superquadracity made by Attila Gillányi*) The definition of superquadratic functions says that a function $f : I \to \mathbb{R}$ where I is $[0, \infty)$ or $[0, L]$ is superquadratic provided that for all $x \in I$ there exists a constant $C(x) \in \mathbb{R}$ such that

$$f(y) \geq f(x) + C(x)(y - x) + f\left(|y - x|\right) \tag{1}$$

for all $y \in I$.
From (1) it follows that for $z, w \in I$, $z \leq w$ a superquadratic function satisfies

$$\frac{f(w) + f(z)}{2} \geq f\left(\frac{w + z}{2}\right) + f\left(\frac{w - z}{2}\right)$$

which is equivalent to

$$f(y + x) + f(y - x) \geq 2f(y) + 2f(x)$$

for $0 \leq x \leq y$, $x + y \in I$.
Also, the proofs of [1] Theorem 2.3 and [2] Lemma 2.1 show that $f(x)$ is superquadratic iff for $0 \leq y_1 < x < y_2$, $y_2 \in I$ the inequality

$$f(x) \leq \frac{x - y_1}{y_2 - y_1}\left(f(y_2) - f(y_2 - x)\right) + \frac{y_2 - x}{y_2 - y_1}\left(f(y_1) - f(x - y_1)\right)$$

holds.

References

[1] S. Abramovich, G. Jameson, G. Sinnamon, *Refining Jensen's inequality*, Bull. Math. Soc. Sc. Math. Roumanie **47 (95)** (2004), 3–14.

[2] S. Banić, *Superquadratic functions*, PhD dissertation, Zagreb, 2007.

Shoshana Abramovich

2. Problem

(*The discrete Sobolev inequality*) Let $n \geq 3$, $h = (h_1, \dots, h_n)$ with $h_i > 0$ for $i = 1, \dots, n$. Consider the discretization $\mathbb{R}^n_h$ of $\mathbb{R}^n$

$$\mathbb{R}^n_h := (h_1\mathbb{Z}) \times \cdots \times (h_n\mathbb{Z}) .$$

The following is a discrete analogue of the Sobolev inequality: there exists a constant $c_s > 0$ such that

$$\sum_{x \in \mathbb{R}^n_h} |u(x)|^{\frac{2n}{n-2}} h_1 \cdot \dots \cdot h_n \leq c_s \left(\sum_{\mathbb{R}^n_h} |\nabla^+_h u(x)|^2 h_1 \cdot \dots \cdot h_n \right)^{\frac{n}{n-2}}$$

for all $u : \mathbb{R}^n_h \to \mathbb{R}$ with compact support. Here

$$\nabla^+_h u(x) := \begin{pmatrix} \frac{u(x+h_1e_1)-u(x)}{h_1} \\ \vdots \\ \frac{u(x+h_ne_n)-u(x)}{h_n} \end{pmatrix}$$

is the discrete gradient and $\{e_1, \dots, e_n\}$ is the standard basis of $\mathbb{R}^n$. The following questions arise:

1. What is the best constant c_s? We know that $c_s \leq \frac{4(n-1)}{n^{3/2}}$.
2. Is it attained in the space

$$\mathcal{D} := \overline{\{u : \mathbb{R}^n_h \to \mathbb{R} \text{ with compact support}\}}^{|||\cdot|||}$$

where

$$|||u||| := \left(\sum_{\mathbb{R}^n_h} |\nabla^+_h u(x)|^2 h_1 \cdot \dots \cdot h_n \right)^{1/2} \quad ?$$

Already the case $h_1 = \dots = h_n > 0$ is interesting.

WOLFGANG REICHEL

3. Problem

It is known that the following statement holds if X is a Banach space: For each $\varepsilon > 0$ and each mapping $f : \mathbb{Z} \to X$ satisfying $f(0) = 0$ and

$$\|f(x + y) + f(x - y) - 2f(x) - 2f(y)\| \leq \varepsilon \quad (x, y \in \mathbb{Z}),$$

there exists a (unique) quadratic mapping $Q : \mathbb{Z} \to X$ such that $\|f(x) - Q(x)\| \leq \varepsilon/2$ for all $x \in \mathbb{Z}$.

PROBLEM. Prove that the converse is true, i.e., every normed space X fulfilling the above assertion is Banach.

MOHAMMAD SAL MOSLEHIAN

4. Problem

Given an operator

$$\mathcal{T}(f) := \frac{7}{15} f(-1) + \frac{64}{45} f\left(\frac{1}{4}\right) + \frac{1}{9} f(1).$$

It is easy to compute that

$$\mathcal{T}(f) = \int_{-1}^{1} f(x)dx$$

for $f(x) = ax^2 + bx + c$. Classical error terms of quadrature operators are of the form

$$\int_{-1}^{1} f(x)dx = Q(f) + \alpha_n f^{(n)}(\xi_f)$$

for some $n \in \mathbb{N}$, $\alpha_n \in \mathbb{R}$, and $\xi_f \in]-1,1[$ (n and α_n are independent of f).

PROBLEM. Find such an error term for the operator $\mathcal{T}$. (The Peano Kernel Theorem is not applicable because Peano Kernel of $\mathcal{T}$ changes sign.)

SZYMON WĄSOWICZ

5. Remark

(*Remark on subquadratic functions, related to Shoshana Abramovich's talk and remark*) At the present meeting, in Shoshana Abramovich's talk, superquadratic functions were investigated, calling a real-valued function f defined on an interval $I = [0, \infty)$ or $I = [0, a]$ with a positive a superquadratic, if, for each $x \in I$, there exists a $C(x) \in \mathbb{R}$ such that

$$f(x) - f(y) \geq f(|y - x|) + C(x)(y - x) \tag{2}$$

for all $x \in I$ (cf. also [1]).
Analogously to the concept of subadditive (and superadditive) functions (cf., e.g., [5] and [4]), we may consider another definition of subquadraticity (and superquadraticity). It is well known, that a real-valued function defined on a group $G = (G, +)$ is called quadratic if it satisfies the functional equation

$$f(x + y) + f(x - y) = 2f(x) + 2f(y) \qquad (x, y \in G).$$

The function f is said to be *subquadratic* if it fulfills

$$f(x + y) + f(x - y) \leq 2f(x) + 2f(y) \qquad (x, y \in G), \tag{3}$$

it is called *superquadratic* if the inequality

$$f(x + y) + f(x - y) \geq 2f(x) + 2f(y) \qquad (x, y \in G) \tag{4}$$

is valid (cf. [3], [6] and [2]). Obviously, a function $f : G \to \mathbb{R}$ is superquadratic if and only if $-f$ is subquadratic, therefore, it is enough to consider one of these concepts.

In Shoshana Abramovich's remark, a connection between inequalities (2) and (4) was established. In the following, we give some examples for subquadratic functions in the sense of the second definition and we present a regularity theorem for them.

Examples.

1. If $B : G \times G \to \mathbb{R}$ is a biadditive symmetric function and b is a nonnegative real number then the function $f : G \to \mathbb{R}$
$$f(x) = B(x,x) + b \qquad (x \in G)$$
satisfies (3).
As a special case of the example above, we obtain that if $a : G \to \mathbb{R}$ is an additive function, c is an arbitrary and b is a nonnegative real constant then the function $f : G \to \mathbb{R}$
$$f(x) = c\,(a(x))^2 + b \qquad (x \in G)$$
solves (3), too.
In the class of continuous real functions, this example gives the subquadratic functions $f : \mathbb{R} \to \mathbb{R}$
$$f(x) = cx^2 + b$$
where c is an arbitrary, b is a nonnegative real constant.
2. A simple calculation yields that $f : G \to \mathbb{R}$,
$$f(x) = c\,|a(x)| + b \qquad (x \in G),$$
where c is an arbitrary, b is a nonnegative real constant, is also a subquadratic function.
3. The function $f : G \to \mathbb{R}$
$$f(x) = \begin{cases} b & \text{if } x \neq 0 \\ d & \text{if } x = 0, \end{cases}$$
where b and d are nonnegative constants such that $d \leq 3b$, is subquadratic.
4. An arbitrary function $f : G \to \mathbb{R}$ satisfying the inequality
$$\sup_{x\in G} f(x) \leq 2 \inf_{x \in G} f(x)$$
is subquadratic.
5. The function $f : \mathbb{R}^n \to \mathbb{R}$
$$f(x) = \begin{cases} 0 & \text{if } x \in \mathbb{Q}^n \\ b & \text{otherwise}, \end{cases}$$
with an arbitrary nonnegative constant b, is subquadratic.

Theorem. *If a real-valued subquadratic function defined on a metric group divisible by 2 is continuous at 0 and its value is 0 there then it is continuous everywhere.*

It is remarkable that if one of the assumptions for the function considered in the theorem above is omitted, it will fail to hold.

- In fact, according to Example 4, there exist (also real) subquadratic functions, which are continuous at 0 but not continuous everywhere.
- Obviously, $f(0) = 0$ does not imply any continuity properties.
- Furthermore, the assumption of the continuity of the function above at a point other than 0 does not imply its continuity everywhere (cf., e.g., Example 3).

References

[1] S. Abramovich, G. Jameson, G. Sinnamon, *Refining Jensen's inequality*, Bull. Math. Soc. Sc. Math. Roumanie **47 (95)** (2004), 3–14.

[2] A. Gilányi, *On subquadratic functions*, manuscript.

[3] Z. Kominek, K. Troczka, *Some remarks on subquadratic functions*, Demonstratio Math. **39** (2006), 751–758.

[4] M. Kuczma, *An Introduction to the Theory of Functional Equations and Inequalities*, Państwowe Wydawnictwo Naukowe – Uniwersytet Śląski, Warszawa – Kraków – Katowice, 1985.

[5] R.A. Rosenbaum, *Subadditive functions*, Duke Math. J. **17** (1950), 227–242.

[6] K. Troczka-Pawelec, *Regular properties of subquadratic functions*, manuscript.

Attila Gilányi

(Compiled by Mihály Bessenyei)

International Series of Numerical Mathematics, Vol. 157, xlvii–xlviii

List of Participants

Abramovich, Shoshana, University of Haifa, Mount Carmel, Haifa, Israel
Aczél, János, University of Waterloo, Waterloo, Ontario, Canada
Adamek, Mirosław, University of Bielsko-Biała, Poland
Baják, Szabolcs, University of Debrecen, Debrecen, Hungary
Bandle, Catherine, Römerstr. 5, Aesch, Switzerland
Barza, Sorina, University of Karlstad, Karlstad, Sweden
Behnke, Henning, Technische Univ. Clausthal, Clausthal-Zellerfeld, Germany
Bessenyei, Mihály, University of Debrecen, Debrecen, Hungary
Boros, Zoltán, University of Debrecen, Debrecen, Hungary
Brown, Malcolm B., University College Cardiff, Cardiff, United Kingdom
Brown, Richard C., Univ. of Alabama, Tuscaloosa, Alabama, United States
Bullen, Peter S., University of British Columbia, Vancouver, BC, Canada
Burai, Pál, University of Debrecen, Debrecen, Hungary
Buşe, Constantin, West University of Timişoara, Timişoara, Romania
Cerone, Pietro, Victoria University, Victoria, Melbourne, Australia
Chudziak, Jacek, University of Rzeszów, Rzeszów, Poland
Daróczy, Zoltán, University of Debrecen, Debrecen, Hungary
Dragomir, S. Sever, Victoria University of Technology, Melbourne, Australia
Fazekas, Borbála, Universität Karlsruhe, Karlsruhe, Germany
Fechner, Włodzimierz, Silesian University, Katowice, Poland
Gavrea, Bogdan, Technical University of Cluj-Napoca, Cluj-Napoca, Romania
Gavrea, Ioan, Technical University of Cluj-Napoca, Cluj-Napoca, Romania
Ger, Roman, Silesian University, Katowice, Poland
Gilányi, Attila, University of Debrecen, Debrecen, Hungary
Goldberg, Moshe, Technion-Israel Institute of Technology, Haifa, Israel
Házy, Attila, University of Miskolc, Miskolc, Hungary
Hoang, Vu, Universität Karlsruhe, Karlsruhe, Germany
Johansson, Maria, Luleå University of Technology, Luleå, Sweden
Kittaneh, Fuad, University of Jordan, Amman, Jordan
Klaričić Bakula, Milica, University of Split, Split, Croatia

KOBAYASHI, KENTA, Kanazawa University, Kanazawa, Japan
LEMMERT, ROLAND, Universität Karlsruhe, Karlsruhe, Germany
LOSONCZI, LÁSZLÓ, University of Debrecen, Debrecen, Hungary
MAKÓ, ZITA, University of Debrecen, Debrecen, Hungary
MAKSA, GYULA, University of Debrecen, Debrecen, Hungary
MATIĆ, MARKO, University of Split, Split, Croatia
MATKOWSKI, JANUSZ, Silesian University, Katowice, Poland
MÉSZÁROS, FRUZSINA, University of Debrecen, Debrecen, Hungary
MITREA, ALEXANDRU, Technical Univ. of Cluj-Napoca, Cluj-Napoca, Romania
MITREA, PAULINA, Technical University of Cluj-Napoca, Cluj-Napoca, Romania
MOHAPATRA, RAM, University of Central Florida, Orlando, Florida, United States
MOSLEHIAN, SAL MOHAMMAD, Ferdowsi University, Mashhad, Iran
MROWIEC, JACEK, University of Bielsko-Biała, Bielsko-Biała, Poland
NAGATOU, KAORI, Kyushu University, Fukuoka, Japan
NAKAO, MITSUHIRO T., Kyushu University, Fukuoka, Japan
NICULESCU, CONSTANTIN, University of Craiova, Craiova, Romania
NIKODEM, KAZIMIERZ, University of Bielsko-Biała, Bielsko-Biała, Poland
OGITA, TAKESHI, CREST, Japan Science and Technology Agency, Tokyo, Japan
OLKIN, INGRAM, Stanford University, Stanford, California, United States
OPIC, BOHUMÍR, Czech Academy of Sciences, Praha, Czech Republic
PÁLES, ZSOLT, University of Debrecen, Debrecen, Hungary
PEARCE, CHARLES EDWARD MILLER, University of Adelaide, Adelaide, Australia
PEČARIĆ, JOSIP, University of Zagreb, Zagreb, Croatia
PERIĆ, IVAN, University of Zagreb, Zagreb, Croatia
PLUM, MICHAEL, Universität Karlsruhe, Karlsruhe, Germany
POPA, DORIAN, Technical University of Cluj-Napoca, Cluj-Napoca, Romania
REICHEL, WOLFGANG, Universität Karlsruhe, Karlsruhe, Germany
SADEGHI, GHADIR, Ferdowsi University, Mashhad, Iran
SPERB, RENÉ P., Eidgen Technische Hochschule Zentrum, Zürich, Switzerland
SZÉKELYHIDI, LÁSZLÓ, University of Debrecen, Debrecen, Hungary
TABOR, JACEK, Jagiellonian University, Kraków, Poland
TABOR, JÓZEF, University of Rzeszów, Rzeszów, Poland
VARGA, ADRIENN, University of Debrecen, Debrecen, Hungary
VINCZE, CSABA, University of Debrecen, Debrecen, Hungary
VOLKMANN, PETER, Universität Karlsruhe, Karlsruhe, Germany
WĄSOWICZ, SZYMON, University of Bielsko-Biała, Bielsko-Biała, Poland

Part I

Inequalities Related to Ordinary and Partial Differential Equations

PHB
'07

International Series of Numerical Mathematics, Vol. 157, 3–12

A Rayleigh-Faber-Krahn Inequality and Some Monotonicity Properties for Eigenvalue Problems with Mixed Boundary Conditions

Catherine Bandle

Abstract. An eigenvalue problem is considered whose eigenvalues appear in the interior and on the boundary. It has been shown in [1] that there exists an infinite sequence of positive and an infinite sequence of negative eigenvalues. The lowest positive and the largest negative eigenvalue λ_1, resp. λ_{-1} can be characterised by means of a Rayleigh principle. It turns out that among all domains of given volume the ball has the smallest λ_1. A partial result in this direction is established for λ_{-1}. The proof uses the isoperimetric inequality of Krahn-Bossel-Daners. Some monotonicity properties similar to those for the elastically supported membrane are included.

Mathematics Subject Classification (2000). 35P15, 47A75, 49R50, 51M16.

Keywords. Comparison theorems for eigenvalues, isoperimetric inequalities, spectrum for elliptic operators with mixed boundary conditions.

1. Introduction

Let $D \subset \mathbb{R}^N$, $N > 1$ be a bounded domain with a Lipschitz boundary and denote by n its outer normal. In this paper we consider the eigenvalue problem

$$\Delta\varphi + \lambda\varphi = 0 \quad \text{in} \quad D, \; \frac{\partial\varphi}{\partial n} = \lambda\sigma\varphi \quad \text{on} \quad \partial D, \tag{1.1}$$

where σ is a *negative* real number. It is obvious that $\lambda_0 = 0$ is an eigenvalue and that φ =const. is the corresponding eigenfunction. It has been shown in [1], see also [2] for the more general case with variable coefficients, that there exists an infinite sequence $\{\lambda_n\}_{n=1}^\infty$ of positive eigenvalues such that $\lambda_n \to \infty$ as $n \to \infty$. In addition there is an infinite sequence $\{\lambda_{-n}\}_{n=1}^\infty$ of negative eigenvalues with the property that $\lambda_{-n} \to -\infty$ as $n \to \infty$. (Here the assumption $N > 1$ is used.) The eigenvalues can be ordered as follows, taking into account their multiplicity,

$$\begin{aligned}\cdots \leq \lambda_{-n-1} \leq \lambda_{-n} \leq \cdots \leq \lambda_{-2} \leq \lambda_{-1} < 0 \\ = \lambda_0 < \lambda_1 \leq \lambda_2 \leq \cdots \leq \lambda_n \leq \lambda_{n+1} \leq \cdots\end{aligned}$$

We are interested in the behaviour of $\lambda_{\pm 1}$. They can be characterised by the following variational principle.

For $u, v \in W^{1,2}(D)$ set

$$a(u,v) := \int_D uv\,dx + \sigma \oint_{\partial D} uv\,ds,$$
$$\langle u, v\rangle := \int_D (\nabla u, \nabla v)\,dx.$$

Define

$$\sigma_0(D) := -\frac{|D|}{|\partial D|}.$$

Then for $\sigma \neq \sigma_0$

$$\begin{aligned}&\frac{1}{\lambda_1(D)} = \sup_{\mathcal{K}} a(v,v), \quad \frac{1}{\lambda_{-1}(D)} = \inf_{\mathcal{K}} a(v,v), \\ &\mathcal{K} := \{v \in W^{1,2}(D), \langle v, v\rangle = 1, a(v,1) = 0\}.\end{aligned} \tag{1.2}$$

It has been observed in [1] that the fact that the eigenfunctions φ_1 and φ_{-1} corresponding to λ_1 and to λ_{-1} are of constant sign depends on the size of σ. More precisely we have:

(i) *If $\sigma < \sigma_0$ then φ_1 is of constant sign and λ_1 is simple, whereas φ_{-1} changes sign.*

(ii) *If $\sigma > \sigma_0$ then φ_{-1} is of constant sign and λ_{-1} is simple, whereas φ_1 changes sign.*

(iii) *If $\sigma = \sigma_0$ both φ_1 and φ_{-1} change sign.*

The main result of this paper is a Rayleigh-Faber-Krahn type inequality.

Theorem 1. *Let D^* be the ball of the same volume as D.*

(i) *For each $\sigma < \sigma_0(D^*)$ we have $\lambda_1(D) \geq \lambda_1(D^*)$. Equality holds only for the ball.*

(ii) *For any domain D there exists a number $\hat{\sigma} \in (\sigma_0(D), 0)$ such that $\lambda_{-1}(D) \geq \lambda_{-1}(D^*)$ whenever $\sigma \in (\hat{\sigma}, \sigma_0(D))$.*

In general $D_1 \subset D_2$ does not imply $\lambda_{\pm 1}(D_1) > \lambda_{\pm 1}(D_2)$. However for special domains it is true. In particular we have

Theorem 2. *Let B be a ball containing D.*

(i) *If $\sigma < \sigma_0(B)$ then $\lambda_1(D) \geq \lambda_1(B)$.*

(ii) *If $\sigma > \sigma_0(D)$ then $\lambda_{-1}(D) \geq \lambda_{-1}(B)$.*

2. Isoperimetric inequality

The first part of the proof of Theorem 1 relies on the isoperimetric inequality of Krahn-Bossel-Daners concerning the elastically supported membrane

$$\Delta\phi + \lambda^+\phi = 0 \quad \text{in} \quad D, \; \frac{\partial\phi}{\partial n} + \alpha\phi = 0 \quad \text{on} \quad \partial D, \; \alpha \in \mathbb{R}^+. \tag{2.1}$$

It is well known that there exists infinitely many positive eigenvalues

$$0 < \lambda_1^+ < \lambda_2^+ \leq \cdots \leq \lambda_n^+ \leq \cdots .$$

The lowest eigenvalue can be obtained from the variational principle

$$\lambda_1^+(D) = \inf_{W^{1,2}(D)} \frac{\int_D |\nabla v|^2 \, dx + \alpha \oint_{\partial D} v^2 \, ds}{\int_D v^2 \, dx}. \tag{2.2}$$

It was conjectured by Krahn, proved by M.-H. Bossel [6] for $N = 2$ and completed by D. Daners [7] for arbitrary N, that $\lambda_1^+(D)$ satisfies the isoperimetric inequality

$$\lambda_1^+(D) \geq \lambda^+(D^*). \tag{2.3}$$

Equality holds only for the ball [8].

For the proof of Theorem 1 we need the following auxiliary lemmas.

Lemma 3. *The lowest positive and the largest negative eigenvalue satisfy the following variational principles.*

(i) *If* $\sigma < \sigma_0(D) < 0$ *then*

$$\frac{1}{\lambda_1(D)} = \sup_{\mathcal{K}_0} a(v,v), \; \mathcal{K}_0 := \{v \in W^{1,2}(D), \langle v, v \rangle = 1\}.$$

(ii) *If* $0 > \sigma > \sigma_0(D)$ *then*

$$\frac{1}{\lambda_{-1}(D)} = \inf_{\mathcal{K}_0} a(v,v).$$

Proof. Let $v \in \mathcal{K}_0$ be fixed and let v_0 be any real number. Clearly $v + v_0$ is also an element of $\mathcal{K}_0$ and

$$a(v + v_0, v + v_0) = a(v,v) + a(1,1)v_0^2 + 2a(v,1)v_0.$$

Since by assumption $a(1,1) < 0$ the function $f(v_0) = a(1,1)v_0^2 + 2a(v,1)v_0$ takes its maximum for $v_0^* = -\frac{a(v,1)}{a(1,1)}$. Hence

$$a(v,v) \leq a(v + v_0^*, v + v_0^*), \quad \text{where} \quad a(1, v + v_0^*) = 0.$$

The assertion (i) now follows from (1.2). The second assertion is proved exactly in the same way. We only have to take into consideration that in this case $a(1,1)$ is positive. □

Lemma 4. *Let $\tilde{D}$ be an arbitrary fixed domain. If $\lambda_1^+(\tilde{D}) \leq \lambda_1^+(D)$ for all positive α, then*

$$\lambda_1(\tilde{D}) \leq \lambda_1(D) \quad \textit{for all} \quad \sigma < \sigma_0(\tilde{D}).$$

Proof. Consider $\lambda_1^+(D)$ with $\alpha = -\lambda_1(D)\sigma > 0$. By assumption

$$\lambda_1^+(D) \geq \lambda_1^+(\tilde{D}) = \frac{\int_{\tilde{D}} |\nabla\tilde{\phi}|^2\, dx - \lambda_1(D)\sigma \oint_{\partial\tilde{D}} \tilde{\phi}^2\, ds}{\int_{\tilde{D}} \tilde{\phi}^2\, dx}, \tag{2.4}$$

where $\tilde{\phi}$ is the eigenfunction corresponding to $\lambda_1^+(\tilde{D})$. Moreover we have by (2.2)

$$\lambda_1(D) = \frac{\int_D |\nabla\varphi_1|^2\, dx - \lambda_1(D)\sigma \oint_{\partial D} \varphi_1^2\, ds}{\int_D \varphi_1^2\, dx} \geq \lambda_1^+(D). \tag{2.5}$$

Here φ_1 is the eigenfunction corresponding to $\lambda_1(D)$. From (2.4) and (2.5) we find

$$\lambda_1(D) \int_{\tilde{D}} \tilde{\phi}^2\, dx \geq \int_{\tilde{D}} |\nabla\tilde{\phi}|^2\, dx - \lambda_1(D)\sigma \oint_{\partial\tilde{D}} \tilde{\phi}^2\, ds,$$

and

$$\frac{\int_{\tilde{D}} \tilde{\phi}^2\, dx + \sigma \oint_{\partial\tilde{D}} \tilde{\phi}^2\, ds}{\int_{\tilde{D}} |\nabla\tilde{\phi}|^2\, dx} \geq \frac{1}{\lambda_1(D)}.$$

By Lemma 3 the left-hand side is bounded from above by $\frac{1}{\lambda_1(\tilde{D})}$. This completes the proof of the lemma. □

As an immediate consequence we have the

Proof of Theorem 1(i). The first assertion is an immediate consequence of the previous lemma and the Krahn-Bossel-Daners inequality (2.3) if we take $\tilde{D} = D^*$ and observe that in view of the classical isoperimetric inequality $\sigma_0(D^*) < \sigma_0(D)$. □

Proof of Theorem 1(ii). In order to prove the second part of Theorem 1 we recall a result derived in [1], namely for $\sigma < 0$ there is a decreasing C^1-curve $\lambda(\sigma)$ such that

$$\lambda(\sigma) = \begin{cases} \lambda_1(\sigma) & \text{if } \sigma < \sigma_0 \\ 0 & \text{if } \sigma = \sigma_0 \\ \lambda_{-1}(\sigma) & \text{if } \sigma_0 < \sigma < 0 \end{cases}$$

Let ν_1 denote the lowest eigenvalue of $\triangle\phi + \nu\phi = 0$ in D, $\phi = 0$ on ∂D. Then

$$\lambda_1(\sigma) \to \nu_1 \quad \text{as} \quad \sigma \to -\infty, \quad \text{and} \quad \lambda_{-1}(\sigma) \to -\infty \quad \text{as} \quad \sigma \to 0.$$

Let $\lambda(\sigma : D)$ and $\lambda(\sigma : D^*)$ be the eigenvalue curves described above corresponding to the domain D and the ball D^* [cf. Fig. 1]. From $\sigma_0(D^*) < \sigma_0(D)$ it follows that $\lambda(\sigma_0(D) : D) = 0$ and $\lambda(\sigma_0(D) : D^*) = \lambda_{-1}(D^*) < 0$. Since both curves are continuous the inequality $\lambda_{-1}(\sigma : D) \geq \lambda_{-1}(\sigma : D^*)$ remains valid in a neighbourhood to the right of $\sigma_0(D)$.

This establishes the second assertion of Theorem 1. □

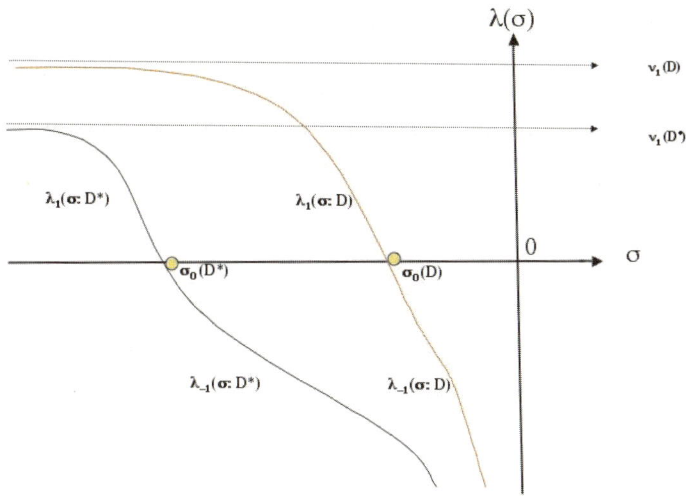

FIGURE 1. Illustration of the curves $\lambda(\sigma)$ corresponding to D and D^*

3. Domain monotonicity

We start with an auxiliary result concerning the eigenvalue problem

$$\triangle u + \lambda u = 0 \quad \text{in} \quad D, \; \frac{\partial u}{\partial n} + \gamma u = 0 \quad \text{on} \quad \partial D. \tag{3.1}$$

Here γ and λ can be positive or negative. We assume that u does not change sign and that $u > 0$ in D. In the ball of radius R the function u is radial (because it is of constant sign) and satisfies the ordinary differential equation

$$w'' + \frac{N-1}{r} w' + \lambda w = 0 \quad \text{in} \quad (0, R), \; u > 0. \tag{3.2}$$

The solutions of (3.2) are of the form

$$w(r) = r^{-\frac{N-2}{2}} \begin{cases} J_{\frac{N-2}{2}}(\sqrt{\lambda} r) & \text{if } \lambda > 0, \\ I_{\frac{N-2}{2}}(\sqrt{-\lambda} r) & \text{if } \lambda < 0. \end{cases} \tag{3.3}$$

Here J_ν and I_ν denote the Bessel, resp. the modified Bessel functions.

Lemma 5. $|\frac{w'(r)}{w(r)}|$ *takes its maximum at* $r = R$. *In view of* (3.1) *we have*

$$\left| \frac{w'(r)}{w(r)} \right| \leq |\gamma|.$$

Proof. From (3.2) we deduce that $v = \frac{w'(r)}{w(r)}$ satisfies

$$v' + v^2 + \frac{N-1}{r} v + \lambda = 0 \quad \text{in} \quad (0, R). \tag{3.4}$$

Let us distinguish between two cases.

(i) $\gamma > 0$. In this case we have $\lambda > 0$. Since $v(0) = 0$ we deduce that $v'(0) = -\lambda/N < 0$. Hence v decreases in a neighbourhood of the origin. Suppose that it attains a local minimum at r_0. Multiplication of (3.4) with r, differentiation with respect to r and evaluation at r_0 implies

$$r_0 v''(r_0) + v^2(r_0) + \lambda = 0.$$

Hence $v''(r_0) < 0$ which is a contradiction. This establishes the lemma in the first case.

(ii) $\gamma < 0$. Then $\lambda < 0$, $v(0) = 0$ and $v'(0) = -\lambda/N > 0$. Suppose that v attains a local maximum at r_1. Differentiation of (3.4) and evaluation at r_1 yields

$$v''(r_1) = \frac{N-1}{r_1^2} v(r_1).$$

This is obviously a contradiction. □

The following lemma is well known (cf. [9] for λ_1^+ and [4] for λ_1^-).

Lemma 6. *Let B_R be a ball of radius R and $D \subset B_R$.*

(i) *If $\gamma > 0$ then*

$$\lambda(D) \geq \lambda(B_R).$$

(ii) *If $\gamma < 0$ then*

$$\lambda(D) \leq \lambda(B_R).$$

Proof. Let u and w be the positive solutions of (3.1) in D, resp. B_R. Then

$$-\oint_{\partial D} \left(\gamma + w^{-1}\frac{\partial w}{\partial n}\right) uw \, ds = \int_D (w\triangle u - u\triangle w)\, dx = (-\lambda(D) + \lambda(B_R)) \int_D uw \, dx. \tag{3.5}$$

From Lemma 5 it follows that $|w^{-1}\frac{\partial w}{\partial n}| \leq |\gamma|$. Therefore the left-hand side of (3.5) is negative (positive) if γ is positive (negative). The proof is now immediate. □

Proof of Theorem 2(i). The comparison of $\lambda_1(D)$ with $\lambda_1(B_R)$ follows immediately from Lemma 6 and Lemma 4. □

The proof for the comparison of λ_{-1} is very similar. We consider the eigenvalue problem

$$\triangle\phi + \lambda^-\phi = 0 \quad \text{in} \quad D, \; \frac{\partial \phi}{\partial n} = \beta\phi \quad \text{on} \quad \partial D, \; \beta > 0. \tag{3.6}$$

From the classical theory of compact operators and the trace inequality it follows that there exist infinitely many eigenvalues

$$\lambda_1^- < \lambda_2^- \leq \cdots, \quad \lambda_n^- \to \infty \quad \text{as} \quad n \to \infty, \; \lambda_1^- < 0.$$

In this case Lemma 4 becomes

Lemma 7. *Let $\tilde{D}$ be an arbitrary fixed domain. If $\lambda_1^-(\tilde{D}) \geq \lambda_1^-(D)$ for all positive β, then*

$$\lambda_{-1}(\tilde{D}) \leq \lambda_{-1}(D) \quad \textit{for all} \quad \sigma > \sigma_0(D).$$

Proof. Consider $\lambda_1^-(D)$ with $\beta = \lambda_{-1}(\tilde{D})\sigma$. If $\tilde{\varphi}_1$ is the eigenfunction corresponding to $\lambda_{-1}(\tilde{D})$ then

$$\lambda_{-1}(\tilde{D}) = \frac{\int_{\tilde{D}} |\nabla \tilde{\varphi}_1|^2\, dx - \lambda_{-1}(\tilde{D})\sigma \oint_{\partial \tilde{D}} \tilde{\varphi}_1^2\, ds}{\int_{\tilde{D}} \tilde{\varphi}_1 2\, dx} \geq \lambda_1^-(\tilde{D}) \geq \lambda_1^-(D).$$

Let ϕ denote the eigenfunction corresponding to $\lambda_1^-(D)$. Then

$$\begin{aligned} \int_D |\nabla \phi|^2\, dx &= \lambda_1^-(D) \int_D \phi^2\, dx + \lambda_{-1}(\tilde{D})\sigma \oint_{\partial D} \phi^2\, ds \\ &\leq \lambda_{-1}(\tilde{D}) \left[\int_D \phi^2\, dx + \sigma \oint_{\partial D} \phi^2\, ds \right] \end{aligned}$$

This inequality together with Lemma 3 implies

$$\frac{1}{\lambda_{-1}(\tilde{D})} \geq \frac{\int_D \phi^2\, dx + \sigma \oint_{\partial D} \phi^2\, ds}{\int_D |\nabla \phi|^2\, dx} \geq \frac{1}{\lambda_{-1}(D)}.$$

This completes the proof. □

Proof of Theorem 2(ii). Theorem 2(ii) is a consequence of the Lemmas 6 and 7.

4. Complements and open problems

4.1. Let D be a fixed and consider λ_1 and λ_{-1} as functions of σ.

Lemma 8. *If $\sigma < \sigma_0$ the function $1/\lambda_1(\sigma)$ is convex whereas for $\sigma > \sigma_0$ the function $1/\lambda_{-1}(\sigma)$ is concave.*

Proof. From Lemma 3 we get for $\sigma_1, \sigma_2 < \sigma_0$

$$\begin{aligned} \lambda_1^{-1}\left(\frac{\sigma_1 + \sigma_2}{2}\right) &= \frac{\int_D u^2\, dx + \frac{\sigma_1 + \sigma_2}{2} \oint_{\partial D} u^2\, ds}{\langle u, u \rangle} \\ &= \frac{1}{2} \frac{\int_D u^2\, dx + \sigma_1 \oint_{\partial D} u^2\, ds}{\langle u, u \rangle} + \frac{1}{2} \frac{\int_D u^2\, dx + \sigma_2 \oint_{\partial D} u^2\, ds}{\langle u, u \rangle} \\ &\leq \frac{1}{2} [\lambda_1^{-1}(\sigma_1) + \lambda_1^{-1}(\sigma_2)], \end{aligned}$$

which proves the first statement. In the same way we prove that $1/\lambda_{-1}(\sigma)$ is concave. □

Notice that the behaviour of $\lambda_{\pm 1}(\sigma)$ differs from the behaviour of $\lambda_1^+(\alpha)$ and $\lambda_1^-(\beta)$. In fact $\lambda_1^+(\alpha)$ is concave and $\lambda_1^-(\beta)$ is convex.

4.2. The question arises whether the inequality $\lambda_{-1}(D) \geq \lambda_{-1}(D^*)$ is valid for all $\sigma > \sigma_0$. For simple examples such as the square

$$S := \{-a < x_1 < a, -a < x_2 < a\}$$

this inequality can be checked directly. In this case the eigenvalues of (1.1) can be expressed in terms of algebraic equations. In fact we have

$$\varphi_{-1} = \cosh(\sqrt{-\lambda/2}x_1)\cosh(\sqrt{-\lambda/2}x_2).$$

The boundary condition yields

$$-\tanh(\sqrt{-\lambda/2}a) = \sqrt{-2\lambda}\sigma.$$

Then $\lambda_{-1}(S)$ is the largest nontrivial negative root of the above equation. For the unit circle $B_1 \subset \mathbb{R}^2$ we have (cf. (3.3))

$$\varphi_{-1}(r) = I_0(\sqrt{-\lambda_{-1}}r)$$

and λ_{-1} is the largest negative root of

$$\frac{I_0'(\sqrt{-\lambda})}{I_0(\sqrt{-\lambda})} = -\sqrt{-\lambda}\sigma.$$

Let us compare $\lambda_{-1}(B_1)$ with $\lambda_{-1}(S)$ where $|B_1| = |S|$, i.e., $a = \sqrt{\pi}/2$. Put for short

$$x := \sqrt{-\lambda},\ F(x) := \frac{I_1(x)}{I_0(x)},\quad f(x) := 2^{-1/2}\tanh\left[\frac{\sqrt{\pi}x}{2^{3/2}}\right] \quad \text{and} \quad g(x) = -\sigma x.$$

We have, cf. Figure 2,

$$f(x) \leq F(x).$$

Hence for all negative $\sigma > \sigma_0$ the inequality $\lambda_{-1}(B_1) < \lambda_{-1}(S)$ holds. This observation together with Theorem 2 leads to the conjecture that λ_{-1} also satisfies a Rayleigh-Faber-Krahn inequality if $\sigma_0 < \sigma < 0$.

The intersection of the straight line $g(x)$ with the upper curve $F(x)$ corresponds to $\sqrt{-\lambda_{-1}(B_1)}$ and the one with the lower curve to $\sqrt{\lambda_{-1}(S)}$.

Problem 1. Prove or disprove Theorem 1(ii) for all $\sigma \in (\sigma_0, 0)$.

M. Bareket derived in [5] a Rayleigh-Faber-Krahn inequality for λ_1^- for nearly circular domains and for small β. She proved that $\lambda_1^-(D) \leq \lambda_1^-(D^*)$. Notice that this is not in contradiction with Theorem 2(ii).

4.3. It is in general not possible to extend Theorem 2(i) to the case $\sigma > \sigma_0$. Indeed if σ vanishes, λ_1, as defined in (1.2), coincides with the first nontrivial eigenvalue ν of the free membrane. According to the classical result of Szegö [10] and Weinberger [11], the opposite inequality $\lambda_1(D) \leq \lambda_1(D^*)$ holds. By continuity this is true in a neighbourhood of $\sigma = 0$. In the case of a square we have $\lambda_1(S) > \lambda_1(S^*)$ for $\sigma_0 < \sigma < \sigma* < 0$.

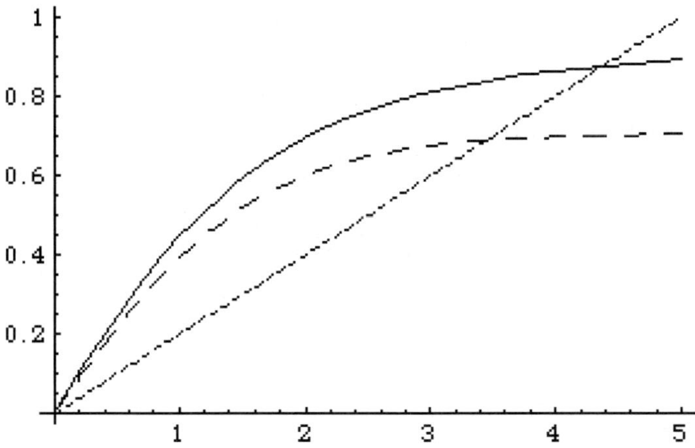

FIGURE 2. upper curve - - - $F(x)$; lower curve $\cdots f(x)$

Problem 2. Prove that there exists a number $\sigma_0 < \sigma^* < 0$ such that

$$\lambda_1(D) \begin{cases} \leq \lambda_1(D^*), & \text{if } \sigma > \sigma^* \\ \geq \lambda_1(D^*), & \text{if } \sigma < \sigma^*. \end{cases}$$

4.4. The monotonicity result in Theorem 2 holds also if B is replaced by a N-cell $C := \{a_i < x_i < b_i, i = 1, \ldots, N\}$. The proof is similar and uses the fact that $\varphi_{\pm 1}$ are of the form $\prod_{i=1}^N h_i(x_i)$.

Problem 3. Describe all domains $\mathcal{B}$ for which Theorem 2 holds.

References

[1] C. Bandle, J. v. Below and W. Reichel, *Parabolic problems with dynamical boundary conditions: eigenvalue expansions and blow up*, Rend. Lincei Mat. Appl. **17** (2006), 35–67.

[2] C. Bandle and W. Reichel, *A linear parabolic problem with non-dissipative dynamical boundary conditions*, Recent advances in elliptic and parabolic problems, Proceedings of the 2004 Swiss-Japanese Seminar in Zürich, World Scientific, 2006.

[3] C. Bandle, J. v. Below and W. Reichel, *Positivity and anti-maximum principles for elliptic operators with mixed boundary conditions*, JEMS **10** (2007), 73–104.

[4] M. Bareket, *On the domain monotonicity of the first eigenvalue of a boundary value problem*, ZAMP **27** (1976), 487–491.

[5] M. Bareket, *On an isoperimetric inequality for the first eigenvalue of a boundary value problem*, SIAM J. Math. Anal. **8** (1977), 280–287.

[6] M.-H. Bossel, *Membranes 'élastiquement' liées: extension du théorème de Rayleigh-Faber-Krahn et de l'inégalité de Cheeger* , C. R. Acad. Sci. Paris Série I Math. **302**(1) (1986), 47–50.

[7] D. Daners, *A Faber-Krahn inequality for Robin problems in any space dimension*, Math. Ann. **335** (2006), 767–785.

[8] D. Daners and J. Kennedy, *Uniqueness in the Faber-Krahn inequality for Robin problems*, to appear.

[9] L. E. Payne and H. Weinberger, *Lower bounds for vibration frequencies of elastically supported membranes and plates*, SIAM J. Appl. Math. **5** (1957), 171–182.

[10] G. Szegö, *Inequalities for certain eigenvalues of a membrane of given area*, J. Rat. Mech. Anal. **3** (1954), 343–356.

[11] H. Weinberger, *An isoperimetric inequality for the N-dimensional free membrane problem*, J. Rat. Mech. Anal. **5** (1956), 533–536.

Catherine Bandle
Mathematisches Institut
Universität Basel
Rheinsprung 21
CH-4051 Basel, Switzerland
e-mail: `catherine.bandle@unibas.ch`

International Series of Numerical Mathematics, Vol. 157, 13–22

Lower and Upper Bounds for Sloshing Frequencies

Henning Behnke

Abstract. The calculation of the frequencies ω for small oscillations of an ideal liquid in a container results in a Steckloff eigenvalue problem. A procedure for calculating lower and upper bounds to the smallest eigenvalues is proposed. For the lower bound computation Goerisch's generalization of Lehmann's method is applied, trial functions are constructed with finite elements. Rounding errors are controlled with interval arithmetic.

Mathematics Subject Classification (2000). Primary 65N25; Secondary 35P15.

Keywords. Sloshing problem, bounds to eigenvalues, Rayleigh-Ritz, Lehmann-Goerisch, finite elements, verified bounds.

1. Introduction

We study an eigenvalue problem with a spectral parameter in a boundary condition.

This problem for the two-dimensional Laplace equation is relevant to sloshing frequencies that describe free oscillations of an inviscid, incompressible, heavy fluid in a canal having uniform cross-section and bounded from above by a horizontal free surface.

It is demonstrated that accurate bounds for the smallest eigenvalues can be computed using inclusion theorems (based on variational principles) and interval arithmetic.

The calculation of the frequencies ω for small oscillations of an ideal liquid in a container results in the eigenvalue problem:

$$\begin{aligned} -\Delta\varphi &= 0 && \text{in } \Omega && \text{(liquid)}, \\ \partial\varphi/\partial\mathbf{n} &= \lambda\varphi && \text{on } \partial_1\Omega && \text{(free surface)}, \\ \partial\varphi/\partial\mathbf{n} &= 0 && \text{on } \partial_2\Omega && \text{(container wall)}, \\ \lambda &> 0; \end{aligned} \tag{1}$$

$\mathbf{n}$ denotes the outward normal to the boundary $\partial\Omega$ of Ω, the relation between λ and ω is $\lambda = \omega^2/g$, g is the acceleration due to gravity.

As an example for the diversity of possible domains with polygonal boundaries, let Ω be defined as

$$\Omega := \{(x,y) \in \mathbb{R}^2 \ : \ -1 < y < 0, -1 - \frac{y}{2} < x < 0\},$$
$$\partial_1\Omega := \{(x,y) \in \mathbb{R}^2 \ : \ -1 \le x \le 0, y = 0\},$$

and $\partial_2\Omega := \partial\Omega \backslash \partial_1\Omega$. The two-dimensional problem is a model for an infinitely long canal with cross section Ω.

Besides numerical examples the rest of the paper is dedicated to a setup for the Goerisch method which is suitable for finite element computations. Since the Goerisch method permits to choose certain quantities in many different ways, this is one of the critical questions.

2. Setting for the problem

Let $\mathbf{H}_a$ and $\mathbf{H}_b$ be two separable, complex Hilbert spaces with inner products $a(\,.\,,\,.\,)$ and $b(\,.\,,\,.\,)$, respectively. Suppose $\mathbf{H}_a$ is a dense subspace of $\mathbf{H}_b$ continuously embedded in $\mathbf{H}_b$ such that for $\kappa > 0$

$$\kappa\, b(u,u) \le a(u,u) \text{ for all } u \in \mathbf{H}_a$$

holds true. The following variationally posed eigenvalue problem is considered:

$$\left.\begin{array}{l} \text{Find eigenpairs } (\lambda, u) \in \mathbb{R} \times \mathbf{H}_a\,,\ u \neq 0\,, \text{ such} \\ \text{that } a(u,v) = \lambda b(u,v) \text{ holds for all } v \in \mathbf{H}_a. \end{array}\right\} \tag{2}$$

Denote by $B \in \mathcal{L}(\mathbf{H}_a)$ the bounded self-adjoint operator that satisfies

$$a(Bu, v) = b(u,v) \text{ for all } u, v \in \mathbf{H}_a\,.$$

By assumption B possesses a self-adjoint inverse $A = B^{-1} \ : \ \mathbf{H}_a \supset \mathcal{D}(A) \longrightarrow \mathbf{H}_a$ and (2) is equivalent to the eigenvalue problem for A. Hence, $\sigma(A)$ and $\sigma_e(A)$ represent the spectrum σ and the essential spectrum σ_e of (2), respectively.

We suppose that for some $N \in \mathbb{N}$ the lower part of σ consists of at least $N+1$ isolated eigenvalues of finite multiplicity

$$0 < \kappa \le \lambda_1 \le \lambda_2 \le \cdots \le \lambda_{N+1} < \inf \sigma_e\,.$$

These eigenvalues are characterized by the variational principle

$$\lambda_j = \inf_{\substack{V \subset \mathbf{H}_a \\ \dim V = j}} \ \max_{0 \neq v \in V} \frac{a(v,v)}{b(v,v)}\,. \tag{3}$$

This formula can be obtained (see [14, Chapter 3]) from Poincaré's principle for the eigenvalues $\mu_j = 1/\lambda_j$ of the operator $B \in \mathcal{L}(\mathbf{H}_a)$.

A discretization of (3) gives the famous Rayleigh-Ritz method for a straightforward and efficient computation of upper bounds to the eigenvalues below the essential spectrum.

In order to compute lower bounds we have to establish an other variationally posed characterization of the eigenvalues as it is given in [15]. Since, the method acts on the Hilbert space $\mathbf{H}_a$ and the inner product $a(\,.\,,\,.\,)$ stands on the left side of (2) this procedure is called the left definite case.

Assume that $\mathbf{X}$ is a further complex Hilbert space with inner product $s(\,.\,,\,.\,)$ and equipped with an isometric embedding $\mathbf{T} : \mathbf{H}_a \longrightarrow \mathbf{X}$ such that

$$s(\mathbf{T}u, \mathbf{T}v) = a(u,v) \text{ for all } u, v \in \mathbf{H}_a\,. \tag{4}$$

Additionally, the method makes use of a separating parameter $\rho \in \mathbb{R}$ with

$$\lambda_N < \rho < \lambda_{N+1}\,,$$

where this lower bound for the $(N+1)$st eigenvalue is known a priori. Sometimes it is hard to determine such a separating parameter ρ, then a homotopy method can be used [1, 12].

Now, if the eigenvalues λ_j are represented in the form

$$\lambda_j = \rho + \frac{\rho}{\tau_j - 1} \quad \text{with} \quad \tau_j = \frac{\lambda_j}{\lambda_j - \rho}\,, \qquad j = 1, \ldots, N\,, \tag{5}$$

we have (see [15, Corollary 2.1]) for $j = 1, \ldots, N$ the variational characterization

$$\tau_j = \inf_{\substack{V \subset \mathbf{H}_a \\ \dim V = j}} \max_{0 \neq v \in V} \min_{w} \frac{a(v,v) - \rho b(v,v)}{a(v,v) - 2\rho b(v,v) + \rho^2 s(w,w)}\,, \tag{6}$$

where the minimum is taken over $w \in \mathbf{X}$ such that

$$s(w, \mathbf{T}u) = b(v,u) \text{ for all } u \in \mathbf{H}_a. \tag{7}$$

Note, that by the spectral mapping theorem $\tau_N \leq \cdots \leq \tau_1 < 0$ are the eigenvalues of the bounded and self-adjoint operator-function $T = A(A - \rho I)^{-1} \in \mathcal{L}(\mathbf{H}_a)$ giving the lower part of its spectrum. This relation is used in equation (5).

A discretization of (6) using a Rayleigh-Ritz procedure to the operator $T \in \mathcal{L}(\mathbf{H}_a)$ combined with a complementary variational principle gives upper bounds to τ_j and hence, by the transformation $\tau \longmapsto \rho + \frac{\rho}{\tau - 1}$ lower bounds to the eigenvalues $\lambda_j\,,\ j = 1, \ldots, N\,,$ of (2).

3. Calculation of bounds

For a discretization of (3) and (6) let $n \in \mathbb{N}\,,\ m \in \mathbb{N}_0$ and suppose the following:

L1. $v_1, \ldots, v_n \in \mathbf{H}_a$ are linearly independent trial functions.
L2. $w_1^\star, \ldots, w_n^\star \in \mathbf{X}$ satisfy $s(w_i^\star, \mathbf{T}u) = b(v_i, u)$ for all $u \in \mathbf{H}_a\,,\quad i = 1, \ldots, n.$
L3. $w_0^\circ, \ldots, w_m^\circ \in \mathbf{X}^\circ = \{w \in \mathbf{X} : s(w, \mathbf{T}u) = 0 \text{ for all } u \in \mathbf{H}_a\}$ where $w_0^\circ = 0$ and $w_1^\circ, \ldots, w_m^\circ$ are linearly independent.

Then, we construct matrices $A_1 = (a_{ik}^{(1)})$, $A_2 = (a_{ik}^{(2)})$ by

$$\begin{aligned} a_{ik}^{(1)} &= a(v_k, v_i) \\ a_{ik}^{(2)} &= b(v_k, v_i) \end{aligned} \qquad \text{for } i, k = 1, \dots, n$$

and matrices $C_{11} = (c_{ik}^{(11)})$, $C_{12} = (c_{ik}^{(12)})$, $C_{22} = (c_{ik}^{(22)})$ by

$$c_{ik}^{(11)} = s(w_k^\star, w_i^\star) \quad \text{for } i, k = 1, \dots, n\,,$$

$$c_{ik}^{(12)} = \begin{cases} s(w_k^\circ, w_i^\star) & \text{if } m > 0 \\ 0 & \text{if } m = 0 \end{cases} \qquad \text{for } i = 1, \dots, n;\ k = 1, \dots, \max\{1, m\}\,,$$

$$c_{ik}^{(22)} = \begin{cases} s(w_k^\circ, w_i^\circ) & \text{if } m > 0 \\ 1 & \text{if } m = 0 \end{cases} \qquad \text{for } i, k = 1, \dots, \max\{1, m\}\,,$$

as well as

$$A_3 = C_{11} - C_{12} C_{22}^{-1} C_{12}^H\,.$$

Now, we consider the following matrix eigenvalue problems

$$(\Lambda^{[n]}, x) \in \mathbb{R} \times \mathbb{C}^n \quad : \quad A_1 x = \Lambda^{[n]} A_2 x\,, \tag{8}$$

$$(\tau^{\rho[n,m]}, x) \in \mathbb{R} \times \mathbb{C}^n \quad : \quad (A_1 - \rho A_2) x = \tau^{\rho[n,m]} (A_1 - 2\rho A_2 + \rho^2 (A_3)) x \tag{9}$$

and arrange the N lowest eigenvalues of both problems in the order

$$\tau_N^{\rho[n,m]} \le \dots \le \tau_2^{\rho[n,m]} \le \tau_1^{\rho[n,m]} < 0 < \Lambda_1^{[n]} \le \Lambda_2^{[n]} \le \dots \le \Lambda_N^{[n]}\,. \tag{10}$$

Such an arrangement is always possible if the trial functions fulfill the assumption

$$\bigcup_{n \in \mathbb{N}} \operatorname{span}\{v_1, \dots, v_n\} \text{ is dense in } \Big(\mathbf{H}_a, a(\,.\,,\,.\,)\Big).$$

Then, for sufficiently large $n \in \mathbb{N}$ and arbitrary $m \in \mathbb{N}_0$ the problem (9) gives exactly N negative eigenvalues (see [11, Theorem 3.4]).

Obviously, the values $\Lambda_j^{[n]}$ are upper Rayleigh-Ritz bounds for the eigenvalues λ_j of (2). Whereas the values $\tau_j^{\rho[n,0]}$ are upper Rayleigh-Ritz bounds for the eigenvalues τ_j of the operator $T \in \mathcal{L}(\mathbf{H}_a)$ with respect to the trial functions $(I - \rho B)v_1, \dots, (I - \rho B)v_n \in \mathbf{H}_a$ (see [11, Theorem 2.6]). A complementary variational principle provides $\tau_j \le \tau_j^{\rho[n,0]} \le \tau_j^{\rho[n,m]}$ for $m \in \mathbb{N}$ (see [11, Lemma 3.3]). Hence, setting

$$\Lambda_j^{\rho[n]} = \rho + \frac{\rho}{\tau_j^{\rho[n,m]} - 1}\,,\ j = 1, \dots, N\,,$$

we arrive at

Theorem 3.1. *Lower and upper bounds to the eigenvalues of* (2) *are given by*

$$\Lambda_j^{\rho[n]} \le \lambda_j \le \Lambda_j^{[n]}\,,\ j = 1, \dots, N\,.$$

For convergence results see [9, 10, 11].

Remark 3.2. The choice $m = 0$ yields the procedure of Lehmann [6] and Maehly [8] (cf. also [15, Remark 2.2]) given by

$$\mathbf{X} = \mathbf{H}_a, \quad s(\,.\,,\,.\,) = a(\,.\,,\,.\,), \quad \mathbf{T} = I \quad \text{and} \quad \mathbf{X}^\circ = \{0\}\,.$$

Generalizing the procedure of Lehmann and Maehly the discretization (9) was originally given by Goerisch [1, 3, 4]. Therefore, we refer to Goerisch method and to Goerisch bounds $\Lambda_j^{\rho[n]}$.

4. Application to the sloshing problem

In order to treat the sloshing problem (1) we define

$$\mathbf{H}_a := \{f \in H^1(\Omega) \;:\; \int_{\partial_1\Omega} f\,ds = 0\}$$

and

$$\begin{aligned} a(f,g) &:= \int_\Omega (\mathrm{grad} f)' \cdot \mathrm{grad} g\,dx\,dy, \\ b(f,g) &:= \int_{\partial_1\Omega} f\,g\,ds \quad \text{for all } f,g \in \mathbf{H}_a; \end{aligned}$$

now the weak form of (1) is:

$$\begin{aligned} &\text{Determine } \varphi \in \mathbf{H}_a, \varphi \neq 0, \lambda \in \mathbb{R} \text{ such that} \\ &a(f,\varphi) = \lambda\, b(f,\varphi) \text{ for all } f \in \mathbf{H}_a. \end{aligned} \tag{11}$$

For an application of the finite element method Ω is divided into subtriangles $\Omega_1, \ldots, \Omega_l$. The trial functions v_i are to be constructed using Lagrange elements. The coupling condition (7) for Lehmann's procedure reads as

$$a(w_i^\star, f) = b(v_i, f) \quad \text{for all } f \in \mathbf{H}_a, i = 1, \ldots, n$$

If this equation holds for $f \in \mathbf{H}_a$, it holds for $f \in C_0^\infty(\Omega)$ as well, hence

$$\begin{aligned} a(w_i^\star, f) &= \int_\Omega (\mathrm{grad}\, w_i^\star)' \cdot \mathrm{grad}\, f\,dx\,dy \\ &= \int_\Omega -w_i^\star\,\Delta f\,dx\,dy = 0 \\ &= \int_{\partial_1\Omega} v_i\,f\,ds \\ &= b(v_i, f) \quad \text{for all} \quad f \in C_0^\infty(\Omega) \end{aligned}$$

According to Weyl's Lemma ([7, 13]) $w_i^\star$ is equivalent to a harmonic function $\tilde{w}_i \in C^\infty(\Omega')$ for each $\Omega' \subset \overline{\Omega'} \subset \Omega$ and can not be constructed with finite elements.

In order to compute lower bounds using finite elements, a more sophisticated definition of the quantities $X, s(.,.)$ and T is required:

$$\mathbf{X} := \prod_{j=1}^{l} (L_2(\Omega_j))^2,$$

$$T : \mathbf{H}_a \to \mathbf{X},\ T\,f := \begin{pmatrix} \operatorname{grad} f|_{\Omega_1} \\ \vdots \\ \operatorname{grad} f|_{\Omega_l} \end{pmatrix},$$

$$s\left(\begin{pmatrix} f_{1,1} \\ f_{1,2} \\ \vdots \\ f_{l,1} \\ f_{l,2} \end{pmatrix}, \begin{pmatrix} g_{1,1} \\ g_{1,2} \\ \vdots \\ g_{l,1} \\ g_{l,2} \end{pmatrix}\right) := \sum_{j=1}^{l} \int_{\Omega_j} (f_{j,1}g_{j,1} + f_{j,2}g_{j,2})\, dx\, dy\,.$$

Now we have $s(T\,f, T\,g) = \sum_{j=1}^{l} \int_{\Omega_j} (\operatorname{grad} f|_{\Omega_j}) \cdot (\operatorname{grad} g|_{\Omega_j})\, dx\, dy = a(f,g)$ for all $f, g \in \mathbf{H}_a$. Assume that we have $w_i^\star \in \mathbf{X}$,

$$w_i^\star = \begin{pmatrix} w_{i,1}^\star \\ \vdots \\ w_{i,l}^\star \end{pmatrix} \quad \text{with} \quad w_{i,j}^\star \in (C^1(\Omega_j))^2,\ i = 1,\ldots,n,\ j = 1,\ldots,l,$$

such that

$$-\operatorname{div} w_{i,j}^\star = 0 \quad \text{in } \Omega_j \quad \text{for } j = 1,\ldots,l \tag{12}$$

$$w_{i,j}^\star \cdot \mathbf{n}_j = -w_{i,k}^\star \cdot \mathbf{n}_k \quad \text{on } \partial\Omega_j \cap \partial\Omega_k \quad \text{for } j,k = 1,\ldots,l \tag{13}$$

$$w_{i,j}^\star \cdot \mathbf{n}_j = 0 \quad \text{on } \partial_2\Omega \cap \partial\Omega_j \quad \text{for } j = 1,\ldots,l \tag{14}$$

$$w_{i,j}^\star \cdot \mathbf{n}_j = v_i \quad \text{on } \partial_1\Omega \cap \partial\Omega_j \quad \text{for } j = 1,\ldots,l \tag{15}$$

A short computation gives

$$\begin{aligned} s(T\,f, w_i^\star) &= \sum_{j=1}^{l} \int_{\Omega_j} (\operatorname{grad} f|_{\Omega_j}) \cdot w_{i,j}^\star\, dx\, dy \\ &= \sum_{j=1}^{l} \Big(-\int_{\Omega_j} f \operatorname{div} w_{i,j}^\star\, dx\, dy + \int_{\partial\Omega_j} f\ (w_{i,j}^\star \cdot \mathbf{n}_j)\, ds\Big) = \int_{\partial_1\Omega} f\, v_i\, ds \\ &= b(f, v_i) \quad \text{for all } f \in \mathbf{H}_a,\ i = 1,\ldots,n\,, \end{aligned}$$

that is (7). The same construction is possible for w_i° if equation (15) is replaced by $w_{i,j}^\circ \cdot \mathbf{n}_j = 0$ on $\partial_1\Omega \cap \partial\Omega_j$ for $j = 1,\ldots,l$.

We use $\hat{w}_i \in C(\overline{\Omega})$, $\hat{w}_i|_{\Omega_j} \in C^2(\Omega_j)$ for $j = 1,\ldots,l$ to construct $w_i^\star$ by

$$w_{i,j}^\star := \begin{pmatrix} \frac{\partial \hat{w}_i}{\partial y}|_{\Omega_j} \\ -\frac{\partial \hat{w}_i}{\partial x}|_{\Omega_j} \end{pmatrix}, \quad j = 1,\ldots,l\,.$$

By this definition (12) and (13) are satisfied.

If we require

$$\hat{w}_i|_{\partial_2\Omega} = 0 \quad \text{and} \tag{16}$$

$$\hat{w}_i(x,0) = -\int_{-1}^{x} v_i(\xi,0)\,d\xi \quad \text{for} \quad -1 \le x \le 0 \tag{17}$$

we have (14) and (15).

The trial functions are constructed using Lagrange elements. For v_i we use polynomials of degree p, for $\hat{w}_i$ polynomials of degree $p+1$. The results are computed with interval arithmetic. For details see [2].

The spectral parameter ρ can be determined due to domain monotonicity. If we have two domains Ω and $\hat{\Omega}$, such that $\hat{\Omega} \subset \Omega$, with the same free surface (that is $\partial_1\hat{\Omega} = \partial_1\Omega$), then the the smaller domain has the smaller eigenvalues, i.e.,

$$\hat{\lambda}_j \le \lambda_j \quad j = 1, 2, \dots .$$

The sloshing problem can easily be solved for an right angled triangle.

5. Results

We give numerical results for the triangulation in Figure 1.

The mesh consists of 694 elements and 376 nodes. The table shows enclosures for the smallest eigenvalues for different degrees of polynomials for the Lagrange elements. dimRR is the dimension of the matrix eigenvalue problem for the Rayleigh-Ritz computation, dimLG is the dimension of the Lehmann-Goerisch matrix eigenvalue problem. For $p = 1$ the dimension is not 376 since we require orthogonality to the constant function.

	$p = 1, \quad \rho = 14.9$	$p = 2, \quad \rho = 14.9$
λ_i	dimRR = 375 ; dimLG = 1708	dimRR = 1444 ; dimLG = 4484
1	$2.^{803}_{787}$	2.7905^{160}_{099}
2	$^{6.067}_{5.874}$	5.95^{209}_{172}
3	$^{9.522}_{7.576}$	9.09^{714}_{227}
4	$^{13.254}_{8.458}$	$12.^{246}_{199}$

	$p = 3, \quad \rho = 14.9$	$p = 4, \quad \rho = 14.9$
λ_i	dimRR = 3207 ; dimLG = 8648	dimRR = 5664 ; dimLG = 14200
1	2.7905110^{909}_{772}	2.79051108^{281}_{033}
2	5.95186^{840}_{785}	5.95186807^{614}_{008}
3	9.0952^{405}_{265}	9.09523^{50595}_{49169}
4	12.237^{242}_{016}	12.237^{20113}_{19944}

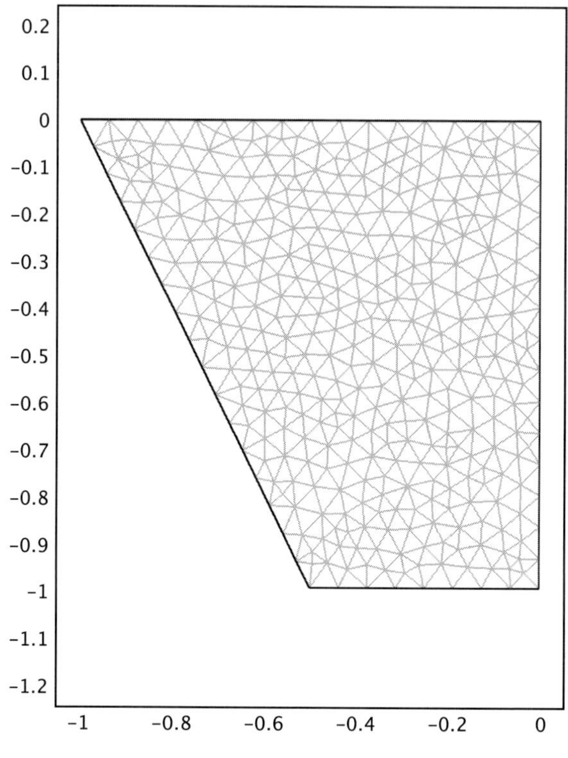

FIGURE 1

Using a finer mesh with 7925 elements, the bounds become sharper:

	$p = 3\,, \quad \rho = 14.9$	$p = 4\,, \quad \rho = 14.9$
λ_i	dimRR = 35934 ; dimLG = 98973	dimRR = 63762 ; dimLG = 162373
1	2.79051108^{205}_{059}	2.7905110^{830}_{795}
2	5.95186807^{379}_{159}	5.9518680^{741}_{698}
3	9.0952350^{341}_{172}	9.09523502^{942}_{136}
4	12.237200^{959}_{729}	12.237200^{920}_{893}

For $p = 4$ the enclosure for the smallest eigenvalue is worse than that for $p = 3$. This is an effect of the rounding errors which are controlled by interval arithmetic. Even though the discretization error is smaller for $p = 4$ the effect of rounding errors becomes larger, since the matrix eigenvalue problems are considerably larger.

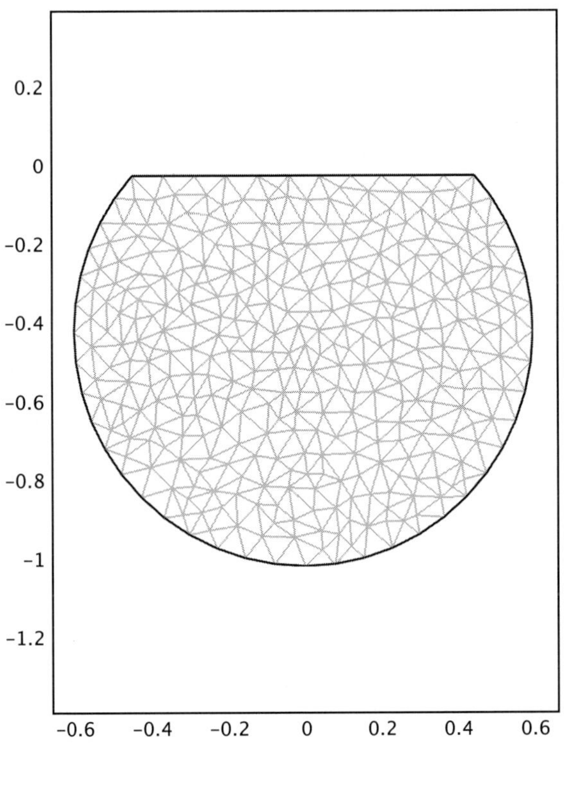

FIGURE 2

Figure 2 shows the triangulation for an other domain. It consists of 659 elements.

Some numerical results are given in the following table:

	$p = 5\,,\quad \rho = 16$
λ_i	dimRR = 35934 ; dimLG = 98973
1	2.88865_{530}^{634}
2	6.1559_{770}^{878}
3	9.466_{476}^{537}
4	12.56_{766}^{808}

The results have been computed on a Fujitsu Siemens SCENIC P computer with 3 GHz clock speed an 1024 MB memory. For the verified computations we used the PROFIL/BIAS library [5].

References

[1] H. Behnke and F. Goerisch, *Inclusions for eigenvalues of selfadjoint problems*, J. Herzberger, editor, Topics in validated computations, Elsevier: Amsterdam, Lausanne, New York, Oxford, Shannon, Tokyo, 1994, 277–322.

[2] H. Behnke and U. Mertins, *Bounds for eigenvalues with the use of finite elements*, Perspectives on enclosure methods, Karlsruhe (2000), Springer, Vienna, 2001, 119–131.

[3] F. Goerisch, *Eine Verallgemeinerung eines Verfahrens von N.J. Lehmann zur Einschließung von Eigenwerten*, Wiss. Z. Tech. Univ. Dresden, **29** (1980), 429–431.

[4] F. Goerisch and H. Haunhorst, *Eigenwertschranken für Eigenwertaufgaben mit partiellen Differentialgleichungen*, ZAMM, **65**(3) (1985), 129–135.

[5] O. Knüppel, *PROFIL/BIAS – a fast interval library*, Computing, **53** (3-4) (1994), 277–287. International Symposium on Scientific Computing, Computer Arithmetic and Validated Numerics, Vienna, 1993.

[6] N.J. Lehmann, *Beiträge zur Lösung linearer Eigenwertprobleme I*, ZAMM **29** (1949), 341–356.

[7] R. Lemmert, *A short proof of Weyl's lemma*, Math. Nachr. **135** (1988), 237–238.

[8] H.J. Maehly, *Ein neues Variationsverfahren zur genäherten Berechnung der Eigenwerte hermitescher Operatoren*, Helv. Phys. Acta, 1952.

[9] U. Mertins, *Zur Konvergenz des Rayleigh-Ritz-Verfahrens bei Eigenwertaufgaben*, Numer. Math. **59** (1991), 667–682.

[10] U. Mertins, *Asymptotische Fehlerschranken für Rayleigh-Ritz-Approximationen selbstadjungierter Eigenwertaufgaben*, Numer. Math. **63** (1992), 227–241.

[11] U. Mertins, *On the convergence of the Goerisch method for self-adjoint eigenvalue problems with arbitrary spectrum*, Z. Anal. Anwendungen, **15** (1996), 661–686.

[12] Michael Plum, *Eigenvalue inclusions for second-order ordinary differential operators by a numerical homotopy method*, Z. Angew. Math. Phys., **41**(2), 1990, 205–226.

[13] W. Velte, *Direkte Methoden der Variationsrechnung*, Teubner: Stuttgart, 1976.

[14] H.F. Weinberger, *Variational Methods for Eigenvalue Approximation*, Regional Conference Series in Applied Mathematics, Band 15. Society for Industrial and Applied Mathematics: Philadelphia, 1974.

[15] S. Zimmermann and U. Mertins. *Variational bounds to eigenvalues of self-adjoint problems with arbitrary spectrum*, Z. Anal. Anwend., **14** (1995), 327–345.

Henning Behnke
Institut für Mathematik
TU Clausthal
Erzstraße 1
D-38678 Clausthal-Zellerfeld, Germany
e-mail: `behnke@math.tu-clausthal.de`

International Series of Numerical Mathematics, Vol. 157, 23–30

On Spectral Bounds for Photonic Crystal Waveguides

B. Malcolm Brown, Vu Hoang, Michael Plum and Ian G. Wood

Abstract. For a $(d+1)$-dimensional photonic crystal with a linear defect strip (waveguide), we calculate real intervals containing spectrum of the associated spectral problem. If such an interval falls completely into a spectral gap of the unperturbed problem (without defect), this will prove the existence of additional spectrum induced by the waveguide.

1. Introduction

Waves in periodic structures may propagate without attenuation except at certain energy levels. This fact is the basic tool used in most integrated electronics, and the idea has been extended to an optical setting, with the notion of photonic band-gap (PBG) materials where bands of forbidden wavelengths are encountered. These ideas have generated a rapidly developing topic in optics and nano-technology and are an important area of engineering practice.

We consider a $(d+1)$-dimensional periodic structure, which has a linear defect introduced. This will create a waveguide which will allow light frequencies to be transmitted that are forbidden in the bulk. The waveguide itself may also exhibit a (possibly different) periodic structure, and indeed waveguides may intersect without causing interference in the transmission. There is much numerical and experimental evidence to show that such waveguides can be effectively created (see [1] and the references contained therein).

We calculate real intervals containing spectrum of the waveguide problem. If such an interval falls completely into a spectral gap of the unperturbed problem (without waveguide), this proves additional spectrum induced by the waveguide. We note that a similar approach, but using L^2-estimates (and not H^{-1}-estimates as we do) has been developed in [1].

2. Preliminaries

In order to model our problem we take a photonic crystal covering the whole of $\mathbb{R}^{d+1}$ (in practice $d = 1, 2$), with a "defect" strip

$$S_l = \{x = (x_1, x') \in \mathbb{R}^{d+1} | x_1 \in \mathbb{R}, x' \in l\Omega\}$$

where $\Omega \subset \mathbb{R}^d$ is a bounded, convex domain containing the origin and $l > 0$. We suppose that the electric permittivity is given as

$$\epsilon(x) = \epsilon > 0 \ (x \in S_l), \quad \epsilon(x) \ \text{ periodic and positive } (x \notin S_l).$$

We shall work in the Hilbert space $H^{-1}(\mathbb{R}^{d+1})$ (the space of bounded linear functionals on $H_0^1(\mathbb{R}^{d+1})$) and take as scalar product on $H_0^1(\mathbb{R}^{d+1})$

$$\langle u, v \rangle_{H_0^1} = \int_{\mathbb{R}^{d+1}} \frac{1}{\epsilon} \nabla u \cdot \overline{\nabla v} dx + \sigma \int_{\mathbb{R}^{d+1}} u \overline{v} dx \tag{1}$$

where $\sigma > 0$ is a parameter at our disposal. Our photonics problem is modelled by the following spectral problem

$$A[u] \equiv -\text{div}\left(\frac{1}{\epsilon} \nabla u\right) = \lambda u.$$

We work with the shifted operator $\Phi : D(\Phi) = H_0^1(\mathbb{R}^{d+1}) \subset H^{-1}(\mathbb{R}^{d+1}) \to H^{-1}(\mathbb{R}^{d+1})$

$$\Phi \equiv A + \sigma = -\text{div}\left(\frac{1}{\epsilon} \nabla\right) + \sigma. \tag{2}$$

Note that

$$(\Phi u)[v] = \langle u, \bar{v} \rangle_{H_0^1} \quad (u, v \in H_0^1)$$

and the scalar product in $H^{-1}(\mathbb{R}^{d+1})$ is $\langle x, y \rangle_{H^{-1}} = \langle \Phi^{-1} x, \Phi^{-1} y \rangle_{H_0^1}$.

We now discuss the spectral problem $Au = \lambda u$, or equivalently

$$\Phi u = (\lambda + \sigma) u. \tag{3}$$

Then, for $u, v \in H_0^1(\mathbb{R}^{d+1})$,

$$\langle \Phi u, v \rangle_{H^{-1}} = \langle u, \Phi^{-1} v \rangle_{H_0^1} = \overline{\langle \Phi^{-1} v, u \rangle_{H_0^1}} = \overline{v[\overline{u}]} = \int_{\mathbb{R}^{d+1}} u \overline{v} dx = \langle u, v \rangle_{L^2}, \tag{4}$$

where $\langle u, v \rangle_{L^2}$ denotes the usual scalar product in $L^2(\mathbb{R}^n)$. Thus Φ is symmetric, and since we know that the range of Φ is $H^{-1}(\mathbb{R}^{d+1})$, we have that $\Phi^{-1} : H^{-1}(\mathbb{R}^{d+1}) \to H^{-1}(\mathbb{R}^{d+1})$ is selfadjoint, which implies that Φ is also selfadjoint.

3. Construction of an enclosing interval

In this section we seek to show that in addition to the spectrum of the unperturbed problem (where $\epsilon(x)$ is periodic throughout $\mathbb{R}^{d+1}$) the perturbation will induce new spectrum which we intend to localize in form of an enclosing interval. We do this by constructing an approximate eigenpair for Φ^{-1} which we denote by

$(\tilde{x}, \tilde{\mu}) \in H^{-1}(\mathbb{R}^{d+1}) \times (0, \infty), \tilde{x} \neq 0$. We then let $\delta > 0$ be an upper bound for the defect

$$\frac{\|\Phi^{-1}\tilde{x} - \tilde{\mu}\tilde{x}\|_{H^{-1}}}{\|\tilde{x}\|_{H^{-1}}} \leq \delta,$$

and the interval $[\tilde{\mu} - \delta, \tilde{\mu} + \delta]$ contains at least one spectral point of Φ^{-1} due to the classical theorem of D. Weinstein (see [2], Theorem 5.9). If $\tilde{\mu} - \delta > 0$, then by the spectral mapping theorem a spectral point of A is in the interval

$$\frac{1}{[\tilde{\mu} - \delta, \tilde{\mu} + \delta]} - \sigma = \left[\tilde{\lambda} - \frac{\delta(\tilde{\lambda} + \sigma)^2}{1 + \delta(\tilde{\lambda} + \sigma)}, \tilde{\lambda} + \frac{\delta(\tilde{\lambda} + \sigma)^2}{1 - \delta(\tilde{\lambda} + \sigma)}\right] \tag{5}$$

where we have written $\tilde{\mu} = \frac{1}{\tilde{\lambda} + \sigma}$.

3.1. Estimate for δ

In this section we seek an upper bound for δ.

Let $\tilde{u} := \Phi^{-1}\tilde{x} \in H_0^1(\mathbb{R}^{d+1})\backslash\{0\}$. Then

$$A\tilde{u} \approx \tilde{\lambda}\tilde{u} \Leftrightarrow \Phi\tilde{u} \approx (\tilde{\lambda} + \sigma)\tilde{u} \Leftrightarrow \tilde{x} \approx \frac{1}{\tilde{\mu}}\Phi^{-1}\tilde{x} \Leftrightarrow \Phi^{-1}\tilde{x} \approx \tilde{\mu}\tilde{x},$$

i.e., $(\tilde{u}, \tilde{\lambda})$ is an approximate eigenpair for A. Moreover, since

$$\|\Phi^{-1}\tilde{x} - \tilde{\mu}\tilde{x}\|_{H^{-1}} = \sup\{\langle\Phi^{-1}\tilde{x} - \tilde{\mu}\tilde{x}, y\rangle_{H^{-1}} : y \in H^{-1}(\mathbb{R}^{d+1}), \|y\|_{H^{-1}} = 1\}$$

we get, using the fact that $\langle\tilde{u}, y\rangle_{H^{-1}} = \langle\Phi^{-1}\tilde{u}, \Phi^{-1}y\rangle_{H_0^1} = \langle\tilde{u}, \Phi^{-1}y\rangle_{L^2}$ by (4),

$$\begin{aligned} \|\Phi^{-1}\tilde{x} - \tilde{\mu}\tilde{x}\|_{H^{-1}} &= \sup_{\|y\|_{H^{-1}}=1} \{\langle\tilde{u}, \Phi^{-1}y\rangle_{L^2} - \tilde{\mu}\langle\tilde{u}, \ \Phi^{-1}y\rangle_{H_0^1}\} \\ &= \sup_{\|v\|_{H_0^1}=1} \{\langle\tilde{u}, v\rangle_{L^2} - \tilde{\mu}\langle\tilde{u}, v\rangle_{H_0^1}\}. \end{aligned}$$

Next we note that, for any $\tilde{w} \in H(div, \mathbb{R}^{d+1})$,

$$\begin{aligned} \langle\tilde{u}, v\rangle_{H_0^1} &= \int_{\mathbb{R}^{d+1}} (\frac{1}{\epsilon}\nabla\tilde{u} - \tilde{w}) \cdot \overline{\nabla v} dx + \int_{\mathbb{R}^{d+1}} \tilde{w} \cdot \overline{\nabla v} dx + \sigma \int_{\mathbb{R}^{d+1}} \tilde{u}\bar{v} dx \\ &= \left\langle \frac{1}{\epsilon}\nabla\tilde{u} - \tilde{w}, \nabla v \right\rangle_{L^2} - \langle\text{div}\tilde{w}, v\rangle_{L^2} + \sigma\langle\tilde{u}, v\rangle_{L^2}, \end{aligned}$$

whence we obtain

$$\begin{aligned} \|\Phi^{-1}\tilde{x} - \tilde{\mu}\tilde{x}\|_{H^{-1}} &= \sup_{\|v\|_{H_0^1}=1} \left\{ \langle\tilde{u} + \tilde{\mu}(\text{div } \tilde{w} - \sigma\tilde{u}), v\rangle_{L^2} - \tilde{\mu} \left\langle \frac{1}{\epsilon}\nabla\tilde{u} - \tilde{w}, \nabla v \right\rangle_{L^2} \right\} \\ &\leq \sigma^{-1/2}\|\tilde{u} + \tilde{\mu}(\text{div } \tilde{w} - \sigma\tilde{u})\|_{L^2} + \tilde{\mu} \left\| \frac{1}{\sqrt{\epsilon}}\nabla\tilde{u} - \sqrt{\epsilon}\tilde{w} \right\|_{L^2} \\ &= \frac{\sigma^{-1/2}}{\tilde{\lambda} + \sigma}\|\tilde{\lambda}\tilde{u} + \text{div } \tilde{w}\|_{L^2} + \tilde{\mu} \left\| \frac{1}{\sqrt{\epsilon}}\nabla\tilde{u} - \sqrt{\epsilon}\tilde{w} \right\|_{L^2}. \end{aligned}$$

For the defect we therefore have

$$\frac{\|\Phi^{-1}\tilde{x} - \tilde{\mu}\tilde{x}\|_{H^{-1}}}{\|\tilde{x}\|_{H^{-1}}} \leq \frac{1}{\|\tilde{u}\|_{H_0^1}}\Big\{\frac{1}{\sqrt{\sigma}(\tilde{\lambda}+\sigma)}\|\mathrm{div}\ \tilde{w} + \tilde{\lambda}\tilde{u}\|_{L^2} + \tilde{\mu}\|\frac{1}{\sqrt{\epsilon}}\nabla\tilde{u} - \sqrt{\epsilon}\tilde{w}\|_{L^2}\Big\}. \tag{6}$$

We have $\tilde{w} \approx \frac{1}{\epsilon}\nabla\tilde{u}$ in mind, in order to make both defect terms "small". Thus, when $\tilde{u}$ is sufficiently smooth and supported in the waveguide (where ϵ is constant), as we will assume from now on, we can choose

$$\tilde{w} = \frac{1}{\epsilon}\nabla\tilde{u}.$$

The last component of (6) is now zero. So the square of the defect has an upper bound given by

$$\delta^2 = \frac{1}{\sigma(\tilde{\lambda}+\sigma)^2}\frac{\|\frac{1}{\epsilon}\Delta\tilde{u} + \tilde{\lambda}\tilde{u}\|_{L^2}^2}{\|\tilde{u}\|_{H_0^1}^2}. \tag{7}$$

We now quote the construction in [1] of the function $\tilde{u}$ which relates our expression to computable objects. Let $\phi(x') \in H_0^2(\Omega)\backslash\{0\}$ and $\phi_l(x') = l^{-d/2}\phi(x'/l)$. Let moreover $\psi(x_1) \in C_0^\infty(\mathbb{R})$ with unit L^2-norm and define

$$\psi_n(x_1) = n^{-1/2}\psi(x_1/n), \quad n > 0.$$

Define $k = \sqrt{\tilde{\lambda}\epsilon}$ and

$$\tilde{u}(x) = \tilde{u}_{l,n}(x) = \phi_l(x')\psi_n(x_1)e^{ikx_1}.$$

We calculate the denominator of (7):

$$\int_{\mathbb{R}^{d+1}}|\nabla\tilde{u}_{l,n}|^2dx = \int_{\mathbb{R}^{d+1}}[|\phi_l(x')[\psi_n'(x_1) + ik\psi_n(x_1)]|^2 + |\nabla\phi_l(x')\psi_n(x_1)|^2]dx.$$

As $n \to \infty$ this tends to $k^2\|\phi\|_{L^2(\Omega)}^2 + \int_{l\Omega}|\nabla\phi_l|^2dx'$, and

$$\begin{aligned}\lim_{n\to\infty}\|\tilde{u}_{l,n}\|_{H_0^1}^2 &= \lim_{n\to\infty}\int_{\mathbb{R}^{d+1}}\frac{1}{\epsilon}|\nabla\tilde{u}_{l,n}|^2dx + \sigma\int_{\mathbb{R}^{d+1}}|\tilde{u}_{l,n}|^2dx \\ &= \tilde{\lambda}\|\phi\|_{L^2(\Omega)}^2 + \frac{1}{\epsilon}\int_{l\Omega}|\nabla\phi_l|^2dx' + \sigma\|\phi\|_{L^2(\Omega)}^2.\end{aligned}$$

Also as $\nabla\phi_l(x') = l^{-(d+2)/2}\nabla\phi(x'/l)$, we get

$$\int_{l\Omega}|\nabla\phi_l|^2dx' = \frac{1}{l^2}\int_\Omega|\nabla\phi(y)|^2dy$$

and thus

$$\lim_{n\to\infty}\|\tilde{u}_{l,n}\|_{H_0^1}^2 = \frac{1}{\epsilon l^2}\|\nabla\phi\|_{L^2(\Omega)}^2 + (\sigma+\tilde{\lambda})\|\phi\|_{L^2(\Omega)}^2.$$

Using the calculations from [1] for the numerator in (7) we finally arrive at

$$\delta^2(\tilde{\lambda}+\sigma)^2 = \frac{\frac{1}{\sigma\epsilon^2l^4}\|\Delta\phi\|_{L^2(\Omega)}^2}{(\tilde{\lambda}+\sigma)\|\phi\|_{L^2(\Omega)}^2 + \frac{1}{\epsilon l^2}\|\nabla\phi\|_{L^2(\Omega)}^2}, \tag{8}$$

where of course we can choose any $\phi \in H_0^2(\Omega)\backslash\{0\}$. Minimizing δ with respect to ϕ, we are led to the first eigenvalue of the problem

$$\Delta^2\phi = \kappa(\epsilon l^2(\tilde{\lambda}+\sigma)\phi - \Delta\phi). \tag{9}$$

Slightly more explicit results are obtained by an obvious comparison of (9) with the clamped plate problem ($\Delta^2\phi = \kappa\phi$) or the buckling plate problem ($\Delta^2\phi = -\kappa\Delta\phi$). In practice, it might however be more useful to choose a specific test function $\phi \in H_0^2(\Omega)$ in (8), as we do in Section 4.

Using (8) and (5) we obtain the enclosing interval, where both $\tilde{\lambda}$ and σ are at our disposal. We have however to make sure that $\tilde{\mu} - \delta > 0$, i.e., that $1 - \delta^2(\tilde{\lambda}+\sigma)^2 > 0$, for example by choosing σ sufficiently large. We note that as $\sigma \to \infty$ we get the result in [1] (developing the theory in L^2 instead of H^{-1}).

4. An example

We consider the case $d = 2$, $\Omega = (-\frac{1}{2}, \frac{1}{2})^2$. Since we cannot solve the eigenvalue problem (9) we use the test function

$$\phi(x,y) = (x-\frac{1}{2})^2(x+\frac{1}{2})^2(y-\frac{1}{2})^2(y+\frac{1}{2})^2$$

which lies in $H_0^2(\Omega)$. The results are illustrated in Figure 1 for some selected values of $\tilde{\lambda}$. Upper and lower bound of the enclosure interval are shown as functions of $\beta = \epsilon l^2$. In these examples it turns out that the condition $\tilde{\mu} - \delta > 0$ mentioned in Section 3 is satisfied if $\beta > 0.026$.

We remark that the length of the enclosing interval is given by

$$\frac{72\sqrt{\beta\sigma(\tilde{\lambda}+\sigma)^2(\beta(\tilde{\lambda}+\sigma)+24)}}{(\beta\sigma(\beta(\tilde{\lambda}+\sigma)+24)-1296)},$$

while the length of the interval in [1] is

$$l_k = \frac{2\sqrt{1295}}{\beta},$$

where we have used the upper bound 1295 for the first eigenvalue of the biharmonic problem on the unit square.

The length of our enclosing interval is essentially the same as in [1], for some ranges of the parameters β and $\tilde{\lambda}$ we get marginal improvements.

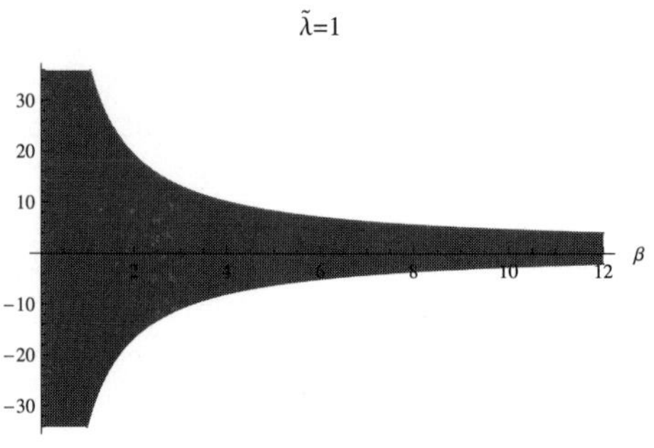

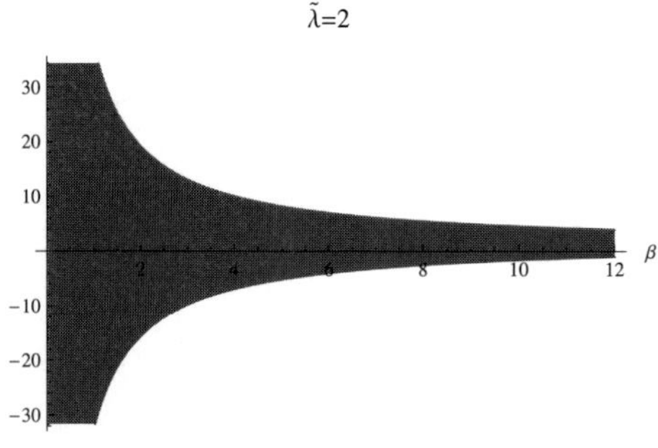

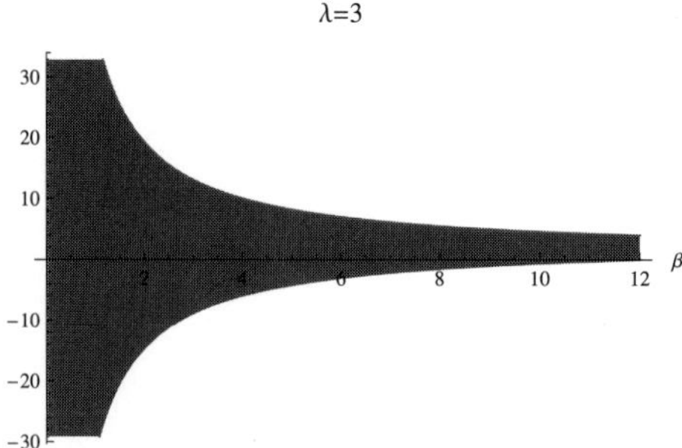

FIGURE 1. Enclosing interval for $\tilde{\lambda} = 1, 2, 3$ plotted as a function of β $(\sigma = 1000)$

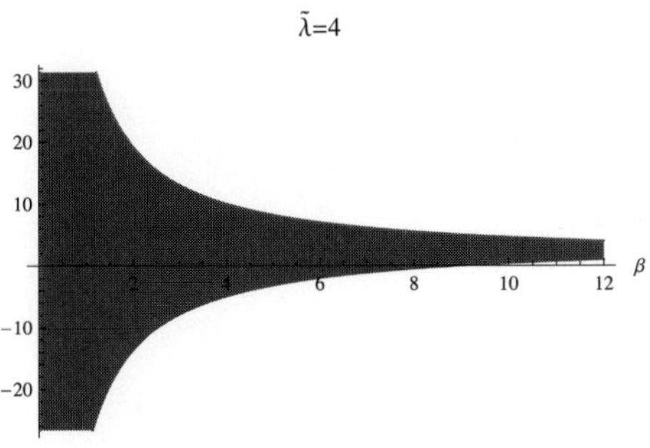

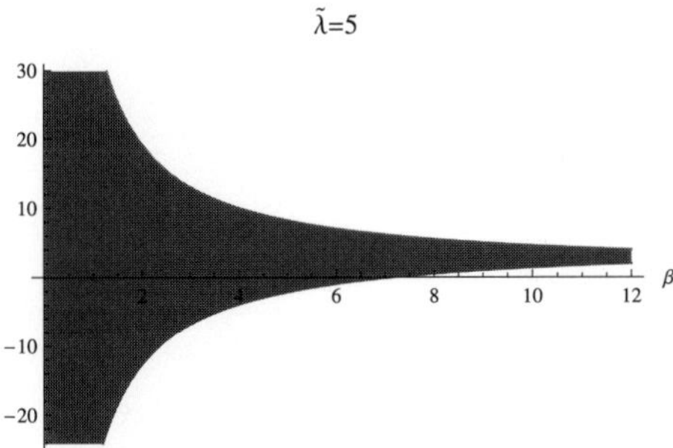

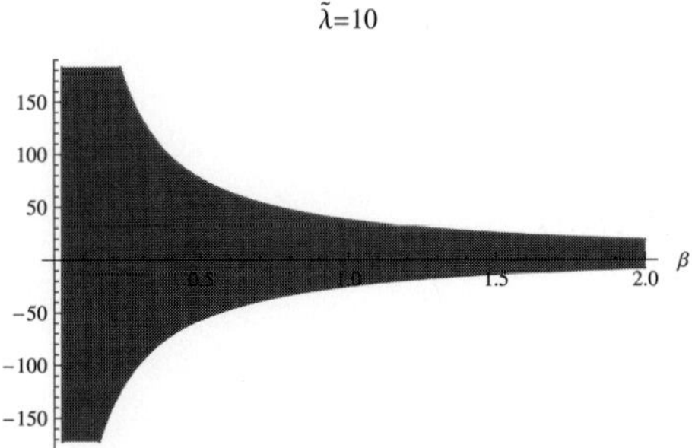

FIGURE 2. Enclosing interval for $\tilde{\lambda} = 4, 5, 10$ plotted as a function of β ($\sigma = 1000$)

References

[1] P. Kuchment, B. Ong, *On guided waves in photonic crystal waveguides*, Waves in periodic and random media, South Hadley, MA (2002), 105–115, Contemp. Math. **339**, Amer. Math. Soc., Providence, RI, 2003.

[2] P.D. Hislop, I.M. Sigal, *Introduction to spectral theory with applications to Schrödinger operators*, Springer, New York, Berlin, Heidelberg, 1996.

B. Malcolm Brown
Cardiff School of Computer Science
Cardiff University
Queen's Buildings, 5 The Parade
Roath, Cardiff CF24 3AA, UK
e-mail: malcolm@cs.cf.ac.uk

Vu Hoang and Michael Plum
Institut für Analysis
Universität Karlsruhe
Englerstrasse 2
D-76128 Karlsruhe, Germany
e-mail: hoang@math.uni-karlsruhe.de
e-mail: michael.plum@math.uni-karlsruhe.de

Ian G. Wood
Institute of Mathematical and Physical Sciences
Aberystwyth University
Penglais
Aberystwyth, Ceredigion, SY23 3BZ, UK
e-mail: ian.wood@aber.ac.uk

International Series of Numerical Mathematics, Vol. 157, 31–42

Real Integrability Conditions for the Nonuniform Exponential Stability of Evolution Families on Banach Spaces

Constantin Buşe

Abstract. Let J be either $\mathbb{R}$ or $\mathbb{R}_+ := [0,\infty)$. We prove that an evolution family $\mathbf{U} = \{U(t,s)\}_{t\geq s\in J}$ which satisfies some natural assumptions is nonuniformly exponentially stable if there exist a positive real number α and a nondecreasing function $\phi : \mathbb{R}_+ \to \mathbb{R}_+$ with $\phi(t)$ positive for all positive t and such that for each $s \in J$, the following inequality

$$\sup_{t>s} \int_0^{t-s} \phi(e^{\alpha u}||U(s+u,s)x||)du = M_\phi(s) < \infty$$

holds true for all $x \in X$ with $||x|| \leq 1$. We arrive at the same conclusion under the assumption that there exist three positive real numbers α, β and K such that for each $t \in J$ the inequality

$$\left(\int_J \chi_{(-\infty,t]}(\tau)e^{-q\alpha\tau}||U(t,\tau)^*x^*||)d\tau\right)^{\frac{1}{q}} \leq Ke^{-\beta t}$$

holds true, for all $x^* \in X^*$ with $||x^*|| \leq 1$ and for some $q \geq 1$.

Mathematics Subject Classification (2000). Primary 47D06, Secondary 35B35.

Keywords. Operator semigroups, evolution families of bounded linear operators, uniform and nonuniform exponential stability.

1. Introduction

The uniform exponential stability of an evolution family $\{U(t,s)\}_{t\geq s}$ of bounded linear operators acting on a Banach space X has been intensively investigated and several important characterizations in terms of integrability conditions with respect to the first variable are known, see [9], [10], [13], [14], [7], [15] and the references therein.

Part of this article was done while the author visited the School of Mathematical Sciences, Governmental College University, Lahore, Pakistan.

On the other hand a reformulation of an old result of E.A. Barbashin [1, Theorem 5.1] reads as follows:

Let $\mathbf{U} = \{U(t,s) : t \geq s \geq 0\}$ *be an exponentially bounded evolution family of bounded linear operators acting on a Banach space* X *such that for each* $t > 0$ *the map* $s \mapsto ||U(t,s)|| : [0,t] \to \mathbb{R}_+$ *is measurable. The following two statements are equivalent:*

(i) *The family* $\mathbf{U}$ *is uniformly exponentially stable.*

(ii) *There exists* $1 \leq p < \infty$ *such that*

$$\sup_{t\geq 0} \left(\int_0^t ||U(t,s)||^p ds \right)^{\frac{1}{p}} < \infty.$$

A similar result for families on the entire real line, can be found in [[6], Theorem 4.1], and reads as follows:

Let $\phi : \mathbb{R}_+ \to \mathbb{R}_+$ *be a nondecreasing function such that* $\phi(t) > 0$ *for all* $t > 0$ *and* $\mathbf{U} = \{U(t,s) : t \geq s\}$ *be an exponentially bounded evolution family of bounded linear operators on* X. *We assume that the function*

$$s \mapsto ||U(t,s)|| : (-\infty, t] \to \mathbb{R}_+,$$

is measurable for all $t \in \mathbb{R}$. *If*

$$\sup_{t\in\mathbb{R}} \int_{-\infty}^t \phi(||U(t,s)||)ds < \infty,$$

then the family $\mathbf{U}$ *is uniformly exponentially stable.*

Among the real integrability conditions which imply nonuniform exponential stability we recall the following:

Let $\phi : \mathbb{R}_+ \to \mathbb{R}_+$ *be a nondecreasing function such that* $\phi(t) > 0$ *for all* $t > 0$ *and let* $\mathbf{U} = \{U(t,s)\}_{t\geq s\geq 0}$ *and* $\mathbf{V} = \{V(t,s)\}_{t\geq s}$ *be exponentially bounded evolution families of bounded linear operator acting on* X *such that for each* $s \geq 0$ *each* $t \in \mathbb{R}$ *and each* $x \in X$ *the maps* $\tau \mapsto ||U(\tau,s)x|| : [s,\infty) \to \mathbb{R}_+$ *and* $\tau \mapsto ||V(t,\tau)|| : (-\infty,t] \to \mathbb{R}_+$ *are measurable. If there exists a positive real number* α *such that*

$$\int_s^\infty \phi(e^{\alpha t}||U(t,s)x||)dt = M_\phi(s) < \infty, \quad \forall s \geq 0, \forall x \in X, ||x|| \leq 1$$

and if for each $t \in \mathbb{R}$

$$\int_{-\infty}^t \phi(e^{-\alpha\tau}||V(t,\tau)||)d\tau < \infty \tag{1.1}$$

then the families $\mathbf{U}$ *and* $\mathbf{V}$ *are non-uniformly exponentially stable.*

We refer to [4], [6] for further details and proofs.

The variant of (1.1), in the case $J = \mathbb{R}_+$, says that for each $t \geq 0$ the inequality

$$\int_0^t \phi(e^{-\alpha\tau}||V(t,\tau)||)d\tau < \infty,$$

holds true. However this condition is trivially verified when the map $\tau \mapsto ||V(t,\tau)||$ is continuous. Therefore, it cannot imply the nonuniform exponential stability of the family V. Moreover, in the literature we did not find characterizations of nonuniform exponential stability in terms of the real integrability conditions along the norm of the trajectories of the evolution family V, similar to (1.1). The aim of this paper is to provide such characterizations.

2. Notations and preliminary results

Let X be a real or complex Banach space and X^* its dual space. By $\mathcal{B}(X)$ will denote the Banach algebra of all linear and bounded operators acting on X. The norms on X, X^* and $\mathcal{B}(X)$ will be denoted by the symbol $||\cdot||$. Let $\mathbb{R}_+ := [0,\infty)$ and let J either $\mathbb{R}$ or $\mathbb{R}_+$. By Δ_J will denote the set of all pairs $(t,s) \in J \times J$ with $t \geq s$. By an *evolution family* of bounded linear operators acting on X we mean a family $\mathbf{U} = \{U(t,s) : (t,s) \in \Delta_J\} \subset \mathcal{B}(X)$ satisfying the following two conditions:

1. $U(t,t) = I$-the identity of $\mathcal{B}(X)$- for all $t \in J$, and
2. $U(t,s)U(s,r) = U(t,r)$ for all $t,s,r \in J$ with $t \geq s \geq r$. If the latter condition holds for all $t,s,r \in J$ we say that $\mathbf{U}$ is a *reversible evolution family* on X.

Let $q > 0$. We say that an evolution family is *q-periodic* if

$$U(t,s) = U(t+q, s+q) \quad \text{for all } (t,s) \in \Delta_J.$$

An evolution family is called *strongly continuous* if for each $x \in X$ the maps

$$\tau \mapsto U(\tau,s)x : [s,t] \to X \text{ and } s \mapsto U(t,s)x : [s,t] \to X$$

are continuous for any pair $(t,s) \in \Delta_J$. An evolution family is *exponentially bounded* if there exist $\omega \in \mathbb{R}$ and $M_\omega \geq 1$ such that

$$||U(t,s)|| \leq M_\omega e^{\omega(t-s)} \text{ for all } (t,s) \in \Delta_J. \tag{2.1}$$

We say that a reversible evolution family is exponentially bounded if there exist $\omega \in \mathbb{R}$ and $M_\omega \geq 1$ such that

$$||U(t,s)|| \leq M_\omega e^{\omega|t-s|} \quad \forall t,s \in J. \tag{2.2}$$

An important example of reversible exponentially bounded evolution family is given by:

Example 1. Let $J = \mathbb{R}_+$ and $t \mapsto A(t) : J \to \mathcal{B}(X)$ be a locally integrable function such that $\sup_{t\geq 0} \int_t^{t+1} ||A(\tau)d\tau < \infty$, and let $U(t)$ be the solution of the Cauchy problem

$$\dot{V}(t) + A(t)V(t) = 0 \quad t \geq 0, \quad V(0) = I. \tag{$A(t)$}$$

Put $U(t,s) := U(t)U^{-1}(s)$ for $t \geq 0$ and $s \geq 0$. The family $\mathbf{U} = \{U(t,s), t \geq 0, s \geq 0\}$ is an exponentially bounded reversible evolution family on X, see e. g. [9].

In order to introduce the concept of nonuniform exponential stability and show that it is quite natural we prove the following:

Proposition 1. *Let $J = \mathbb{R}_+$ and $\mathbf{U}$ be an exponentially bounded reversible evolution family of bounded linear operators acting on the Banach space X. The following four statements are equivalent:*

1. *There exist two maps $N : [0,\infty) \to (0,\infty)$ and $\nu : [0,\infty) \to (0,\infty)$ such that*
$$||U(t,s)|| \leq N(s)e^{-\nu(s)(t-s)} \quad \text{for all } t \geq s \geq 0. \tag{2.3}$$
2. *There exist a function $M : [0,\infty) \to (0,\infty)$ and a positive real number μ such that*
$$||U(t,s)|| \leq M(s)e^{-\mu(t-s)} \quad \forall t \geq s \geq 0. \tag{2.4}$$
3. *There exist two positive constants K and ρ such that*
$$||U(t,0)|| \leq Ke^{-\rho t} \text{ for all } t \geq 0. \tag{2.5}$$
4. *There exist three constants $L > 0$, $a > 0$ and $b \in \mathbb{R}$ such that*
$$||U(t,s)|| \leq Le^{-at}e^{bs} \text{ for all } t \geq s \geq 0. \tag{2.6}$$

Proof. The inequality (2.3) with $s = 0$ yields the inequality (2.5). The inequality (2.6) may be rewritten as

$$||U(t,s)|| \leq Le^{(b-a)s}e^{-a(t-s)} \text{ for all } t \geq s \geq 0,$$

i.e., (2. 4) holds true. Assume that (2.5) fulfils and obtain:

$$||U(t,s)|| = ||U(t,0)U(0,s)|| \leq ||U(t,0)||||U(0,s)|| \leq Ke^{-\rho t}M_\omega e^{\omega s}.$$

This is a variant of (2.6). □

An evolution family (reversible or not) is called:

(i) *uniformly bounded* if
$$\sup_{(t,s)\in\Delta_J} ||U(t,s)|| \leq M_U < \infty.$$

(ii) *non-uniformly strongly stable* if for each $x \in X$ and each $s \in J$ we have that $\lim_{t\to\infty} U(t,s)x = 0$.

(iii) *uniformly exponentially stable* if there exist two positive constants N and ν such that $||U(t,s)|| \leq Ne^{-\nu(t-s)}$ for all $t, s \in J$ with $t \geq s$.

(iv) *non-uniformly exponentially stable* if there exist a function $N : [0,\infty) \to (0,\infty)$ and a positive real number ν such that
$$||U(t,s)|| \leq N(s)e^{-\nu(t-s)} \text{ for all } t, s \in J \text{ with } t \geq s.$$

The next example shows that there exist exponentially bounded reversible evolution families which are uniformly bounded and non-uniformly exponentially stable and are not uniformly exponentially stable.

Example 2. By $\mathbf{C}$ we denote the set of all complex numbers. Let us consider the map

$$t \mapsto A(t) := \sqrt{2} - \sin \ln(t+1) - \cos \ln(t+1) : \mathbb{R}_+ \to \mathbf{C}.$$

The solution $P(\cdot)$ of the Cauchy problem $(A(t))$ with the initial condition $U(0) = 1$ is $P(t) = \exp(-\int_0^t A(\tau)d\tau), t \geq 0$. It is clear that the map $A(\cdot)$ is bounded, so the family $\mathbf{U} = \{U(t,s) := P(t)P^{-1}(s) : t \geq 0, s \geq 0\}$ of bounded linear operators on $\mathbf{C}$ is a reversible and exponentially bounded evolution family. Moreover, for each $t \geq s \geq 0$ we have that

$$|P(t)P^{-1}(s)| = \exp(-\int_s^t A(\tau)d\tau) \leq 1,$$

that is $\mathbf{U}$ is uniformly bounded.

On the other hand

$$\begin{aligned} |P(t)P^{-1}(0)| &= \exp(-\int_0^t A(\tau)d\tau) = e^{-\sqrt{2}t+(t+1)\sin\ln(t+1)} \\ &= e^{\sqrt{2}}e^{-(t+1)[\sqrt{2}-\sin\ln(t+1)]} \leq e^{\sqrt{2}-0.4}e^{-0.4t} \end{aligned}$$

for all $t \geq 0$, then the evolution family $\mathbf{U}$ is non-uniformly exponentially stable such as been stated in the Proposition 1, above.

In order to prove that the family $\mathbf{U}$ is not uniformly exponentially stable first remark that the following two statements are equivalent. We refer to [11] for theorems of this type in the infinite dimensional case.

(j) There exist constants $a > 0$ and $b \geq 0$ such that

$$\int_s^t A(\tau)d\tau \geq a(t-s) - b \text{ for all } t \geq s \geq 0.$$

(jj) The family $\mathbf{U}$ is uniformly exponentially stable.

Indeed the implication **(j)** $\Rightarrow$ **(jj)** follows by the inequality

$$P(t)P^{-1}(s) = \exp(-\int_s^t A(\tau)d\tau) \leq e^b e^{-a(t-s)} \text{ for all } t \geq s \geq 0,$$

while that the inequality

$$\exp(-\int_s^t A(\tau)d\tau) \leq Ne^{-\nu(t-s)} = e^b e^{-\nu(t-s)}$$

where $N = e^b \geq 1$ (i.e., $b \geq 0$) and $t \geq s \geq 0$, implies that $\int_s^t A(\tau)d\tau \geq -b+\nu(t-s)$, i.e., **(j)** holds with $a = \nu$ and $b = \ln N$. We suppose for a contradiction that the family $\mathbf{U}$ is uniformly exponentially stable. Then there exist $a > 0$ and $b \geq 0$ such that

$$\sqrt{2}(t-s) - [(t+1)\sin\ln(t+1) - (s+1)\sin\ln(s+1)] \geq a(t-s) - b \qquad (2.7)$$

holds for any $t \geq s \geq 0$. Let $g(\xi) := \sqrt{2}\xi - \xi\sin(\ln\xi), \xi \geq 1$. The above inequality (2.7) may be written as

$$g(u) - g(v) \geq a(u-v) - b \text{ for all } u \geq v \geq 1. \qquad (2.8)$$

Let $u_n := e^{2n\pi+\frac{\pi}{4}}$ and $v_n := u_n(1+\frac{1}{n})$. Using the mean value theorem follows that there exists $\xi_n \in (u_n, v_n)$ such that

$$g(u_n) - g(v_n) = (u_n - v_n)\sqrt{2}(1 - \cos\left(\ln\left(\xi_n - \frac{\pi}{4}\right)\right)$$

Then from (2.8) we get

$$\sqrt{2}\left(1 - \cos\left(\ln\left(\xi_n - \frac{\pi}{4}\right)\right)\right) \geq a - \frac{nb}{u_n} \quad \forall n = 1, 2, \ldots \qquad (2.9).$$

But $1 > \cos(\ln(\xi_n - \frac{\pi}{4})) > \cos(\ln(1+\frac{1}{n}))$, so the left side of (2.9) tends to 0 when $n \to \infty$. This leads with the inequality $0 \geq a$, which it is a contradiction.

For a more technical treatment of this example see [5].

Recall that if an evolution family $\mathbf{U}$ satisfies the convolution condition

$$U(t,s) = U(t-s,0) \text{ for all } (t,s) \in \Delta_J$$

then the one parameter evolution family $\mathbf{T} := \{U(t,0), t \geq 0\}$ is a semigroup of operators on X. If $\mathbf{T}$ is strongly continuous then it is exponentially bounded, that is there exist a real numbers ω and $M_\omega \geq 1$ such that $||T(t)|| \leq M_\omega e^{\omega t}$ for all $t \geq 0$. The converse statement is not true but exponentially bounded semigroups possess a certain type of measurability. More precisely we use the following proposition.

Proposition 2. *If a one parameter semigroup $\{T(t)\}_{t\geq 0}$ is exponentially bounded then for each $x \in X$, the map $t \mapsto ||T(t)x||$ is measurable.*

Let $\{T(t)\}_{t\geq 0}$ be a strongly continuous one parameter semigroup on a Banach space X and $\mathbf{T}^* = \{T(t)^*\}_{t\geq 0}$ the associated one parameter dual semigroup on X^*. It is known that the dual semigroup $\mathbf{T}^*$ may be not strongly continuous but it is exponentially bounded because $||T(t)|| = ||T(t)^*||$ for all $t \geq 0$. Then for each $x^* \in X^*$ the map $t \mapsto ||T(t)^* x^*||$ is measurable. We say that an evolution family $\{U(t,s)\} : (t,s) \in \Delta_J$ of bounded linear operators acting on X satisfies the measurability condition on Δ_J, if for each pair $(t,s) \in \Delta_J$, each $x \in X$ and each $x^* \in X^*$ the maps $U(\cdot, s)x$ and $U(t,\cdot)^*)x^*$ are measurable.

Remark.

1. For q-periodic evolution families and for semigroups the notions of uniform exponential stability and nonuniform exponential stability are equivalent, such as been stated in [3], [5].

2. Let $\mathcal{X} := C_{00}(\mathbb{R}_+, X)$ the Banach space of all continuous X-valued functions f on $\mathbb{R}_+$ with the property that $f(0) = f(\infty) = 0$, endowed with the "sup" norm and let $\mathbf{U} = \{U(t,s)\}_{t\geq s\geq 0}$ be an exponentially bounded evolution family on X.

The evolution family $\mathbf{U}$ is uniformly exponentially stable if and only if its associated evolution semigroup $\mathbf{T} = \{T(t)\}_{t\geq 0}$ on $\mathcal{X}$, defined by

$$(T(t)f)(\xi) = \begin{cases} U(\xi, \xi - t)f(\xi - t) & \text{if} \quad \xi \geq t \\ 0, & \text{if} \quad 0 \leq \xi < t \end{cases}$$

is uniformly exponentially stable. See [12] and [8] for further details.

Also it is known that a uniformly stable evolution family $\mathbf{U}$ is non-uniformly strongly stable if and only if its associated evolution semigroup on $\mathcal{X}$ is strongly stable. We refer to [2] for further details and proofs.

3. Real integrability conditions for nonuniform exponential stability

Theorem 1. *Let $p, q \in (1, \infty)$ with $\frac{1}{p} + \frac{1}{q} = 1$ and let $\mathbf{U} = \{U(t,s) : (t,s) \in \Delta_J\}$ be an evolution family of bounded linear operator acting on X which satisfies the measurability condition on Δ_J. The following two statements are equivalent:*

(i) *The evolution family $\mathbf{U}$ is non-uniformly exponentially stable.*

(ii) *There exist a positive real number α and two positive functions $M_p(\cdot)$ and $M_q^*(\cdot)$ on $\mathbb{R}_+$ such that*

$$\sup_{t>s} \left(\frac{1}{t-s} \int_s^t e^{p\alpha(\tau - s)} ||U(\tau, s)x||^p d\tau \right)^{\frac{1}{p}} \leq M_p(s)||x||, \quad \forall s \in J \quad \forall x \in X$$

and

$$\sup_{t>s} \left(\frac{1}{t-s} \int_s^t e^{q\alpha(t-\tau)} ||U(t,\tau)^* x^*||^q \right)^{\frac{1}{q}} \leq M_q^*(s)||x^*||, \quad \forall s \in J \quad \forall x^* \in X^*.$$

Proof. The implication **(i)** $\Rightarrow$ **(ii)** is trivial. We shall prove the implication **(ii)** $\Rightarrow$ **(i)**. Let $x \in X, x^* \in X^*$ and $(t,s) \in \Delta_J$ with $t > s$. Using the Hölder inequality we get:

$$(t-s)|\langle x^*, e^{\alpha(t-s)}U(t,s)x\rangle| = \int_s^t e^{\alpha(t-\tau)} e^{\alpha(\tau-s)} |\langle x^*, U(t,\tau)U(\tau,s)x\rangle| d\tau$$

$$\leq \left(\frac{1}{t-s} \int_s^t e^{q\alpha(t-\tau)} ||U(t,\tau)^* x^*||^q \right)^{\frac{1}{q}} \left(\frac{1}{t-s} \int_s^t e^{p\alpha(\tau-s)} ||U(\tau,s)x||^p d\tau \right)^{\frac{1}{p}} (t-s)$$

$$\leq M_p(s) M_q^*(s) ||x|| ||x^*|| (t-s).$$

Finally we have

$$||U(t,s)|| \leq M_p(s)M_q^*(s)e^{-\alpha(t-s)}, \text{ for all } (t,s) \in \Delta_J,$$

that is the family $\mathbf{U}$ is non-uniformly exponentially stable. □

Corollary 1. *Let $\mathbf{T} = \{T(t)\}_{t\geq 0}$ be a semigroup of bounded linear operators acting on X, and let $p, q \in (1,\infty)$ with $1/p + 1/q = 1$. We assume that for each $x \in X$ and each $x^* \in X^*$ the maps $t \mapsto ||T(t)x||$ and $t \mapsto ||T(t)^*x^*||$ are measurable and moreover there exists a positive real number α such that*

$$\sup_{t>0} \frac{1}{t} \int_0^t e^{p\alpha s}||T(s)x||^p ds < \infty \tag{3.1}$$

and

$$\sup_{t>0} \frac{1}{t} \int_0^t e^{q\alpha s}||T(s)^*x^*||^q ds < \infty. \tag{3.2}$$

Then the semigroup $\mathbf{T}$ is uniformly exponentially stable.

It is worth to mention that the previous corollary is stated under the minimal assumption that the maps $t \mapsto ||T(t)x||$ and $t \mapsto ||T(t)^*x^*||$ are measurable. If the semigroup $\mathbf{T}$ is exponentially bounded the above maps are measurable and moreover either one of the conditions (3.1) or (3.2) can be dropped. We leave open the question whether the measurability conditions imply the exponential boundedness of $\mathbf{T}$.

Theorem 2. *Let $\phi : \mathbb{R}_+ \to \mathbb{R}_+$ be a nondecreasing function such that $\phi(t) > 0$ for all $t > 0$ and let $\mathbf{U} = \{U(t,s)\}_{(t,s)\in\Delta_J}$ be an exponentially bounded evolution family on a Banach space X which satisfies the measurability condition on Δ_J. If there exists $\alpha > 0$ such that for each $s \in J$ the following inequality*

$$\sup_{t>s} \int_{-\infty}^{\infty} \phi(\chi_{[s,t]}(\tau)e^{\alpha(\tau-s)}||U(\tau,s)x||)d\tau := M_\phi(s) < \infty$$

holds true for every $x \in X$ with $||x|| \leq 1$, then the family $\mathbf{U}$ is non-uniformly exponentially stable.

Proof. Let $\omega > 0$ and $M_\omega \geq 1$ such that $||U(t,s)|| \leq M_\omega e^{\omega(t-s)}$ for every $t \geq s$. Let $x \in X$, as above, and let $x^* \in X^*$ be a non-zero vector. We may suppose that $\phi(1) = 1$ otherwise we replace ϕ be an appropriate multiple of itself. Let $N = N(s) > M_\phi(s)$, $t \geq s + N(s)$ and $t - N \leq \tau \leq t$. We have successively:

$$\begin{aligned}
&\chi_{[t-N,t]}(\tau)e^{-(\omega+\alpha)N(s)}|\langle x^*, e^{\alpha(t-s)}U(t,s)x\rangle| \\
&\leq \chi_{[t-N,t]}(\tau)e^{-\omega(t-\tau)}||U(t,\tau)^*||e^{\alpha(\tau-s)}||U(t,s)x||||x^*|| \\
&\leq M_\omega||x^*||\chi_{[s,t]}(\tau)e^{\alpha(\tau-s)}||U(\tau,s)x||.
\end{aligned}$$

Using the properties of ϕ we get:

$$N\phi(M_\omega^{-1}||x^*||^{-1}e^{-(\omega+\alpha)N}|\langle x^*, e^{\alpha(t-s)}U(t,s)x\rangle|)$$
$$= \int_{-\infty}^{\infty} \phi(\chi_{[t-N,t]}(\tau)M_\omega^{-1}||x^*||^{-1}e^{-(\omega+\alpha)N}|\langle x^*, e^{\alpha(t-s)}U(t,s)x\rangle|)d\tau$$
$$\leq \int_{-\infty}^{\infty} \phi(e^{\alpha(\tau-s)}\chi_{[s,t]}(\tau)||U(\tau,s)x||)d\tau \leq M_\phi(s) < N(s).$$

Now it is easy to see that:

$$||U(t,s)|| \leq M_\omega e^{(\omega+\alpha)N(s)}e^{-\alpha(t-s)}$$

for all $t \geq s$, that is the family **U** is non-uniformly exponentially stable. □

A simple calculations shows that:

$$\int_s^t \phi(e^{\alpha(\tau-s)}||U(\tau,s)x||)d\tau = \int_0^{t-s} \phi(e^{\alpha u}||U(s+u,s)x||)du.$$

Thus we have

Corollary 2. *Let* **U** *be an exponentially bounded evolution family on* X *which satisfies the measurability condition on* Δ_J*. If there exist* ϕ*, as above, and a positive real number* α *such that for each* $s \in J$ *the inequality*

$$\int_s^\infty \phi(e^{\alpha t}e^{-\alpha s}||U(t,s)x||)dt = \int_0^\infty \phi(e^{\alpha u}||U(s+u,s)x||)du < \infty, \tag{3.3}$$

holds true for every $x \in X$ *with* $||x|| \leq 1$*, then the family* **U** *is non-uniformly exponentially stable.*

In the case $J = \mathbb{R}_+$ from (3.3) follows also the following known result: *Let* **U**, ϕ *and* α *be as above. If for each* $s \geq 0$ *and each* $x \in X$ *we have that*

$$\int_s^\infty \phi(e^{\alpha t}||U(t,s)x||)dt < \infty$$

then the family $\mathbf{U} = \{U(t,s)\}_{t\geq s\geq 0}$ *is non-uniformly exponentially stable.* See [3].

Theorem 3. *Let* $\mathbf{U} = \{U(t,s)\}_{t\geq s\in J}$ *be an evolution family on a Banach space* X *which satisfies the measurability condition on* Δ_J *and is non-uniformly exponentially bounded, that is there exist two positive maps* ω *and* M_ω *on* J *such that*

$$||U(t,s)|| \leq M_\omega(s)e^{\omega(s)(t-s)}, \quad \forall (t,s) \in \Delta_J.$$

The following two statements are equivalent:

(i) *The family* **U** *is non-uniformly stable.*

(ii) *There exist three positive real numbers α, β and K such that for some $q \geq 1$ and each $t \in J$ the inequality*

$$\left(\int_J \chi_{(-\infty,t]}(\tau) e^{-q\alpha\tau} ||U(t,\tau)^* x^*||^q d\tau \right)^{\frac{1}{q}} \leq K e^{-\beta t}$$

holds true, for all $x^* \in X^*$ with $||x^*|| \leq 1$.

Proof. We consider the case $q = 1$ and $q > 1$ separately.

Case $q = 1$. Let $s \in J$, $N > 0$ and let $s \leq \tau \leq s + N \leq t$. Then for each $x \in X$ and each $x^* \in X^*$ with $||x^*|| \leq 1$, have that:

$$\begin{aligned} & N e^{-\omega(s)+\alpha)N} |\langle x^*, e^{\alpha(t-s)} U(t,s)x \rangle| \\ & \quad \leq \int_J \chi_{[s,s+N]}(\tau) e^{-(\omega(s)+\alpha)(\tau - s)} e^{\alpha(\tau-s)} ||U(\tau,s)x|| e^{\alpha(-\tau)} ||U(t,\tau)^* x^*|| d\tau \\ & \quad \leq M_\omega(s) \int_J \chi_{(-\infty,t]}(\tau) ||x|| e^{\alpha(-\tau)} ||U(t,\tau)^* x^*|| d\tau \leq M_\omega(s) K e^{-\beta t}, \end{aligned}$$

$$||U(t,s)|| \leq K \frac{M_\omega(s)}{N} e^{(\omega(s)+\alpha)N} e^{\alpha s} e^{-\beta t}.$$

Case $q > 1$. Let t, s, τ and N as above and let $p > 1$ such that $\frac{1}{p} + \frac{1}{q} = 1$. Using the Hölder inequality we have that:

$$\begin{aligned} & N e^{-(\omega(s)+\alpha)N} |\langle x^*, e^{\alpha(-s)} U(t,s)x \rangle| \\ & \leq \int_J \chi_{[s,s+N]}(\tau) e^{-(\omega(s)+\alpha)(\tau-s)} e^{\alpha(\tau-s)} ||U(\tau,s)x|| e^{\alpha(-\tau)} ||U(t,\tau)^* x^* d\tau \\ & \leq \left(\int_J M_\omega^p(s) \chi_{[s,s+N]}(\tau) ||x||^p d\tau \right)^{\frac{1}{p}} \left(\int_J \chi_{(-\infty,t]}(\tau) e^{\alpha(-q\alpha\tau)} ||U(t,\tau)^* x^*||^q d\tau \right)^{\frac{1}{q}} \\ & \leq N^{\frac{1}{p}} K M_\omega(s) ||x|| e^{-\beta t}. \end{aligned}$$

This yields

$$||U(t,s)|| \leq N^{-\frac{1}{q}} K M_\omega(s) e^{(\omega(s)+\alpha)N} e^{\alpha s} e^{-\beta t}.$$

A similar inequality can be easily obtained, in the both cases, for $t \in [s, s+N]$. □

Corollary 3. *Let $\mathbf{U} = \{U(t,s)\}_{t \geq s \geq 0}$ and q be as in the above Theorem 3 with $J = \mathbb{R}_+$. If there exist three positive real numbers α, β and K such that for each*

$t \geq 0$ the inequality

$$\int_0^t e^{-q\alpha\tau}||U(t,\tau)^*x^*||^q d\tau \leq Ke^{-q\beta t}$$

*holds true for every $x^* \in X^*$ with $||x^*|| \leq 1$, then the family **U** is non-uniformly exponentially stable.*

A similar corollary can be formulated for $J = \mathbb{R}$. For semigroups the above Theorem 3 yields the following interesting variant of the Datko-Pazy Theorem:

Corollary 4. *An exponentially bounded semigroup $\mathbf{T} = \{T(t)\}_{t\geq 0}$ on a Banach space X is uniformly exponentially stable if and only if there exist three positive real numbers α, β and K such that for each $t \geq 0$ the inequality:*

$$\int_0^t e^{q\alpha\rho}||T(\rho)^*x^*||^q d\rho \leq Ke^{q(\alpha-\beta)t}$$

holds true for all $x^* \in X^*$ with $||x^*|| \leq 1$ and some $q \geq 1$.

Proof. The measurability of the function $\rho \mapsto ||T(\rho)^*x^*|| : \mathbb{R}_+ \to \mathbb{R}_+$ is a consequence of the Proposition 1 stated above. □

Acknowledgement

This paper was completed while the author was a Research Assistant of the "Mathematics and Applications" research unit of the University of Aveiro, supported by the "Ciência 2007" program of FCT-the Portuguese Foundation for Science and Technology.

References

[1] E.A. Barbashin, *Introduction in the theory of stability*, Izd. Nauka, Moscow, 1967 (Russian).

[2] C. Batty, J.K. Chill, Y. Tomilov, *Strong stability of bounded evolution families and semigroups*, J. Funct. Anal. **193** (2002), 116–139.

[3] C. Buşe, *On nonuniform exponential stability of evolutionary processes*, Rend. Sem. Univ. Pol. Torino, **52**(4) (1994), 395–406.

[4] C. Buşe, *Nonuniform exponential stability and Orlicz functions*, Comm. Math. Prace Matematyczne, **36** (1996), 34–97.

[5] C. Buşe, *Asymptotic stability of evolutors and normed functions spaces*, Rend. Sem. Mat. Univ. Pol. Torino **55** (2) (1997), 109–122.

[6] C. Buşe, S.S. Dragomir, *A theorem of Rolewicz's type in solid function spaces*, Glasgow Math. Journal **44** (2002), 125–135.

[7] C. Chicone, Yuri Latushkin, *Evolution Semigroups in Dynamical Systems and Differential Equations*, Amer. Math. Soc., Math. Surv. and Monographs **70** (1999).

[8] S. Clark, Yu. Latushkin, S. Montgomery-Smith, and T. Randolph, *Stability radius and internal versus external stability in Banach spaces: An evolution semigroup approach*, SIAM J. Control and Optimization **38**(200), 1757–1793.

[9] Y.L. Daletkii, M.G. Krein, *Stability of Solutions of Differential Equations in Banach Spaces*, Izd. Nauka, Moskwa, 1970.

[10] R. Datko, *Uniform asymptotic stability of evolutionary processes in Banach space*, SIAM J. Math. Analysis **3** (1973), 428–445.

[11] E. Nakaguchi, S. Saito, A. Yagi, *Exponential decay of solutions of linear evolution equations in a Banach space*, Math. Japon. **44**(1996), 469–481.

[12] Van Minh, Nguyen, F. Räbiger, and R. Schnaubelt, *Exponential stability, exponential expansiveness, and exponential dichotomy of evolution equations on the half-line*, Integral Equations Operator Theory **32** (1998), 332–353.

[13] A. Pazy, *Semigroups of linear operators and applications to partial differential equations*, Springer-Verlag, 1983.

[14] S. Rolewicz, *On uniform N-equistability,* Journal of Math. Anal. Appl. **115** (1986), 434–441.

[15] J.M.A.M. van Neerven, *Exponential stability of operators and operator semigroups*, Journal of Functional Analysis **130** (1995), 293–309.

Constantin Buşe
West University of Timisoara
Department of Mathematics
Bd. V. Parvan, No. 4
RO-300223 Timisoara, Romania
e-mail: constantin@ua.pt

Current address: University of Aveiro
Research unit of "Mathematics and Applications"
Department of Mathematics
3810-193 AVEIRO, Portugal
e-mail: buse@math.uvt.ro

International Series of Numerical Mathematics, Vol. 157, 43–50

Validated Computations for Fundamental Solutions of Linear Ordinary Differential Equations

Kaori Nagatou

Abstract. We present a method to enclose fundamental solutions of linear ordinary differential equations, especially for a one dimensional Schrödinger equation which has a periodic potential. Our method is based on Floquet theory and Nakao's verification method for nonlinear equations. We show how to enclose fundamental solutions together with characteristic exponents and give a numerical example.

Mathematics Subject Classification (2000). 34B05, 34B15, 65L10.

Keywords. Fundamental solution, Floquet theory, verification method.

1. Introduction

We consider to compute fundamental solutions for the following equation

$$L\psi \equiv -\psi'' + q(x)\psi = 0, \ x \in \mathbb{R}, \tag{1}$$

where we assume that q is a bounded, continuous and periodic function with a period $r > 0$.

By Floquet Theory there exist fundamental solutions $\psi_1(x), \psi_2(x)$ of $L\psi = 0$ s.t.

$$\psi_1(x) = e^{\mu x} p_1(x), \ \psi_2(x) = e^{-\mu x} p_2(x), \tag{2}$$

or

$$\psi_1(x) = e^{\mu x} p_1(x), \ \psi_2(x) = e^{\mu x}(x p_1(x) + p_2(x)), \tag{3}$$

where μ is the characteristic exponent and p_1, p_2 are $r-$periodic functions. In this paper we assume that $|\phi_1(r) + \phi_2'(r)| > 2$ for the solutions ϕ_1, ϕ_2 of (6) and (7), and note that this excludes the case (3), and allows to choose μ in (2) with positive real part. Our aim is to compute ψ_1, ψ_2 with guaranteed accuracy.

Once we obtained enclosures for such fundamental solutions, we may use them for another problem. For example, using these guaranteed enclosures for the fundamental solutions we can enclose the Green's function $G(x,y)$ defined by [1]

$$G(x,y) = \begin{cases} \psi_1(x)\psi_2(y)/W(\psi_1,\psi_2)(x) & (x \le y) \\ \psi_2(x)\psi_1(y)/W(\psi_1,\psi_2)(x) & (x \ge y) \end{cases} \tag{4}$$

for $-\infty < x, y < +\infty$, where $W(\psi_1,\psi_2)(x) \equiv \psi_1(x)\psi_2'(x) - \psi_1'(x)\psi_2(x)$ stands for the Wronskian. Since $Re(\mu) > 0$, G decays exponentially as $y \to -\infty$ and $y \to +\infty$, and we have [1]

$$-(L^{-1}f)(x) = \int_{\mathbb{R}} G(x,y)f(y)dy. \tag{5}$$

This kind of expression for L^{-1} is useful to execute another verification method (e.g., for a spectral problem of a perturbed Schrödinger operator) related to the selfadjoint operator $L : \mathcal{D}(L) \subset L^2(\mathbb{R}) \to L^2(\mathbb{R})$ defined on a suitable dense subspace $\mathcal{D}(L) \subset L^2(\mathbb{R})$; see [1] for details on the construction of this operator realizing the differential expression (1).

2. Verification for fundamental solutions

In order to verify the fundamental solutions ψ_1 and ψ_2 in (1) satisfying (2), it is sufficient, as explained at the end of this section, to enclose the functions ϕ_1 and ϕ_2 which are solutions for the following equations:

$$\begin{cases} -\phi_1''(x) + q(x)\phi_1(x) = 0 & \text{in } [0,r] \\ \phi_1(0) = 1,\ \phi_1'(0) = 0 \end{cases} \tag{6}$$

$$\begin{cases} -\phi_2''(x) + q(x)\phi_2(x) = 0 & \text{in } [0,r] \\ \phi_2(0) = 0,\ \phi_2'(0) = 1 \end{cases} \tag{7}$$

Let S_h denote the set of continuous and piecewise linear functions on $[0,r]$ with uniform mesh $0 = x_0 < x_1 < \cdots < x_{N+1} = r$ and mesh size h. We define a function space

$$V \equiv W^1_{\infty,0}(0,r) \cap \left(\bigwedge_{i=0}^{N} C^1[x_i, x_{i+1}] \right),$$

and we define the norm for $(w,\mu) \in V \times \mathbb{R}$ by

$$\|(w,\mu)\|_{V\times\mathbb{R}} \equiv \max\{\|w\|_{W^1_{\infty,0}},\ |\mu|\},$$

where $W^1_{\infty,0}(0,r)$ is a usual Sobolev space defined by

$$W^1_{\infty,0}(0,r) \equiv \{\phi \in W^1_\infty(0,r) \mid \phi(0) = \phi(r) = 0\}$$

Setting $\phi_1(r) = \kappa$, $\phi_2(r) = \tau$ and transforming

$$\widetilde{\phi_1}(x) \equiv \phi_1(x) + \frac{1-\kappa}{r}x - 1, \quad \widetilde{\phi_2}(x) \equiv \phi_2(x) - \frac{\tau}{r}x,$$

we consider the following equivalent problems:

Find $(\widetilde{\phi}_1, \kappa) \in V \times \mathbb{R}$ **s.t.**

$$\begin{cases} -\widetilde{\phi}_1''(x) + q(x)\left(\widetilde{\phi}_1(x) + \dfrac{\kappa - 1}{r}x + 1\right) & = \quad 0 \quad \text{in } [0, r] \\ \widetilde{\phi}_1(0) = \widetilde{\phi}_1(r) = 0 \\ \widetilde{\phi}_1'(0) = \dfrac{1-\kappa}{r} \end{cases} \tag{8}$$

Find $(\widetilde{\phi}_2, \tau) \in V \times \mathbb{R}$ **s.t.**

$$\begin{cases} -\widetilde{\phi}_2''(x) + q(x)\left(\widetilde{\phi}_2(x) + \dfrac{\tau}{r}x\right) & = \quad 0 \quad \text{in } [0, r] \\ \widetilde{\phi}_2(0) = \widetilde{\phi}_2(r) = 0 \\ \widetilde{\phi}_2'(0) = 1 - \dfrac{\tau}{r} \end{cases} \tag{9}$$

Below we describe how to enclose $(\widetilde{\phi}_1, \kappa) \in V \times \mathbb{R}$ in (8). The enclosure for $(\widetilde{\phi}_2, \tau) \in V \times \mathbb{R}$ is analogous.

Let $P_{h0} : V \to S_h$ denote the H_0^1-projection defined by

$$(\nabla(u - P_{h0}u), \nabla v)_{L^2} = 0 \text{ for all } v \in S_h,$$

and define the projection $P_h : V \times \mathbb{R} \to S_h \times \mathbb{R}$ by

$$P_h(u, \lambda) \equiv (P_{h0}u, \lambda).$$

Now, let $(\widetilde{\phi}_{1,h}, \kappa_h) \in S_h \times \mathbb{R}$ be the exact finite element solution of (8). (Here we assume that there exists a unique finite element solution. This assumption can be checked in the actual verified computation.) We will verify the solution $(\widetilde{\phi}_1, \kappa)$ in the neighborhood of $(\bar{\phi}_1, \kappa_h)$ satisfying

$$\begin{cases} -\bar{\phi}_1{}''(x) & = & -q(x)\left(\widetilde{\phi}_{1,h}(x) + \dfrac{\kappa_h - 1}{r}x + 1\right) & \text{in } (0, r), \\ \bar{\phi}_1(x) & = & 0 & \text{for } x = 0, r. \end{cases} \tag{10}$$

Notice that $\bar{\phi}_1 \in W_{\infty,0}^1(0, r) \bigcap W_\infty^2(0, r)$, and $\widetilde{\phi}_{1,h} = P_{h0}\bar{\phi}_1$. Defining $w = \widetilde{\phi}_1 - \bar{\phi}_1$, $v_0 = \bar{\phi}_1 - \widetilde{\phi}_{1,h}$, $\mu = \kappa - \kappa_h$, we have

$$\begin{cases} -w''(x) = -q(x)\left(w(x) + v_0(x) + \dfrac{\mu}{r}x\right) & \text{in } (0, r) \\ w(0) = w(r) = 0 \\ w'(0) = \dfrac{1 - \mu - \kappa_h}{r} - v_0'(0) - \widetilde{\phi}_{1,h}'(0) \end{cases} \tag{11}$$

Thus using the following compact map on $V \times \mathbb{R}$

$$\begin{aligned} F(w, \mu) \quad \equiv \quad & \left((-\Delta)^{-1}\left\{-q(x)\left(w(x) + v_0(x) + \frac{\mu}{r}x\right)\right\},\right. \\ & \left.\mu + w'(0) - \frac{1 - \mu - \kappa_h}{r} + v_0'(0) + \widetilde{\phi}_{1,h}'(0)\right), \end{aligned} \tag{12}$$

where $(-\Delta)^{-1}$ means the solution operator for the Poisson equation with homogeneous boundary condition, we have the fixed point equation for $z = (w, \mu)$

$$z = F(z). \tag{13}$$

Now we decompose (13) into finite- and infinite-dimensional parts:

$$\begin{cases} P_h(z) &= P_h F(z), \\ (I - P_h)(z) &= (I - P_h)F(z). \end{cases} \tag{14}$$

We define the operator

$$\mathcal{N}_h(z) \equiv P_h(z) - [I - F']_h^{-1}(P_h(z) - P_h F(z)),$$

where we omitted to express the evaluation point for the Fréchet derivative F' because the operator F is affine, and assumed that the restriction to $S_h \times \mathbb{R}$ of the operator $P_h[I - F'] : V \times \mathbb{R} \to S_h \times \mathbb{R}$ has an inverse

$$[I - F']_h^{-1} : S_h \times \mathbb{R} \to S_h \times \mathbb{R}.$$

This assumption can be numerically checked in the actual computation.

We next define the operator $T : V \times \mathbb{R} \to V \times \mathbb{R}$ as

$$T(z) \equiv \mathcal{N}_h(z) + (I - P_h)F(z). \tag{15}$$

Then T becomes a compact map on $V \times \mathbb{R}$, and

$$z = T(z) \Leftrightarrow z = F(z) \tag{16}$$

holds.

Our purpose is to find a fixed point of T in a certain set $Z \subset V \times \mathbb{R}$, which is called a 'candidate set'. Given positive real numbers α, γ and γ_R we define the corresponding candidate set Z by

$$Z \equiv Z_h + [\alpha], \tag{17}$$

where

$$Z_h \equiv \{z_1 \in S_h \mid \|z_1\|_{W^1_{\infty,0}} \leq \gamma\} \times \{z_2 \in \mathbb{R} \mid |z_2| \leq \gamma_R\}, \tag{18}$$

$$[\alpha] \equiv \{z_\perp \in S_h^\perp \times \{0\} \mid \|w_\perp\|_{W^1_{\infty,0}} \leq \alpha\}. \tag{19}$$

Here $S_h^\perp$ denotes the orthogonal complement of S_h in $H_0^1(0, r)$. We define the two operators $\mathcal{N}_{h,1}$ and $\mathcal{N}_{h,1}$ by

$$\mathcal{N}_{h,1} : V \times \mathbb{R} \to V, \ \mathcal{N}_{h,2} : V \times \mathbb{R} \to \mathbb{R}$$

such that $\mathcal{N}_h = (\mathcal{N}_{h,1}, \mathcal{N}_{h,2})$. If the relation

$$T(Z) \subset Z \tag{20}$$

holds, by Schauder's fixed point theorem, there exists a fixed point of T in Z. Decomposing $T(Z) \subset Z$ into finite- and infinite-dimensional parts we have a sufficient conditions for it as follows:

$$\begin{cases} \sup_{z\in Z} \|\mathcal{N}_{h,1}(z)\|_V \le \gamma, \\ \sup_{z\in Z} |\mathcal{N}_{h,2}(z)| \le \gamma_R, \\ \sup_{z\in Z} \|(I-P_h)F(z)\|_{W^1_{\infty,0}\times\{0\}} \le \alpha. \end{cases} \tag{21}$$

We find γ, γ_R and α which satisfy the conditions (21) by an iteration method. Details (on a more general level) can be found, e.g., in [4, 6, 7].

After enclosing $\phi_1(x)$ and $\phi_2(x)$ by the method mentioned above, we evaluate $\phi_1(r)$ and $\phi_2'(r)$ rigorously. Then we can calculate the real values ρ_1 and ρ_2 which are solutions of the quadratic equation:

$$\rho^2 - \{\phi_1(r) + \phi_2'(r)\}\rho + 1 = 0. \tag{22}$$

Here ρ_1 and ρ_2 are obtained as real values by our assumption that $|\phi_1(r)+\phi_2'(r)| > 2$. Note that ρ_1 and ρ_2 are the characteristic multipliers for $L\psi = 0$ and the characteristic exponents μ_1 and μ_2 are calculated by $\mu_1 \equiv r^{-1}\log\rho_1$, $\mu_2 \equiv -\mu_1$. (Note that $e^{r\mu_i} = \rho_i$ $(i = 1, 2)$ holds.)

Here we mention about the relation between ϕ_1, ϕ_2 and ψ_1, ψ_2. We define the matrix A by

$$A = \begin{pmatrix} \phi_1(r) & \phi_1'(r) \\ \phi_2(r) & \phi_2'(r) \end{pmatrix}.$$

Since we enclose $\phi_1(r)$, $\phi_1'(r)$, $\phi_2(r)$ and $\phi_2'(r)$ by intervals, the matrix A is usually an interval matrix. Then clearly ρ_1 and ρ_2 are the eigenvalues of A. (Note that $\rho_1 \neq \rho_2$ holds by our assumption that $|\phi_1(r) + \phi_2'(r)| > 2$.) Let v_1 and v_2 be the corresponding eigenvectors for ρ_1 and ρ_2, respectively. Then we can define ψ_1 and ψ_2 by

$$\begin{pmatrix} \psi_1 \\ \psi_2 \end{pmatrix} \equiv \begin{pmatrix} v_1 & v_2 \end{pmatrix}^{-1} \begin{pmatrix} \phi_1 \\ \phi_2 \end{pmatrix}. \tag{23}$$

Now we define p_1 and p_2 by

$$\begin{pmatrix} p_1 \\ p_2 \end{pmatrix} \equiv \begin{pmatrix} e^{-\mu x}\psi_1 \\ e^{\mu x}\psi_2 \end{pmatrix}, \tag{24}$$

where $\mu \in \{\mu_1, \mu_2\}$ is chosen such that $\mathrm{Re}\mu > 0$, which is possible if $|\phi_1(r) + \phi_2'(r)| > 2$. Then we can observe that $p_i(x + r) = p_i(x)$ $(i = 1, 2)$ and ψ_1 and ψ_2 defined by (23) satisfy the relation (2).

3. Numerical examples

We consider the case $q(x) = 5\cos 2\pi x - 9.0$, $r = 3$ and $N = 2000$ as an example.

The computations were carried out on the DELL Precision WorkStation 340 (Intel Pentium4 2.4GHz) using MATLAB (Ver. 7.0.1). The verification results for

$\widetilde{\phi}_k$ $(k=1,2)$ are shown in Table 1 and 2, and the solutions $(\widetilde{\phi}_1,\kappa)$, $(\widetilde{\phi}_2,\tau)$ are enclosed as

$$\|\widetilde{\phi}_k-\widetilde{\phi}_{k,h}\|_V \le \|v_0\|_{W^1_{\infty,0}}+\alpha+\gamma\ (k=1,2),$$
$$|\kappa-\kappa_h|\le\gamma_R,\ |\tau-\tau_h|\le\tau_R.$$

TABLE 1. Verification Results for $(\widetilde{\phi}_1,\kappa)$

$\|v_0\|_{W^1_{\infty,0}}$	γ	γ_R	α
0.1472	0.4842	2.101×10^{-7}	0.0204

TABLE 2. Verification Results for $(\widetilde{\phi}_2,\tau)$

$\|v_0\|_{W^1_{\infty,0}}$	γ	τ_R	α
0.0061	0.0037	1.56×10^{-9}	1.5188×10^{-4}

From these verified results we could obtain

$$\begin{aligned}\kappa &\in [-9.4586\times10^{-6}, -9.0384\times10^{-6}]\\ \tau &\in [2.5080\times10^{-7}, 2.5392\times10^{-7}]\end{aligned}$$

and finally

$$\mu\in[0.73995\mathbf{353},\ 0.73995\mathbf{485}],$$

which gives the desired fundamental solutions ψ_1 and ψ_2.

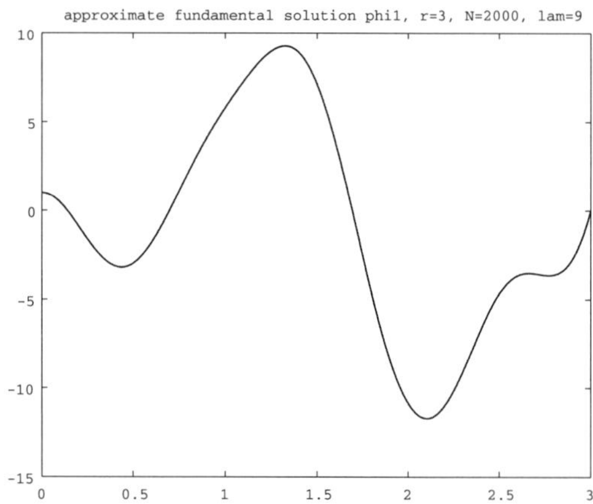

FIGURE 1. Approximate solution for ϕ_1

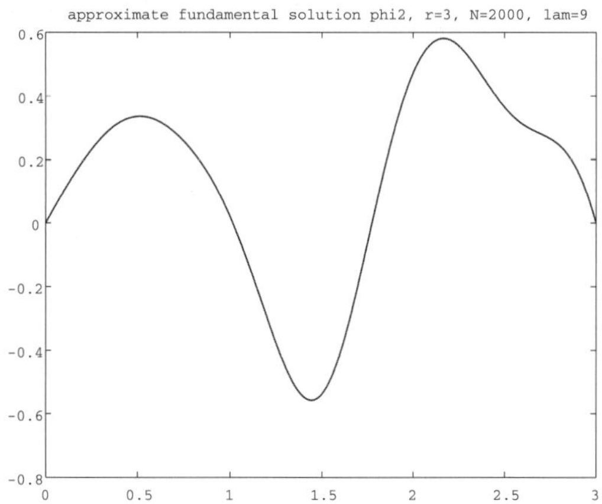

FIGURE 2. Approximate solution for ϕ_2

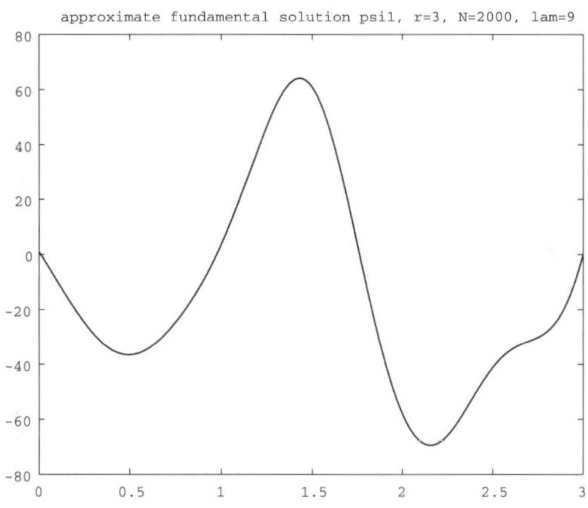

FIGURE 3. Approximate solution for ψ_1

Acknowledgment

The author would like to express many thanks to referees for their important and useful suggestions to improve this paper.

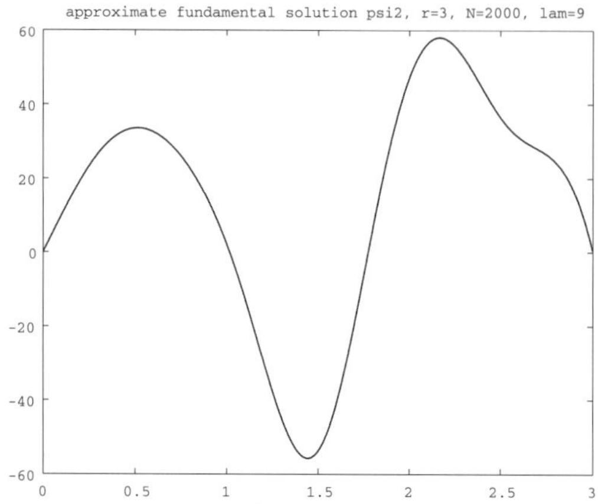

FIGURE 4. Approximate solution for ψ_2

References

[1] M.S.P. Eastham, *The Spectral Theory of Periodic Differential Equations*, Scottish Academic Press, 1973.

[2] K. Nagatou, *A numerical method to verify the elliptic eigenvalue problems including a uniqueness property*, Computing **63** (1999), 109–130.

[3] M.T. Nakao and Y. Watanabe, *An efficient approach to the numerical verification for solutions of the elliptic differential equations*, Numerical Algorithms **37** (2004), 311–323.

[4] M.T. Nakao, K. Hashimoto and Y. Watanabe, *A numerical method to verify the invertibility of linear elliptic operators with applications to nonlinear problems*, Computing **75**(1) (2005), 1–14.

[5] M.T. Nakao, N. Yamamoto and K. Nagatou, *Numerical Verifications for eigenvalues of second-order elliptic operators*, Japan Journal of Industrial and Applied Mathematics, **16**(3) (1999), 307–320.

[6] K. Nagatou, N. Yamamoto and M.T. Nakao, *An approach to the numerical verification of solutions for nonlinear elliptic problems with local uniqueness*, Numerical Functional Analysis and Optimization **20**(5 & 6) (1999), 543–565.

[7] K. Nagatou, K. Hashimoto, and M.T. Nakao, *Numerical verification of stationary solutions for Navier-Stokes problems*, Journal of Computational and Applied Mathematics **199** (2007) 445–451.

[8] M.H. Schultz, *Spline Analysis*, Prentice-Hall, London, 1973.

[9] K. Yosida, *Functional Analysis*, Springer-Verlag, Berlin, 1995.

Kaori Nagatou
Faculty of Mathematics, Kyushu University, Japan, *and*
PRESTO, Japan Science and Technology Agency

Part II
Integral Inequalities

RHB
'07

International Series of Numerical Mathematics, Vol. 157, 53–59

Equivalence of Modular Inequalities of Hardy Type on Non-negative Respective Non-increasing Functions

Sorina Barza and Lars-Erik Persson

Abstract. Some weighted modular integrals inequalities with Volterra type operators are considered. The equivalence of such inequalities on the cones on non-negative respective non-increasing functions is established.

Mathematics Subject Classification (2000). Primary 26D07; 26D15.

Keywords. Weighted inequalities, modular inequalities, Volterra type operators, Orlicz spaces.

1. Introduction

The aim of this paper is to extend some weighted inequalities, proved in [6] and [7], from the Lebesgue setting to more general "modular inequality" settings equipped with Orlicz spaces. Let w be a weight, i.e., a positive and measurable function on $\mathbb{R}$. The Orlicz spaces $L_\Phi(\mathbb{R}, w(x)dx)$ are defined by

$$\|f\|_{L_\Phi(\mathbb{R},w(x)dx)} = \inf_{\lambda>0}\left\{\int_{\mathbb{R}} \Phi\left(\frac{|f(x)|}{\lambda}\right) w(x)dx \leq 1\right\} < \infty \tag{1}$$

where Φ is a Young function. For standard notations and details concerning Orlicz spaces and modular functions we refer to [1]. Let now

$$E := \{f(x) \geq 0, x \in \mathbb{R}_+\},$$

$$E\downarrow := \{f(x) \geq 0, f(x) \text{ is non-increasing for } x \in \mathbb{R}_+\}$$

be two standard cones in the space of Lebesgue measurable functions. A Volterra type integral operator is an operator K defined by

$$Kf(x) = \int_0^x k(x,y)f(y)dy, \tag{2}$$

where the kernel k is nonnegative, nondecreasing in x and nonincreasing in y (such kernels are sometimes called Oinarov kernels). These include the Hardy operator if $k \equiv 1$ and the Riemann-Liouville fractional integral operators if $k(x,y) = (x-y)^\alpha, \quad \alpha > 0$.
In [6, Theorem 2.3] it was proved, in particular, that, for $1 \le p < \infty$ and $0 < q < \infty$, the inequalities

$$\left(\int_{\mathbb{R}_+} (Kf(x))^q\, u(x)dx\right)^{1/q} \le C_1(p,q)\left(\int_{\mathbb{R}_+} (f(x))^p\, dx\right)^{1/p}, \quad f \in E,$$

and

$$\left(\int_{\mathbb{R}_+} (Kf(x))^q\, u(x)dx\right)^{1/q} \le C_2(p,q)\left(\int_{\mathbb{R}_+} (f(x))^p\, dx\right)^{1/p}, \quad f \in E\downarrow,$$

are equivalent and $C_1(p,q) = C_2(p,q)$. In this paper we will generalize this and other similar results to more general modular inequalities.
Here and in the sequel functions are assumed to be measurable, weight functions, denoted by u, v, and w, are locally integrable and the left-hand sides of inequalities exist if the right-hand sides do. Also constants are assumed to be positive and denoted by C or c (sometimes with different subscripts to pronounce that they can be different from each other).

2. Main results

We consider first some auxiliary results that will be used to prove the main results. The notion of level function was first introduced by Halperin [3]. We use the extension of this notion given by Lorentz [2] and based on the following result:

Theorem 1. *Assume that f is nonnegative measurable function on $\mathbb{R}_+$ such that*

$$\lim_{x\to\infty} \frac{\int_0^x f(t)dt}{x} = 0. \tag{3}$$

Then there exists a unique nonnegative decreasing function f° on $\mathbb{R}_+$ satisfying the following conditions:

1. $\int_0^x f(t)dt \le \int_0^x f^\circ(t)dt$
2. *up to a set of measure zero, the set $\{t \in \mathbb{R}_+ : f(t) \ne f^\circ(t)\}$ is the union of bounded disjoint intervals I_k such that*

$$\int_{I_k} f(t)dt = \int_{I_k} f^\circ(t)dt$$

and $f^\circ(t)$ is constant on I_k.

Remark 2. This theorem is a slight modification of the results of Chapter 3.6 of [2]; the proof is similar to the one given in [2] so we omit the details. The function f° is called the level function of f.

Lemma 3. *Let $P \geq 0$ be a convex function on $[0,\infty)$. Let $f \geq 0$ be a bounded function with compact support $\subset R_+$ and let f° be its level function. Then*

$$\int_0^\infty P(f^\circ(x))dx \leq \int_0^\infty P(f(x))dx. \tag{4}$$

Proof. Since f has compact support the condition (3) is satisfied. By Theorem 1 we have that $f^\circ(t) = \lambda_k$ for any $t \in I_k$ and $\lambda_k = \frac{1}{|I_k|}\int_{I_k} f(t)dt$. Hence, by Jensen's inequality, we get that

$$\int_{I_k} P(f^\circ(x))dx = P(\lambda_k)\,|I_k| = P\left(\frac{1}{|I_k|}\int_{I_k} f(t)dt\right)|I_k| \leq \int_{I_k} P(f(x))dx.$$

Since on $[0,\infty) \setminus \cup_{k=1}^\infty I_k$, $f(x) = f^\circ(t)$, by summing up we obtain the inequality (4) and the proof is complete. □

Remark 4. If $P(x) = x^p$, $p \geq 1$, then we get the result of Proposition 2.1 in [6].

Our first main result reads:

Theorem 5. *Let $P \geq 0$ be a convex and increasing function on $[0,\infty)$ and $Q \geq 0$ be an increasing function on $[0,\infty)$ such that $\lim_{x\to 0} Q(x) = 0$ and $\lim_{x\to\infty} Q(x) = \infty$. Let the kernel $k(x,y) : \mathbb{R}_+ \times \mathbb{R}_+ \to \mathbb{R}_+$ be nonincreasing in $y \in [0,x]$ for every x and let K be defined by (2). Then the inequalities*

$$Q^{-1}\left(\int_0^\infty Q(Kf(x)u(x))v(x)dx\right) \leq P^{-1}\left(\int_0^\infty P(C_1 f(x))dx\right), \quad f \in E, \tag{5}$$

and

$$Q^{-1}\left(\int_0^\infty Q(Kf(x)u(x))v(x)dx\right) \leq P^{-1}\left(\int_0^\infty P(C_2 f(x))dx\right), \quad f \in E\downarrow, \tag{6}$$

are equivalent and $C_1 = C_2$.

Remark 6. Note that by choosing $Q(u) = u^q$, $0 < q < \infty$, $P(u) = u^p$, $1 \leq p < \infty$, we obtain exactly the previous mentioned result by Persson-Stepanov-Ushakova [6, Theorem 2.3].

Proof. The implication (5)$\Longrightarrow$ (6) and the inequality $C_2 \leq C_1$ are trivial because $E\downarrow \subset E$. Now we will show the reversed implication and assume that (6) holds. By monotonicity of the kernel $k(x,y)$ we have in the sense of Stieltjes' integral that

$$k(x,y) = k(x,x) + \int_y^x d_z(-k(x,z)),$$

for all $0 \leq y < x < \infty$. Using the above representation and the fact that $\int_0^x f(t)dt \leq \int_0^x f^\circ(t)dt$ it was proved in [6, Theorem 2.3] that

$$Kf(x) \leq Kf^\circ(x)$$

for all bounded $f \in E$ with compact support. Since Q is increasing we also have that

$$Q(Kf(x)u(x))v(x) \leq Q(Kf^\circ(x)u(x))v(x), \quad x \geq 0,$$

and, hence, by our assumption and by Lemma 3 (see inequality (4)) we get that

$$Q^{-1}\left(\int_0^\infty Q(Kf(x)u(x))v(x)dx\right) \le Q^{-1}\left(\int_0^\infty Q(Kf^\circ(x)u(x))v(x)dx\right)$$
$$\le P^{-1}\left(\int_0^\infty P(C_2 f^\circ(x))dx\right) \le P^{-1}\left(\int_0^\infty P(C_2 f(x))dx\right).$$

Hence (5) holds and we also have that $C_1 \le C_2$. For arbitrary $f \in E$ the implication follows by a standard approximation argument. The proof is complete. □

As remarked in [4, Proposition 3, p. 115] the integral modular inequalities are equivalent to the uniform boundedness of a family of norm inequalities. More precisely we have the following general result:

Lemma 7. *Let P and Q be two Young functions, $f \in E$, u, v, w be weight functions and K an operator. Then the modular inequality*

$$Q^{-1}\left(\int_0^\infty Q(Kf(x)u(x))v(x)dx\right) \le P^{-1}\left(\int_0^\infty P(Cf(x))w(x)dx\right), \quad f \in E$$

holds if and only if

$$\|uKf\|_{L_Q(\mathbb{R}_+,\varepsilon v(x)dx)} \le C\|f\|_{L_P(\mathbb{R}_+,\eta(\varepsilon)w(x)dx)}$$

holds for all $\varepsilon > 0$, with $\eta(\varepsilon) = \frac{1}{Q(P^{-1}(\frac{1}{\varepsilon}))}$ and C independent of ε.

(Here $\|f\|_{L_P(\mathbb{R}_+,w(x)dx)}$ is the Orlicz norm as defined in (1).)

Using the above lemma with $\varepsilon = 1$ we can state the following corollary of Theorem 5:

Corollary 8. *Let P and Q be two Young functions and let the integral operator K be defined by (2). The inequalities*

$$\|uKf\|_{L_Q(\mathbb{R}_+,v(x)dx)} \le C_1\|f\|_{L_P(\mathbb{R}_+,dx)}, \quad f \in E$$

and

$$\|uKf\|_{L_Q(\mathbb{R}_+,v(x)dx)} \le C_2\|f\|_{L_P(\mathbb{R}_+,dx)}, \quad f \in E\downarrow$$

are equivalent and $C_1 = C_2$.

Let now $Hf(x) = \frac{1}{x}\int_0^x f(t)dt$ be the classical Hardy operator and denote by C the best constant in the inequality

$$\|Hf\|_{L_P(\mathbb{R}_+,dx)} \le C\|f\|_{L_P(\mathbb{R}_+,dx)}. \tag{7}$$

It is well known that (7) always holds if the complementary function $\widetilde{P}$ to the Young function P satisfies the Δ_2 condition, see, e.g., [5].

The following result gives a generalization of Corollary 2.4 in [6]:

Corollary 9. *Let P and Q be two Young functions, where $\widetilde{P}$ satisfies the Δ_2 condition. Then the inequalities*

$$\|Hf\|_{L_Q(\mathbb{R}_+,u(x)dx)} \le C_1\|f\|_{L_P(\mathbb{R}_+,dx)}, \quad f \in E, \tag{8}$$

$$\|Hf\|_{L_Q(\mathbb{R}_+,u(x)dx)} \le C_2\|f\|_{L_P(\mathbb{R}_+,dx)}, \quad f \in E\downarrow, \tag{9}$$

and

$$\|f\|_{L_Q(\mathbb{R}_+,u(x)dx)} \le C_3\|f\|_{L_P(\mathbb{R}_+,dx)}, \quad f \in E\downarrow \tag{10}$$

are equivalent. Moreover, $C_3 \le C_1 = C_2 \le CC_3$, where C is the best constant in (7)*.*

Proof. The first equivalence follows from Corollary 8. Now we show that (9) implies (10). Since $f(x) \le Hf(x))$ for any $f \in E\downarrow$ we have by the inequality (9) that

$$\|f\|_{L_Q(\mathbb{R}_+,u(x)dx)} \le \|Hf\|_{L_Q(\mathbb{R}_+,u(x)dx)} \le C_2\|f\|_{L_P(\mathbb{R}_+,dx)}, f \in E\downarrow$$

and $C_3 \le C_2$. To show that (10) implies (9) we first note that if f is decreasing, then Hf is decreasing. Hence, by (10) and (7), we have that

$$\|Hf\|_{L_Q(\mathbb{R}_+,u(x)dx)} \le C_3\|Hf\|_{L_P(\mathbb{R}_+,dx)} \le CC_3\|f\|_{L_P(\mathbb{R}_+,dx)}, f \in E\downarrow,$$

i.e., (9) holds with $C_2 \le CC_3$ and this completes the proof. □

Remark 10. There exist several results concerning the characterization of the above inequalities. For a comprehensive study see, e.g., [4]. By using our corollary some complementary information to these characterizations can be given.

Let $\phi : (0,\infty) \to \mathbb{R}$ be an invertible monotone function such that either

(a) ϕ is concave and increasing or

(b) ϕ is convex and decreasing.

Put $H_\phi f(x) = \phi^{-1}\left(\frac{1}{x}\int_0^x \phi(f)(y)dy\right)$. Our last main result is a generalization of results which can be found in [7] and [6].

Theorem 11. *Suppose that ϕ satisfies the condition* **(a)** *or* **(b)***. Let P be a positive, convex and increasing function and Q be a positive, increasing function on $[0,\infty)$ such that the inequality*

$$\left(\int_0^\infty P(Hf(x))dx\right) \le \left(\int_0^\infty P(Cf(x))dx\right), \quad f \in E,$$

holds. Then the following inequalities are equivalent:

$$Q^{-1}\left(\int_0^\infty Q(f(x)u(x))v(x)dx\right) \le P^{-1}\left(\int_0^\infty P(c_1f(x))dx\right), \quad f \in E\downarrow, \tag{11}$$

$$Q^{-1}\left(\int_0^\infty Q(Hf(x)u(x))v(x)dx\right) \le P^{-1}\left(\int_0^\infty P(c_2f(x))dx\right), \quad f \in E, \tag{12}$$

$$Q^{-1}\left(\int_0^\infty Q(Hf(x)u(x))v(x)dx\right) \leq P^{-1}\left(\int_0^\infty P(c_3 f(x))dx\right), \quad f \in E\downarrow, \tag{13}$$

$$Q^{-1}\left(\int_0^\infty Q(H_\phi f(x)u(x))v(x)dx\right) \leq P^{-1}\left(\int_0^\infty P(c_4 f(x))dx\right), \quad f \in E, \tag{14}$$

$$Q^{-1}\left(\int_0^\infty Q(H_\phi f(x)u(x))v(x)dx\right) \leq P^{-1}\left(\int_0^\infty P(c_5 f(x))dx\right), \quad f \in E\downarrow. \tag{15}$$

Moreover, the least possible constants in the inequalities (11)–(15) *are pairwise equivalent.*

Proof. The proof of the implication (11)$\Longrightarrow$ (13) follows in a similar way as the proof of the implication (10)$\Longrightarrow$ (9) in Corollary 9 and the equivalence (12)$\Longleftrightarrow$ (13) follows from Theorem 5. Moreover, by applying Jensen's inequality for the case (b) and the reversed Jensen's inequality for the case (a) we find that $H_\phi f(x) \leq Hf(x)$ so that (13)$\Longrightarrow$ (15). Also the equivalence (14)$\Longleftrightarrow$ (15) follows from Theorem 5. It remains to prove that (15)$\Longrightarrow$ (11). Let $f \in E\downarrow$ and ϕ satisfies condition (a). Then $\phi(f)$ is decreasing. This implies that $\phi(f(x)) \leq H(\phi(f(x)))$. And since ϕ^{-1} is increasing we conclude that $f(x) \leq H_\phi f(x)$ and therefore (15)$\Longrightarrow$ (11). Moreover $c_1 \leq c_5$. If ϕ satisfies the condition (b) and $f \in E\downarrow$, then $\phi(f)$ is increasing. Hence $\phi(f(x)) \geq H(\phi(f(x)))$. Applying the decreasing function ϕ^{-1} to this inequality we obtain that $f(x) \leq H_\phi f(x)$ and we have also in this case that (15)$\Longrightarrow$ (11) and $c_1 \leq c_5$. The proof is complete. □

Finally we state the following direct consequence of Theorem 11:

Corollary 12. *Suppose that the hypothesis of Theorem* 11 *are satisfied. Then the inequalities* (11) *and*

$$Q^{-1}\left(\int_0^\infty Q(\exp(H(\log f(x))u(x))v(x)dx\right) \leq P^{-1}\left(\int_0^\infty P(c_6 f(x))dx\right)$$

are equivalent either for all $f \in E$ or for all $f \in E\downarrow$.

Remark 13. This result is useful when investigating the relations between Hardy-type inequalities and its limit Pólya-Knopp type inequalities both on the cone of positive functions and the cone of positive and decreasing functions.

Acknowledgement

We thank the referee for some useful remarks which have improved the final version of the paper.

References

[1] C. Bennett and R. Sharpley, *Interpolation of Operators*, Academic Press, Inc., 1988.

[2] G.G. Lorentz, *Bernstein Polynomials*, Univ. of Toronto Press, Toronto, 1953.

[3] I. Halperin, *Function spaces*, Canad. J. Math. **5** (1953), 273–288.

[4] A. Kufner, L. Maligranda and L.-E. Persson, *The Hardy Inequality-About its History and Some Related Results*, Pilsen, 2007.

[5] G. Palmieri, *An approach to the theory of some trace spaces related to Orlicz-Sobolev spaces*, Boll. Un. Mat. Ital. B(**5**) **16** (1979), no. 1, 100–119 (Italian).

[6] L.-E. Persson, V.D. Stepanov and E.P. Ushakova, *Equivalence of Hardy type inequalities with general measures on the cones of non-negative respective non-increasing functions*, Proc. Amer. Math. Soc. **134**(8)(2006), 2363–2372.

[7] G. Sinnamon, *Hardy's inequality and monotonicity*, In: Function spaces, Differential Operators and Nonlinear Analysis (Eds.: P. Drabek and J. Rakosnik)), Mathematical Institute of the Academy of Sciences of the Czech Republic, Prague (2005), 292–310.

Sorina Barza
Department of Mathematics
Karlstad University
SE-65188, Karlstad, Sweden
e-mail: sorina.barza@kau.se

Lars-Erik Persson
Department of Mathematics
Luleå University of Technology
SE-97187, Luleå, Sweden
e-mail: larserik@sm.luth.se

International Series of Numerical Mathematics, Vol. 157, 61–76

Some One Variable Weighted Norm Inequalities and Their Applications to Sturm-Liouville and Other Differential Operators

Richard C. Brown and Don B. Hinton

Mathematics Subject Classification (2000). Primary:26D10, 47A30 34B24; Secondary 47E05, 34C10.

Keywords. Weighted sum inequalities, weighted product inequalities, weighted Hardy inequalities, nonoscillation, discrete spectra, strong limit-point and Dirichlet conditions.

1. Introduction

In this paper we are interested in the validity and applications – especially to problems involving Sturm-Liouville operators – of inequalities of the following type:

$$\int_I w|u^{(j)}|^q \leq K_1(\epsilon)\left(\int_I v_0|u|^p\right)^{q/p} + \epsilon\left(\int_I v_1|u^{(m)}|^r\right)^{q/r} \tag{1.1}$$

$$\int_I w|u^{(j)}|^q \leq K_2\left(\int_I v_0|u|^p\right)^{(q/p)\lambda}\left(\int_I v_1|u^{(m)}|^r\right)^{(q/r)(1-\lambda)} \tag{1.2}$$

$$\int_I w|u|^q \leq K_3\left(\int_I v_1|u^{(m)}|^r\right)^{q/r}. \tag{1.3}$$

Here w, v_0, v_1 are positive a.e. measurable functions or "weights", K_1, K_2, K_3 are positive constants independent of u, $I = (a,b)$, $-\infty \leq a < b \leq \infty$, $1 \leq p, q, r \leq \infty$, and ϵ, λ are real numbers such that $\epsilon \in (0, \epsilon_0)$ where $\epsilon_0 \leq \infty$ and $0 \leq \lambda < 1$.

The domains of the "sum" inequality (1.1) or the "product" inequality (1.2) are one of the following linear spaces:

$$\begin{aligned}
\mathcal{D}^{p,r}(v_0,v_1;I) &:= \left\{u : u \in AC^{m-1}(I);\ \int_I v_0|u|^p,\ \int_I v_1|u^{(m)}|^r < \infty\right\}\\
\mathcal{D}^{p,r}_L(v_0,v_1;I) &:= \{u \in \mathcal{D}^{p,r}(v_0,v_1;I) : \lim_{t\to a^+} u^{(i)}(t) = 0\\
&\qquad \text{for } i = 0,\dots,m-1\}\\
\mathcal{D}^{p,r}_R(v_0,v_1;I) &:= \{u \in \mathcal{D}^{p,r}(v_0,v_1;I) : \lim_{t\to b^-} u^{(i)}(t) = 0\\
&\qquad \text{for } i = 0,\dots,m-1\}\\
\mathcal{D}^{p,r}_{LR}(v_0,v_1;I) &:= \{u \in \mathcal{D}^{p,r}_L(v_0,v_1;I) \cap \mathcal{D}^{p,r}_R(v_0,v_1;I)\}
\end{aligned}$$

where $u \in AC^k(I)$ if and only all derivatives of u up to and including $u^{(k)}$ are absolutely continuous on compact subintervals of I. For the Hardy-type inequality (1.3) we require that $\int_I v_1|u^{(m)}|$ be finite and define analogously the subspaces $\mathcal{D}^{q,r}_L(v_1;I)$, $\mathcal{D}^{q,r}_R(v_1;I)$, or $\mathcal{D}^{q,r}_{LR}(v_1;I)$.

There has been much recent interest in the inequalities (1.1)–(1.3) with results due to many mathematicians. We will discuss some of them here. Unfortunately, while a great many facts are known, there is as yet no satisfactory theory of necessary and sufficient conditions for the inequalities (1.1) or (1.2). One topic concerning (1.1)–(1.3) which we will omit is the determination of the optimal values of the constants K_1, K_2, and K_3. For nontrivial weights these values are almost all unknown. In the unweighted case (1.2) includes many special cases due to Landau, Kolgomorov, Hardy, Kwong and Zettl, and others for which the best constants are known. For a survey of results concerning this problem see Kwong and Zettl, [26]. We mention in passing, however, that one new result (Brown and Kwong, [10]) in the unweighted case of (1.2) is the sharp inequality

$$\int_0^\infty |y'|^2 \le (4/3)\sqrt{3}\left(\int_0^\infty |y|\right) \max_{(0,\infty)} |y''|.$$

The best constants for the unweighted (Poincaré-type) inequalities (1.3) on finite intervals when $m = 1$ have also been characterized by Talenti [38].

A few additional remarks concerning notation: Upper case letters (principally K or C) denote an arbitrary constant whose value may vary. We write $K_1, K_2, \dots,$ etc. to distinguish between different constants. $K(\cdot)$ indicates dependence on a parameter, e.g., $K_1(\epsilon)$. If w is a weight, $L^p(w;I)$ signifies the L^p space on the interval I having the norm $||u||_{I;w,p} := \left(\int_I w|u|^p\right)^{1/p}$ where the cases $p = 1$ or $w = 1$ are omitted from the notation. A local property is indicated by the subscript "loc", e.g., $f \in L^p_{\text{loc}}(I)$, etc. Also $AC(I) \equiv AC^0(I)$, $L(w;I) \equiv L^1(w;I)$, and when the context is obvious we sometimes write $\mathcal{D}^{p,r}$, $\mathcal{D}^{p,r}_L$, $\mathcal{D}^{q,r}_R$, etc. Lastly, finite subintervals of I are denoted by J or Δ with $|J|$ or $|\Delta|$ signifying their length.

2. Interpolation inequalities of sum type

The following basic lemma is a one dimensional version of a Sobolev embedding theorem.

Lemma 2.1. *Let J be a finite subinterval of $\mathbb{R}$, and set $L = |J|$. Then for all $u \in AC^{m-1}(J)$ and $0 \le j \le m-1$ the inequality*

$$||u^{(j)}||_{J,\infty} \le K\left\{L^{-(j+1)}\int_J |u| + L^{m-j-1}\int_J |u^{(m)}|\right\} \tag{2.1}$$

holds.

Proof. For simplicity we suppose that $m = 2$, $j = 1$, and $J = [0,1]$. Clearly,

$$|u'(t)| \le \int_0^1 |u''| + |u'(0)|.$$

Let $\mathcal{F}$ denote the class of appropriately smooth functions ψ such that $\psi(0) = -1$, $\psi(1) = \psi'(0) = \psi'(1) = 0$. Then integration by parts and the triangle inequality gives

$$|u'(0)| \le \int_0^1 |u''||\psi| + \int_0^1 |u||\psi''|.$$

(2.1) follows with $K \le \inf_{\psi\in\mathcal{F}} \max\{1 + |\psi|, |\psi''|\}$. A change of variable yields the inequality on general J. The proof is similar for general m and j. For other proofs see [3] or [19]. □

Lemma 2.1 will yield inequalities like (1.1). Let $I = [a,\infty)$ where $a > -\infty$. Suppose f is a positive continuous function. Let $J_{t,\epsilon} := [t, t+\epsilon f(t)]$, and define for $1 < p, q, r < \infty$

$$S_1(\epsilon_0) := \sup_{t\in I,\, 0<\epsilon\le\epsilon_0} f^{-\phi}\left[(\epsilon f)^{-1}\int_{J_{t,\epsilon}} w\right]\left[(\epsilon f)^{-1}\int_{J_{t,\epsilon}} v_0^{-p'/p}\right]^{q/p'} \tag{2.2}$$

$$S_2(\epsilon_0) := \sup_{t\in I,\, 0<\epsilon\le\epsilon_0} f^{\theta}\left[(\epsilon f)^{-1}\int_{J_{t,\epsilon}} w\right]\left[(\epsilon f)^{-1}\int_{J_{t,\epsilon}} v_1^{-r'/r}\right]^{q/r'} \tag{2.3}$$

where $\phi := q(j + 1/p - 1/q)$ and $\theta := q(m - j - 1/r + 1/q)$.

Our first example of inequality (1.1) is the following:

Theorem 2.1. *Suppose $\infty \ge q \ge \max\{p, r\} \ge 1$ and $S_1(\epsilon_0), S_2(\epsilon_0) < \infty$ then the inequality*

$$\int_I w|u^{(j)}|^q \le K_1\left\{\epsilon^{\phi}\left(\int_I v_0|u|^p\right)^{q/p} + \epsilon^{\theta}\left(\int_I v_1|u^{(m)}|^p\right)^{q/r}\right\} \tag{2.4}$$

holds on $\mathcal{D}^{p,r}(v_0, v_1; I)$ for $\epsilon \le \epsilon_0$ and $K_1 \approx \max\{S_1(\epsilon_0), S_2(\epsilon_0)\}$.

Proof. Assume $1 < p, q, r < \infty$. Set $t_0 := a$, $t_{i+1} := t_i + \epsilon f(t_i)$, etc., and $\Delta_i := [t_i, t_{i+i}]$. (The continuity and positivity of f, shows that $\{t_i\}$ has no finite limit point.) A Hölder's inequality calculation applied to (2.1) with $L = |\Delta_i|$ yields that

$$\int_{\Delta_i} w|u^{(j)}|^q \leq K\left\{\epsilon^{-\phi}S_1(\epsilon_0)\left(\int_{\Delta_i} v_0|u|^p\right)^{q/p} + \epsilon^{\theta}S_2(\epsilon_0)\left(\int_{\Delta i} v_1|u^{(m)}|^r\right)^{q/r}\right\}.$$

Adding these inequalities over i and the elementary inequality $\sum A_i^s \leq (\sum A_i)^s$ for $s \geq 1$, gives (2.4). We omit the cases when p, q, or $r = 1$ or ∞. The proof is similar if we make the necessary (and obvious) changes in the definitions of $S_1(\epsilon_0)$, and $S_2(\epsilon_0)$. □

Remark 2.1. Weighted sum inequalities can also be proven in $\mathbb{R}^n$, and conditions involving averages like (2.2) and (2.3) can be stated if $q < \min\{p, r\}$, $p \leq q < r$, or $r \leq q < p$. See [5], [6] for details.

Remark 2.2. If $\infty > q \geq \max\{p, r\}$ (2.4) remains true if the "semi-pointwise" averages

$$R_1(\epsilon_0) := \sup_{t\in I, 0<\epsilon<\infty} f(t)^{-qj}w(t)\left[(\epsilon f)^{-1}\int_{J_{t,\epsilon}} v_0^{-p'/p}\right]^{q/p'}$$

$$R_2(\epsilon_0) := \sup_{t\in I, 0<\epsilon\leq\infty} f(t)^{q(m-j)}w(t)\left[(\epsilon f)^{-1}\int_{J_{t,\epsilon}} v_1^{-r'/r}\right]^{q/r'}$$

are finite and f is nondecreasing (cf. [4, (14)]). Likewise, Theorem 2.1 may be improved by assuming that p, q, r satisfy (3.1) below and $\epsilon_0 = \infty$. This is a consequence of Theorem 3.1 in the next section. See Remark 3.1.

How can conditions like (2.2), (2.3) be verified? Essentially we will want to choose J_t so that the weights are "almost" constant on this interval.

Example 2.1. Let $q \geq \max\{p, r\}$, $w(t) = t^\beta$, $v_0(t) = t^\gamma$, $v_1(t) = t^\alpha$, $f(t) = t^\delta$ where $\delta \leq 1$, and $I = [1, \infty)$. Then

$$1 \leq \sup_{s\in J_{t,\epsilon}} st^{-1} \leq 1 + \epsilon t^{\delta-1} \leq 1 + \epsilon.$$

A calculation shows that $S_1(\epsilon_0)$, $S_2(\epsilon_0) < \infty$ if

$$\beta \leq \min\left\{\delta\phi + qp^{-1}\gamma, -\delta\theta + qr^{-1}\alpha\right\}, \tag{2.5}$$

and ϵ is sufficiently small (say, $\epsilon \in (0, 1]$). If $I = (0, b]$, $b < \infty$, then these inequalities are reversed. If $I = (0, \infty)$, there is equality . In (2.5) β will be as large as possible relative to α and γ if δ is chosen by "equality", i.e.,

$$\delta = (\alpha/r - \gamma/p)(m + 1/p - 1/r)^{-1} \leq 1. \tag{2.6}$$

With this choice of δ

$$\frac{\beta}{q} \leq \left(\frac{\gamma}{p}\right)\left(\frac{\theta}{\phi+\theta}\right) + \left(\frac{\alpha}{r}\right)\left(\frac{\phi}{\phi+\theta}\right). \tag{2.7}$$

In particular if $q = p = r$ we get that

$$\beta m \le (m - j)\gamma + j\alpha. \tag{2.8}$$

It turns out that if (2.6) is assumed, then (2.7) is *necessary* as well as *sufficient* for the sum inequality to hold. (In general the condition $S_1(\epsilon_0) < \infty$ is necessary and sufficient for the inequality provided the weights are chosen so that the two integral averages $S_1(\epsilon_0)$, $S_2(\epsilon_0)$ are equivalent.)

Example 2.2. The analysis works if $w(t) = e^{\beta t}$, $v_0(t) = e^{\gamma t}$, $v_1(t) = e^{\alpha t}$, $f(t) = e^{\delta t}$ where $\delta \le 0$, and $I = [0, \infty)$.

Example 2.3. If $I = [0, \infty)$, $w(t) = e^{\beta t^2}$, $v_0(t) = e^{\gamma t^2}$, $v_1(t) = e^{\alpha t^2}$, and $f(t) = e^{\delta t^2}$ with $\delta \le 0$, then $1 \le e^{s^2} e^{-t^2} \le \mathrm{O}(e^{2\epsilon t})$. In this case a calculation shows that (2.4) holds if there is *strict* inequality in (2.7). Again this condition is *necessary* as well as *sufficient.*

Remark 2.3. As mentioned above, no necessary and sufficient conditions are known for the sum inequality (1.1) which are valid for all p, q, r, j, m and weights w, v_0, and v_1. However in the case $m = 1$, $j = 0$, $p = r$, $q \ge p$, and $\epsilon = 1$, then a necessary and sufficient condition for (1.1) has been derived by Oĭnarov [34]. These conditions are too complicated to state here but they also requires the finiteness of integral averages distantly resembling (2.2), (2.3).

We also note that Kwong and Zettl have shown in the case $q = p = r$ that the sum inequality (1.1) is sometimes valid when the local integrability of $v_0^{-p/p'}$ is replaced by weaker conditions, for example if $v_0 \ge 0$ and has a positive integral on subintervals. Several such examples are given in [24] and [26]. When $p = q = r = 2$, $m = 2$, and $j = 1$ Wojteczek [41] has also given examples of weighted sum inequalities which are not derivable from Theorem 2.1.

3. Product inequalities

Brown and Hinton [4] have shown:

Theorem 3.1. *If* $1 \le p, q, r < \infty$, $I = [a, \infty)$, $a > -\infty$, $S_1(\infty), S_2(\infty) < \infty$ *and*

$$m/q \le (m - j)/p + j/r, \tag{3.1}$$

or $R_1(\infty), R_2(\infty) < \infty$ *with* f *nondecreasing and* $q \ge \max\{p, r\}$, *then*

$$\int_I w|u^{(j)}|^q \le K_2 \left(\int_I v_0|u|^p\right)^{(q/p)\lambda} \left(\int_I v_1|u^{(m)}|^r\right)^{(q/r)(1-\lambda)} \tag{3.2}$$

where

$$\lambda := \frac{m - j - 1/r + 1/q}{m + 1/p - 1/r}, \qquad 1 - \lambda := \frac{j + 1/p - 1/q}{m + 1/p - 1/r} \tag{3.3}$$

holds for all $u \in \mathcal{D}^{p,r}(v_0, v_1; I)$ *such that* $\int_I v_1|u^{(m)}|^r \neq 0$.

Remark 3.1. An arithmetic-geometric mean inequality argument shows that (3.2) implies (2.4) if $K_2 := K_1/(\lambda^\lambda(1-\lambda)^{(1-\lambda)})$. Conversely, if (2.4) holds for all $\epsilon > 0$, then we get (3.2). If p, q, r do not satisfy (3.1), for example if $q \le \min\{p, r\}$ versions of (3.2) still hold if we change the definitions of $S_1(\infty)$ and $S_2(\infty)$. The theory (as in the case for sum inequalities) may also be developed in $\mathbb{R}^n$. See [5] or [6] for details.

Example 3.1. An interesting special case of Theorem 3.1 is to take $w(t) = v_0(t) = v_1(t) = f(t) = 1$. Then

$$\int_I |u^{(j)}|^q \le K_2 \left(\int_I |u|^p\right)^{(q/p)\lambda} \left(\int_I |u^{(m)}|^r\right)^{(q/r)(1-\lambda)}. \tag{3.4}$$

With a little further argument (see [4, p.118]) it can be shown that (3.4) is valid when $I = \mathbb{R}$. This inequality is due to Gabushin [18] and is an $n = 1$ analogue of the interpolation inequality of Nirenberg [33] which holds on $C_0^\infty(\mathbb{R}^n)$.

Example 3.2. If $I = [0, \infty)$ and $q \ge \max\{p, r\}$, then

$$\int_I e^{ct}|u^{(j)}|^q \le K_2 \left(\int_I e^{ct}|u|^p\right)^{(q/p)\lambda} \left(\int_I e^{ct}|u^{(m)}|^r\right)^{(q/r)(1-\lambda)}$$

where $c > 0$ since $R_1(\infty), R_2(\infty) < \infty$. The case $p = q = r$ was proven by Kwong and Zettl in [25].

Other examples of Theorem 3.1 may be obtained by taking $w = v_0 = v_1$, $p = q = r$, and assuming that w satisfies the "A_p condition" (i.e., exactly equivalent to assuming $S_1(\infty), S_2(\infty)$ finite with $f(t) = 1$.[1])

In the case of power weights with $q = p = r$, $m = 2$, and $j = 1$, the following result was shown by Brown, Hinton, and Kwong [11].

Theorem 3.2. *Let $u \in \mathcal{D}^{p,p}(t^\gamma, t^\alpha; (0, \infty))$, $1 \le p < \infty$, then the inequality*

$$\int_0^\infty t^\beta |u'|^p \le K_2 \left(\int_0^\infty t^\gamma |u|^p\right)^{1/2} \left(\int_0^\infty t^\alpha |u''|^p\right)^{1/2}$$

holds if and only if the following conditions are satisfied:

(i) $\{\alpha, \beta, \gamma\} \ne \{p-1, -1, -1-p\}$.
(ii) $\beta = (\alpha + \gamma)/2$.
(iii) $\lim_{t \to 0^+} u'(t) = 0$ *when* $\beta \le -1$ *and* $\beta > \alpha - p$.
(iv) $\lim_{t \to \infty} u'(t) = 0$ *when* $\beta \ge -1$ *and* $\beta < \alpha - p$.

[1] Because of this connection, one might think of (2.2), (2.3) as requiring that the weights satisfy a kind of generalized A_p condition. It may be useful therefore to investigate whether our conditions have the same sort of applications to analysis as A_p conditions.

Also in the exceptional case $\{\alpha,\gamma\}=\{p-1,\,-1-p\}$ *the inequality*

$$\int_0^\infty t^{-1}|u'|^p \le K_3\left\{\left(\int_0^\infty t^{-1-p}|u|^p\right)^{1/2}\left(\int_0^\infty t^{p-1}|u''|\right)^{1/2}+\int_0^\infty t^{-1-p}|u|^p\right\} \tag{3.5}$$

is valid.

This result extends Kwong and Zettl [25],[26] who proved the inequality when $\gamma > -1-p$ and $\beta > -1$. For additional information on the power weight case see [8] and [9, Example 2.1].

If one is willing to allow u to vanish at the end points of I, many more multiplicative inequalities can be derived. One of the most general [9, Theorem 2.1] has the form

$$\int_I M^\beta |M'|^{\beta_0}|u^{(j)}|^q \le K_2\left(\int_I M^\gamma|M'|^{\gamma_0}|u|^p\right)^{(q/p)\lambda}\left(\int_I M^\alpha|M'|^{\alpha_0}|u^{(m)}|^r\right)^{(q/r)(1-\lambda)}$$

on $\mathcal{D}_L^{p,r}$, $\mathcal{D}_R^{p,r}$, or $\mathcal{D}_{LR}^{p,r}$ where λ is given by (3.3), q,p,r,n,j satisfy (3.1), M is positive, strictly monotone, M' is strictly monotone, $M/|M'|$ is bounded below by a positive constant, and is uniformly Lipshitz on I. $\beta,\beta_0,\alpha,\alpha_0$ satisfy certain complicated conditions which will not be given here. But when $p=q=r$ we have that:

$$\beta = \lambda\gamma+(1-\lambda)\alpha$$
$$\beta_0 = \lambda\gamma_0+(1-\lambda)\alpha_0.$$

Example 3.3. (cf. [9, Example 2.2]) If $q=p=r$, $M=e^{-t}$ on $I=(0,\infty)$ we have the inequality

$$\int_I e^{-ct}|u^{(j)}|^p \le K_2\left(\int_I e^{-ct}|u|^p\right)^{(m-j)/m}\left(\int_I e^{-ct}|u^{(m)}|^p\right)^{j/m}$$

on $\mathcal{D}_L^{p,p}$ where $c>0$. Notice that the zero endpoint conditions on u are now necessary for the inequality to hold.

4. Weighted Hardy inequalities

The classical Hardy inequality (Hardy, Littlewood, and Polya [21, Theorem 330] states that if $1<p<\infty$, $c<-1$, and $u\in\mathcal{D}_L^p(t^{c+p};(0,\infty))$, then

$$\int_0^\infty t^c|u(t)|^p\,dt \le [p/|c+1|]^p\int_0^\infty t^{c+p}|u'(t)|^p\,dt. \tag{4.1}$$

If $c>-1$ the inequality is true on $\mathcal{D}_R^p(t^{c+p};(0,\infty))$. There have been many generalizations of this inequality (excellent surveys may be found in Opic and Kufner

[32] and Kufner, Maligranda, and Persson [23]). It has been shown by Stepanov [36] that

$$\int_0^\infty w|u|^q \leq K_3 \left(\int_0^\infty v_1 |u^{(m)}|^r \right)^{q/r}$$

where $q \geq r$ holds on $\mathcal{D}_L^r$ if and only if

$$\sup_{t>0} \left[\int_t^\infty w(x)\, dx \right]^{1/q} \left[\int_0^t (t-x)^{(m-1)r'} v_1(x)^{-r'/r} dx \right]^{1/r'} < \infty$$

$$\sup_{t>0} \left[\int_t^\infty (x-t)^{(m-1)q} w(x)\, dx \right]^{1/q} \left[\int_0^t v_1(x)^{-r'/r} dx \right]^{1/r'} < \infty. \tag{4.2}$$

A slightly different (though equivalent) pair of conditions has been given by Martin-Reyes and Sawyer [29]. When $m = 1$ the conditions (4.2) collapse into a single well-known necessary and sufficient condition discovered independently by Muckenhoupt [30], Chisholm and Everitt [12], Talenti [37], and Tomaselli [39]. It seems hard, however, to find more than very few specific weights (e.g., powers of t) which satisfy these conditions. (For examples in the case $m = 1$ see Love [28].) Another approach is to show that Hardy's inequality holds for a large family of weights having a certain structure. For instance, (Brown and Hinton [7]) one can obtain the following results:

Theorem 4.1. *Let M be a positive locally absolutely continuous function on $I = (a, b)$ such that either $M' > 0$ or $M' < 0$ a.e. on I and:*

(i) *ω is a positive measurable function that is nonincreasing if $\lim_{t\to a^+} u^{(i)}(t) = 0$ and nondecreasing if $\lim_{t\to b^-} u^{(i)}(t) = 0$ for $i = 0, \dots, m-1$;*
(ii) *we suppose that*

$$\beta = (q/r)(\alpha + 1) - mq - 1 \tag{4.3}$$

$$\beta_0 = (q/r)(\alpha_0 - 1) + mq + 1. \tag{4.4}$$

Then if M is increasing and concave up the inequality

$$\int_I M^\beta |M'|^{\beta_0} |u|^q \omega \leq K_3 \left[\int_I M^\alpha |M'|^{\alpha_0} |u^{(m)}|^r \omega^{r/q} \right]^{q/r}$$

holds on $\mathcal{D}_L^r$ if $\beta < -1 - (m-1)q$ and $\beta_0 \leq 1$; and it holds on $\mathcal{D}_R^r$ if $\beta > -1$ and $\beta_0 \geq 1 + (m-1)q$.

For other choices of M, e.g.: M increasing and concave down, M decreasing and concave up or down, the Hardy inequality will hold with other conditions on β_0, β.

Corollary 4.1. *Let M_j, M_j', $j = 1, \dots, l$, and ω be functions satisfying the conditions of Theorem* 4.1. *Define*

$$w := \left(\prod_{j=1}^{l} w_j^{1/\lambda_j} \right) \omega,$$

$$v_1 := \left(\prod_{j=1}^{l} v_{1j}^{1/\lambda_j} \right) \omega$$

where

$$w_j = M_j^{\beta_j} |M_j'|^{\beta_{j0}},$$
$$v_{1j} = M_j^{\alpha_j} |M_j'|^{\alpha_{j0}},$$

the λ_j are positive numbers such that $\sum_{j=1}^{l} \lambda_j^{-1} \geq 1$, and the other parameters are as above. Then the Hardy inequality

$$\int_I w|u|^q \leq K_3 \left(\int_I v_1 |u^{(m)}|^r \right)^{q/r}$$

holds on $\mathcal{D}_R^r$ or $\mathcal{D}_L^r$.

Example 4.1. Let $M(t) = e^{t^2}$, $q = r$, $I = (1, \infty)$, $\omega = t^{-mq-\alpha_0} \equiv t^{-1}$ if $\alpha_0 = 1 - mq$. A calculation based on Theorem 4.1 gives:

$$\int_I e^{-ct^2} |u(t)|^q \leq K_3 \int_I e^{-ct^2} t^{-mq} |u(t)^{(m)}|^q.$$

on $\mathcal{D}_L^q$ where $c = -(\alpha + \alpha_0) > (m-1)q$.

5. Applications

We look at a few examples showing some of the ways the inequalities discussed in the previous sections can be applied to problems in differential equations. Our selection is only intended to give the "flavor" of the theory and is not exhaustive. We have omitted, for instance, the quadratic form perturbation theory of Kato, disconjugacy criteria, bounds of the number of negative eigenvalues, or the determination of lower bounds on the spectrum. In all these areas Hardy or interpolation type inequalities have proven useful.

A THEOREM OF LANDAU. In 1930 Landau [27] proved the following result:

Theorem 5.1. *Consider the differential operator*

$$L(y) := y^{(m)} + p_{m-1} y^{(m-1)} + \cdots + p_0 y$$

defined on $I := [a, \infty)$. Suppose y and $L(y)$ are bounded, and that the coefficients p_i, $i = 0, \dots, m-1$, are bounded. Then the functions $y^{(i)}$, $i = 1, \dots, m$, are also bounded.

Landau's proof was technically difficult. However, our technology allows us to prove:

Theorem 5.2. *Suppose $y, L(y) \in L^r(I)$ for some $1 \le r \le \infty$, the functions $\int_t^{t+\eta} |p_i|^r$ are bounded for some $\eta > 0$ when $r < \infty$, and that the p_i are bounded when $r = \infty$. Then* (i) $y^{(i)} \in L^q(I)$, $i = 0, \dots, m-1$, *for all* $\infty \ge q \ge r$; (ii) $y^{(m)} \in L^r(I)$.

Proof. Assume $r < \infty$. Let $I_{n\epsilon} := [a, a+n\epsilon]$ where $0 < \epsilon < \eta$ will be chosen later, and n be an arbitrary positive integer. The hypothesis on the coefficients p_i, Theorem 2.1, and an extension of the interval of integration for $|y|^r$ gives the sum inequalities

$$\int_{I_{n\epsilon}} |p_i|^r |y^{(i)}|^r \le K_i \left\{ \epsilon^{-r(i+1/r)} \int_I |y|^r + \epsilon^{r(m-i-1/r)} \int_{I_{n\epsilon}} |y^{(m)}|^r \right\} \tag{5.1}$$

for $i = 1, \dots, m-1$. Moreover, the constants K_i do not depend on $n\epsilon$ Next, the triangle inequality, the Minkowski inequality for sums, and extension of the range of integration for $|L(y)|^r$ yields that

$$\int_{I_{n\epsilon}} |y^{(m)}|^r \le (m+1)^{r-1} \left\{ \int_I |L(y)|^r + \sum_{i=0}^{m-1} \int_{I_{n\epsilon}} |p_i|^r |y^{(i)}|^r \right\}. \tag{5.2}$$

Therefore, choosing ϵ small enough, substituting (5.1) into (5.2), and subtracting the terms involving $|y^{(m)}|^r$ from both sides we see that

$$(1/2) \int_{I_{n\epsilon}} |y^{(m)}|^r \le K_1(\epsilon) \left\{ \int_I |L(y)|^r + \int_I |y|^r \right\}.$$

Since n is arbitrary, this proves (ii). Next (i) follows from the interpolation inequalities

$$||y^{(i)}||_{I,q} \le K(i,m,q,r)\{||y||_{I,r} + ||y^{(m)}||_{I,r}\}$$

which are unweighted versions of Theorem 2.1. In the case $r = \infty$ the analogs of (5.1) and (5.2) are

$$||p_i y^{(i)}||_{\infty, I_{n\epsilon}} \le K \left\{ \epsilon^{-i} ||y||_{\infty,I} + \epsilon^{(m-i)} ||y^{(m)}||_{\infty, I_{n\epsilon}} \right\}, \quad i = 1, \dots, m-1, \tag{5.3}$$

$$||y^{(m)}||_{\infty, I_{n\epsilon}} \le \left\{ ||L(y)||_{\infty,I} + \sum_{i=0}^{m-1} ||p_i y^{(i)}||_{\infty, I_{n\epsilon}} \right\}. \tag{5.4}$$

(5.3) is an obvious modification of Lemma 2.1. (5.4) follows from the triangle inequality. The remainder of the argument parallels the case $r < \infty$. □

Example 5.1. Suppose $I = [1, \infty)$, $r = 2$, $m = 2k$, and $L(y) = y^{(m)} - p_0 y$ where $\int_t^{t+\eta} |p_0|^2$ is bounded on I for some $\eta > 0$. Then Theorem 5.1 implies that $y^{(m)} \in L^2(I)$ and that all intermediate derivatives $y^{(i)}$ are bounded and L^q integrable for $q \ge 2$. In particular, all square integrable solutions y of $L(y) = 0$ have these properties. Further, from the triangle inequality $p_0 y \in L^2(I)$; in other words, this symmetric differential operator is *separated.*

Discrete spectra and Non-oscillation criteria. For convenience we again take $I = (a, \infty)$, $a > -\infty$. Recall that the differential expression

$$M[u] := \sum_{i=0}^{m} (-1)^i (p_i u^{(i)})^{(i)}$$

where $p_m > 0$ and p_i is assumed to have a continuous ith derivative is said to be oscillatory at ∞ if for every $n > a$ there is an interval $[c, d] \subset (n, \infty)$ and a nontrivial solution y of $M[y] = 0$ such that

$$y(c) = \cdots = y^{(m-1)}(c) = 0 = y(d) = \cdots = y^{(m-1)}(d).$$

Two basic results in oscillation theory (cf. Glazman [20] or Reid [35]) are:

Theorem 5.3. *Let $\mathcal{Q}(u) := \sum_{i=0}^{m} \int_I p_i |u^{(i)}|^2$, and let $\mathcal{A}_m(c, d) \subset AC^{m-1}(c, d)$ be the class of functions u such that*

$$u^{(i)}(c) = u^{(i)}(d), \quad i = 0, \ldots, m-1, \quad \textit{and} \quad u^{(m)} \in L^2(c, d).$$

Then M is oscillatory at ∞ if and only if for every $n > a$ there is an interval $[c, d] \subset (n, \infty)$ and a nontrivial $y \in \mathcal{A}_m(c, d)$ such that $\mathcal{Q}(\widetilde{y}) \leq 0$ where $\widetilde{y} := y$ on $[c, d]$ and 0 otherwise.

In particular, M will be non-oscillatory at ∞ if for every $n > a$, $\mathcal{Q}(u) > 0$ for all $u \in AC^{m-1}$ having compact support in (n, ∞) and L^2 integrable mth derivative. This concept has a connection to the property "**BD**" of M – that the spectrum of every self-adjoint extension A of the minimal operator $T_0(M)$ in the Hilbert space $L^2(I)$ determined by $M[u]$[2] is discrete and bounded below. The classical result is:

Theorem 5.4. *M has **BD** if and only if for all real numbers λ, $M(u) - \lambda u$ is non-oscillatory at ∞, i.e., for each λ there is an n such that $\mathcal{Q}(u) > \lambda \int_n^\infty |u|^2$ for all compact support functions in $AC^{m-1}(n, \infty)$ with L^2 integrable mth derivative. Further if this is true up to $\lambda = \mu$ the spectrum of A is finite on $(-\infty, \mu)$.*

These theorems together with our inequalities can be used to give conditions that M be *non-oscillatory* at ∞ or that its spectrum of self-adjoint extensions of be finite on $(-\infty, \mu)$ for $\mu \in \mathbb{R}$.

Example 5.2. (cf. Everitt [13]) Consider $M[y] := -y'' + qy$ on $(0, \infty)$. Let $q = q_+ - q_-$ where $q_+ := \max\{q, 0\}$. Then if $q_- \in L^p(0, \infty)$ for some $1 < p < \infty$. The spectrum of self-adjoint extensions is discrete on $(-\infty, 0)$. To see this choose n large enough that $||q_-||_{(n,\infty),p} \leq \epsilon < 1$ and note that

$$\mathcal{Q}(u) \geq \int_n^\infty |u'|^2 - \int_n^\infty q_- |u|^2.$$

---

[2]For the definitions of the minimal and maximal operators determined by M see Naimark [31].

Next let $t_0 = n$, $t_{i+1} = t_i + \epsilon$, $\Delta_i = [t_i, t_{i+1}]$, $i = 0, \ldots,$ on Δ_i. A calculation using the basic sum inequality (2.1) and Hölder's inequality gives

$$\int_{\Delta_i} q_-|u|^2 \leq 2\epsilon^{1/p'} \left\{ \int_{\Delta_i} |u|^2 + \int_{\Delta_i} |u'|^2 \right\}$$

where $1/p' = 1 - 1/p$. By addition the same inequality is true on (n, ∞). Hence

$$\begin{aligned} Q(u) &\geq (1 - 2\epsilon^{1/p'}) \int_n^\infty |u'|^2 - 2\epsilon^{1/p'} \int_n^\infty |u|^2 \\ &\geq -2\epsilon^{1/p'} \int_n^\infty |u|^2. \end{aligned}$$

This shows that the spectrum is finite.[3] on $(-\infty, -\delta)$ for all $\delta > 0$, so that 0 is a limit point of the point spectrum. The same argument works for the weaker assumption on q that $\lim_{t\to\infty} \int_t^{t+\epsilon} |q| = 0$. Indeed, if $\lim_{t\to\infty} \epsilon^{-1} \left(\int_t^{t+\epsilon} |q| \right) = \mu/2$, the spectrum of self-adjoint extensions is finite on $(-\infty, -\mu - \delta)$, $\forall \delta > 0$[4].

Example 5.3. Consider the two term operator

$$M[u] = (-1)^m (e^{ct} u^{(m)})^{(m)} + q(t)u\,.$$

Then by an analogous but more involved argument using a weighted Hardy inequality Brown and Hinton [7] showed that M has **BD** if

$$\int_I |e^{-cx} q(x)|^s < \infty, \quad \text{for } c > 0,$$

and

$$\int_I |e^{|c|x} q(x)|^s < \infty, \quad \text{for } c < 0,$$

and some s, $1 < s < \infty$. Many other results of this type may be found in Ashbaugh, Brown, and Hinton [2].

Dirichlet and Strong Limit-point Results. Let $I = [a, \infty)$, $a > -\infty$ and consider the differential expression

$$M[y] := w^{-1}[-(py')' + qy] \tag{5.5}$$

where $p, q > 0$ and $p^{-1}, q, w \in L^1_{\text{loc}}[a, \infty)$. This means that M is regular at a and and singular at ∞.

The following result is in the spirit of the main result in Kalf [22] but has a simpler proof; it also extends Amos and Everitt [1, Theorem 1].

Theorem 5.5. *Suppose in* (5.5)*:*

(i) $q = q_1 - q_2$ *where* $q_1 + cw \geq q_3$ *for some constant* $c \geq 0$ *on* I.

[3]For the determination of a precise lower bound on the spectrum using the best constant of a product type interpolation inequality see Veiling [40]

[4]On the other hand, [20, Theorem 25] shows that continuous spectrum of these example covers the nonnegative part of the axis

(ii) *For a sufficiently large n, all $t > n$, and fixed positive constants K_1, $\mu < 1$, and $K_2 < 1 - \mu$ the inequalities*

$$\int_n^t |q_2||y|^2 \le K_1 \int_n^t w|y|^2 + \mu \int_n^t p|y'|^2 \tag{5.6}$$

$$\int_n^t |q_3||y|^2 \le K_2 \int_n^t p|y'|^2 \tag{5.7}$$

hold for all $y \in \mathcal{D}(T(M))$ with support in (n, ∞).

(iii) $w \notin L(I)$.

Then the maximal operator $T(M)$ is "Dirichlet" and "strong limit-point" (SLP) at ∞; i.e., $p^{1/2}y', |q|^{1/2}y \in L^2(I)$ and

$$\lim_{t\to\infty} f(t)(p\bar{g}')(t) = 0\,, \quad \forall f, g \in D(T(M)).$$

Proof. Given $f \in \mathcal{D}(T(M))$ using Lemma 2 of [31, §17.3] we can construct $\tilde{f}$ with support in (n, ∞) such that $\tilde{f} \equiv f$ on $[n + 1, \infty)$. Clearly f satisfies the Dirichlet and SLP condition if and only if $\tilde{f}$ does. This observation justifies us in considering only those $f \in D(T(M))$ such that $f(n) = f'(n) = 0$. Let n be such that (5.6) and (5.7) hold. Integrating by parts and an elementary estimate gives that for $t > n$

$$\int_n^t \{p|f'|^2 + (q_1 + cw)|f|^2\} \le (p\bar{f}'f)(t) + \int_n^t (|q_2| + |q_3|)|f|^2 + \int_I wfM[\bar{f}]. \tag{5.8}$$

Consequently we have that

$$\int_n^t \{(1 - \mu - K_2)p|f'|^2 + (q_1 + cw)|f|^2\} \le (p\bar{f}'f)(t) + K_1 \int_I w|f|^2 + \int_I wfM[\bar{f}].$$

Since $1 - \mu - K_2 > 0$, it is clear that if $p^{1/2}f'$ or $(q_1 + cw)^{1/2}f \notin L^2(a, \infty)$ then $\lim_{t\to\infty}(p\bar{f}'f)(t) = \infty$, whence in particular $|f|^2$ is monotone increasing near ∞, contradicting the fact that both $w \notin L(I)$ and $f \in L^2(w; I)$. By (5.6)–(5.7) it follows that $|q_2|^{1/2}f$, $|q_3|^{1/2}f$ and therefore also $|q|^{1/2}f \in L^2(I)$. The argument that M is strong limit-point at ∞ is omitted since it may be found in [15] or [16]. For further information concerning the Dirichlet and SLP properties see [14]. □

References

[1] R.J. Amos and W.N. Everitt, *On integral inequalities and compact embeddings associated with ordinary differential expressions*, Arch. f. Rat. Mech. and Analysis **71** (1979), 15–40.

[2] M. Ashbaugh, R.C. Brown, and D.B. Hinton, *Interpolation inequalities and nonoscillatory differential equations*, International Series of Numerical Mathematics, Vol 103 (Proceedings of the Oberwolfach Conference "General Inequalities 6"), Birkhäuser Verlag, Basel, 1992, 243–255.

[3] R.C. Brown and D.B. Hinton, *Sufficient conditions for weighted inequalities of sum form*, J. Math. Anal. Appl. **112** (1985), 563–578.

[4] R.C. Brown and D.B. Hinton, *Sufficient conditions for weighted Gabushin inequalities*, Casopis Pěst. Mat. **111** (1986), 113–122.

[5] R.C. Brown and D.B. Hinton, *Weighted interpolation inequalities of sum and product form in R^n*, J. London Math Soc. (3) **56** (1988), 261–280.

[6] R.C. Brown and D.B. Hinton, *Weighted interpolation inequalities and embeddings in R^n*, Canad. J. Math. **47** (1990), 959–980.

[7] R.C. Brown and D.B. Hinton, *A Weighted Hardy's inequality and nonoscillatory differential equations*, Quaes. Mathematicae **115** (1992), 197–212.

[8] R.C. Brown and D.B. Hinton, *Interpolation inequalities with power weights for functions of one variable*, J. Math. Anal. Appl. **172** (1993), 233–242.

[9] R.C. Brown and D.B. Hinton, *An interpolation inequality and applications*, Inequalities and Applications (R. P. Agarwal, ed.). World Scientific, Singapore-New Jersey-London-Hong Kong, 1994, 87–101.

[10] R.C. Brown and Man K. Kwong, *On the best constant for the inequality $\int_0^\infty y'^2 \leq K\left(\int_0^\infty |y|\right) \max_{(0,\infty)} |y''|$*, Inequalities and Applications (R.P. Agarwal, ed.), World Scientific, Singapore-New Jersey-London-Hong Kong, 1994, 103–121.

[11] R.C. Brown and Man K. Kwong, D.B. Hinton, and Man K. Kwong, *A remark on a weighted Landau inequality of Kwong and Zettl*, Canad. Math. Bull. **38**(1) (1995), 34–41.

[12] R.S. Chisholm and W.N. Everitt, *On bounded integral operators in the space of integrable square functions*, Proc. R.S.E. (A), **69** (1970/71), 199–204.

[13] W.N. Everitt, *On the spectrum of a second order linear differential equations with a p-integrable coefficient*, Applicable Anal. **2** (1972), 143–160.

[14] W.N. Everitt, *On the strong limit-point condition of second-order differential expressions*, International Conference of Differential Equations (H.A. Antosiewicz, ed.) (Proceedings of an international conference held at the University of Southern California, September 3–7, 1974). Academic Press, New York, 1974, 287–306.

[15] W.N. Everitt, M. Giertz, and J. Weidmann, *Some remarks on a separation and limit-point criterion of second-order ordinary differential expressions*, Math. Ann. **200** (1973), 335–346.

[16] W.N. Everitt, M. Giertz, and J. Weidmann, and J.B. McLeod, *On strong and weak limit-point classification of second-order differential expressions*, Proc. London Math. Soc.(3) **29** (1974), 142–158.

[17] W.N. Everitt, M. Giertz, and J. Weidmann, D.B. Hinton, and J. S. W. Wong, *On strong limit-n classification of linear ordinary differential expressions of order* $2n$, Proc. London Math. Soc.(3) **29** (1974), 351–367.

[18] V.N. Gabushin, *Inequalities for norms of a function and its derivatives in L_p metrics*, Mat. Zam. **1** (1967), 291–298 (Russian).

[19] S. Goldberg, *Unbounded Linear Operators Theory and Applications*, McGraw-Hill Series in Higher Mathematics (E.H. Spanier, ed.), McGraw-Hill Book Company, New York, St. Louis, San Francisco, Toronto, London, Sydney, 1966.

[20] M. Glazman, *Direct Methods of Qualitative Spectral Analysis of Singular Differential Operators*, Israel Program for Scientific Translation, Jerusalem, 1965.

[21] G.H. Hardy, J.E. Littlewood, and G. Pólya, *Inequalities*, Cambridge University Press, Cambridge, 1959.

[22] H. Kalf, *Remarks on some Dirichlet type results for semibounded Sturm-Liouville operators*, Math. Ann. **210** (1974), 197–205.

[23] A. Kufner, L. Maligranda, and L. Persson, *The Hardy Inequality: About its History and Some Related Results*, Pilsen, 2007.

[24] M.K. Kwong and A. Zettl, *Weighted norm inequalities of sum form involving derivatives*, Proc. R.S.E. (A) **88** (1981), 121–134.

[25] M.K. Kwong and A. Zettl, *Norm inequalities of product form in weighted L^p spaces*, Proc. R.S.E. (A) **89** (1981), 293–307.

[26] M.K. Kwong and A. Zettl, *Norm Inequalities for Derivatives and Differences*, Lecture Notes in Mathematics 1536, Springer-Verlag, Berlin, 1992.

[27] E. Landau, *Über einen Satz von Herr Esclangon*, Math. Ann. **102** (1930), 177–188.

[28] E.R. Love, *Generalizations of Hardy's integral inequality*, Proc. R.S.E. (A) **100** (1985), 237–262.

[29] F.J. Martin-Reyes and E. Sawyer, *Weighted inequalities for Riemann-Liouville fractional integrals of order one and higher*, Proc. A.M.S. **106** (1989), 727–733.

[30] B. Muckenhoupt, *Hardy's inequality with weights*, Studia Math. **44** (1972), 31–38.

[31] M.A. Naimark, *Linear Differential Operators, Part II*, Frederick Ungar, New York, 1968.

[32] B. Opic and A. Kufner, *Hardy-type Inequalities*, Longman Scientific & Technical, Harlow, Essex, UK, 1989.

[33] L. Nirenberg, *On elliptic partial differential equations*, Annali della Scuola Norm. Sup. Pisa, Ser III, **16** (1958), 115–162.

[34] R. Oĭnarov, *On weighted norm inequalities with three weights*, J. London Math. Soc. **48**(2) (1993), 103–116.

[35] W.T. Reid, *Sturmian Theory for Ordinary Differential Equations*, Springer-Verlag, Berlin, 1980.

[36] V. Stepanov, *Two weighted estimates for Riemann-Liouville integrals*, Preprint No. 39, Mat. Ustav. Csav, Prague (1988); Math. USSR Izvestiya **36** (1991), 669–681 (Russian).

[37] G. Talenti, *Sopra una diseguaglianza integrale*, Ann. Scuola Norm. Sup. Pisa (3) **31** (1967), 167–188.

[38] G. Talenti, *Best constant in Sobolev inequality*, Istit. Mat. Univ. Firenze **22** (1974–75), 1–32.

[39] G. Tomaselli, *A class of inequalities*, Bol. Un. Mat. Ital. IV Ser. 2 (1969), 622–631.

[40] E.J.M. Veiling, *Optimal lower bounds for the spectrum of a second order linear differential equation with p-integrable coefficient*, Proc. R.S.E. (A) **92** (1982), 95–101.

[41] K. Wojteczek, *On quadratic integral inequalities*, J. Math. Anal. Appl., to appear.

Richard C. Brown
Department of Mathematics
University of Alabama-Tuscaloosa
AL 35487-0350, USA
e-mail: dicbrown@bama.ua.edu

Don B. Hinton
Department of Mathematics
University of Tennessee-Knoxville
TN 37996, USA
e-mail: hinton@math.utk.edu

International Series of Numerical Mathematics, Vol. 157, 77–89

Bounding the Gini Mean Difference

Pietro Cerone

Abstract. Some recent results on bounding and approximating the Gini mean difference in which the author was involved for both general distributions and distributions supported on a finite interval are surveyed. The paper supplements the previous work utilising the Steffensen and Karamata type approaches in approximating and bounding the Gini mean difference.

Mathematics Subject Classification (2000). 62H20, 62G10, 62P20, 26D15.

Keywords. Gini mean difference, continuous distributions, expectation, moments, variance, Steffensen and Karamata type inequalities.

1. Introduction

Let $f : \mathbb{R} \rightarrow [0, \infty)$ be a *probability density function* (pdf), meaning that f is integrable on $\mathbb{R}$ and $\int_{-\infty}^{\infty} f(t)\,dt = 1$, and define

$$F(x) := \int_{-\infty}^{x} f(t)\,dt, \qquad x \in \mathbb{R} \text{ and } E(f) := \int_{-\infty}^{\infty} x f(x)\,dx, \tag{1.1}$$

to be its *cumulative function* and the *expectation* provided that the integrals exist and are finite.

The *mean difference*

$$R_G(f) := \frac{1}{2} \int_{-\infty}^{\infty} \int_{-\infty}^{\infty} |x - y|\,dF(x)\,dF(y) \tag{1.2}$$

was proposed by Gini in 1912 [12], after whom it is usually named, but it was discussed by Helmert and other German writers in the 1870's (cf. H.A. David [10], see also [18, p. 48]). The mean difference has a certain theoretical attraction, being dependent on the spread of the variate-values among themselves rather than on the deviations from some central value ([18, p. 48]). Further, its defining integral (1.2) may converge when the *variance* $\sigma^2(f)$,

$$\sigma^2(f) := \int_{-\infty}^{\infty} (x - E(f))^2\,dF(x), \tag{1.3}$$

does not. It can, however, be more difficult to compute than (1.3).

Another useful concept that will be utilized in the following is the *mean deviation* $M_D(f)$, defined by [18, p. 48]

$$M_D(f) := \int_{-\infty}^{\infty} |x - E(f)|\, dF(x). \tag{1.4}$$

As G.M. Giorgi noted in [13], some of the many reasons for the success and the relevance of the Gini mean difference or *Gini index* $I_G(f)$,

$$I_G(f) = \frac{R_G(f)}{E(f)}, \tag{1.5}$$

are their simplicity, certain interesting properties and useful decomposition possibilities, and these attributes have been analysed in an earlier work by Giorgi [14]. For a bibliographic portrait of the Gini index, see [13] where numerous references are given.

The aim of the present paper is to supplement a survey of some recent inequalities for the mean differences [3] obtained by Cerone and Dragomir in a sequence of four papers ([4]–[7]). In the four papers, the Sonin and Korkine identities were used to point out various bounds for the Gini mean difference in both the general case and in the case of distributions supported on a finite interval.

Bounds for the Gini mean difference are developed in the current article utilising Steffensen and Karamata type inequalities. Specifically, some new or less well-known developments will be used to obtain bounds for $R_G(f)$ which supplements knowledge of this important quantity.

2. Some identities and inequalities

Some identities for the Gini mean difference, $R_G(f)$ will be stated here since they will form the basis for obtaining approximations and bounds. The reader is referred to the book [18], Exercise 2.9, p. 94 or [3].

Define the function $e : \mathbb{R} \to \mathbb{R}$, $e(x) = x$ and with $F(\cdot)$ and $E(\cdot)$ as given by (1.1) then, the covariance of e and F is given by:

$$\mathrm{Cov}(e, F) := E[(e - E(f))(F - E(F))]. \tag{2.1}$$

The following result holds (see for instance [18, p. 54] or [3]).

Theorem 1. *With the above notation then*

$$R_G(f) = 2\,\mathrm{Cov}(e, F) = \int_{-\infty}^{\infty} (1 - F(y))\, F(y)\, dy \tag{2.2}$$

$$= 2\int_{-\infty}^{\infty} x f(x)\, F(x)\, dx - E(f).$$

Utilising Sonin's identity (see for instance [21, p. 246] for the case of univariate real functions) was used in [4], to obtain the following identities for $R_G(f)$.

Theorem 2. *With the above assumptions for f and F, we have for any $\gamma, \delta \in \mathbb{R}$ the identities:*

$$R_G(f) = 2\int_{-\infty}^{\infty} (x - E(f))(F(x) - \gamma) f(x)\, dx \tag{2.3}$$
$$= 2\int_{-\infty}^{\infty} (x - \delta)\left(F(x) - \frac{1}{2}\right) f(x)\, dx.$$

The following result was produced in [5] *using the Korkine identity (see for instance* [21, p. 242]*):*

Theorem 3. *With the above assumptions for f and F, we have the following representation for the Gini mean difference:*

$$R_G(f) = \int_{-\infty}^{\infty}\int_{-\infty}^{\infty} (x - y)(F(x) - F(y)) f(x) f(y)\, dxdy. \tag{2.4}$$

The identities of Theorems 1–3 namely, (2.2), (2.3) and (2.4) provide a means of bounding $R_G(f)$ which will be outlined in the following sections.

2.1. Inequalities for $R_G(f)$

The following result compares the Gini mean difference with the mean deviation defined by (1.4) which was obtained in [4] using (2.3).

Theorem 4. *With the above assumptions, we have the bounds:*

$$\frac{1}{2} M_D(f) \le R_G(f) \le 2\sup_{x\in\mathbb{R}} |F(x) - \gamma|\, M_D(f) \le M_D(f), \tag{2.5}$$

for any $\gamma \in [0, 1]$, where $F(\cdot)$ is the cumulative distribution of f and $M_D(f)$ is the mean deviation defined by (1.4).

It was pointed out by J.L. Gastwirth in [11], using inequality 105 from the book [15] by Hardy, Littlewood and Polya and the fact that F is increasing, that one can state the following results.

Theorem 5. *Assume that F is supported on a finite interval (a, b). Then*

$$0 \le R_G(f) \le \frac{1}{b-a}(b - E(f))(E(f) - a). \tag{2.6}$$

Theorem 6. *If F is concave on (a, b), then*

$$\frac{1}{3}(E(f) - a) \le R_G(f) \le \frac{1}{b-a}(E(f) - a)\left[(b - E(f)) - \frac{1}{3}(E(f) - a)\right]. \tag{2.7}$$

It should be noted that when F is convex on (a, b), then the mean difference $R_G(f)$ is bounded by [11, p. 309]

$$\frac{b - E(f)}{3(b-a)} \le R_G(f) \le \frac{b - E(f)}{3(b-a)}[4(E(f) - a) - (b - a)]. \tag{2.8}$$

Recall that (see for instance [11, p. 309]) a cumulative function F supported on (a, ∞) has the *decreasing hazard rate* (DHR) property if $-\ln[1 - F(x)]$ is concave for $x \geq a$. When the density function $f(x) = F'(x)$ exists, then the function $q(x) = f(x)[1 - F(x)]^{-1}$ is nonincreasing.

Using Lemma 5 from [15], Gastwirth has obtained the following results as well [11, Theorem 3]:

Theorem 7. *If F is a cumulative function defined on (a, ∞) with the DHR property, density f and finite mean $E(f)$, then*

$$\frac{1}{2}(E(f) - a) \leq R_G(f) \leq E(f) - a. \tag{2.9}$$

2.2. Inequalities via Grüss and Sonin type results

The following representation for the Gini mean difference

$$R_G(f) = \int_a^b F(x)(1 - F(x))\,dx, \tag{2.10}$$

holds provided that F is supported on $[a, b]$, a finite interval.

Bounds for the quantity $R_G^*(f)$, involving $R_G(f)$ and defined here for simplicity

$$R_G^*(f) := \frac{1}{b-a}[b - E(f)][E(f) - a] - R_G(f), \tag{2.11}$$

will be obtained below.

Utilising the well-known Grüss inequality the following simple bound for the Gini mean difference was obtained in [6].

Theorem 8. *If f is defined on the finite interval $[a, b]$ and $R_G^*(f)$ is given by (2.11) then*

$$0 \leq R_G^*(f) \leq \frac{1}{4}(b - a). \tag{2.12}$$

The following improvement of Theorem 8 was obtained in [6] using an improvement (see [5] and [9]) of the Grüss inequality .

Theorem 9. *If f is defined on the finite interval $[a, b]$and $R_G^*(f)$ is given by (2.11), then*

$$\begin{aligned} 0 \leq R_G^*(f) &\leq \frac{1}{2} \cdot \int_a^b \left| F(x) - \frac{b - E(f)}{b-a} \right| dx \\ &\leq \frac{1}{2} \cdot \left[\int_a^b \left(F(x) - \frac{b - E(f)}{b-a} \right)^2 dx \right]^{\frac{1}{2}} \leq \frac{1}{4}(b-a). \end{aligned} \tag{2.13}$$

The Sonin identity [21, p. 246] on (2.10) produces the result:

$$R_G^*(f) = \int_a^b \left(F(t) - \frac{b - E(f)}{b-a} \right)(F(t) - \lambda)\,dt. \tag{2.14}$$

which was used in [7] to obtain the following theorem.

Theorem 10. *Assume that f is defined on the finite interval $[a,b]$ and $R_G^*(f)$ is given by* (2.11), *then*

$$(0\le) R_G^*(f) \le \inf_{\lambda\in\mathbb{R}} \|F-\lambda\|_\infty \int_a^b \left| F(t) - \frac{b-E(f)}{b-a} \right| dt \qquad (2.15)$$
$$\le \frac{1}{2}\int_a^b \left| F(t) - \frac{b-E(f)}{b-a} \right| dt.$$

Taking $\lambda = \frac{b-E(f)}{b-a}$ in (2.14), the following simple bound was also obtained in [7],

$$R_G^*(f) \le \frac{1}{b-a}\left[\frac{b-a}{2} + \left| E(f) + \frac{a+b}{2}\right|\right]^2. \qquad (2.16)$$

The following identity was developed in [7] from using the Korkine identity [21, p. 242] on (2.10)

$$R_G^*(f) = \frac{1}{2(b-a)}\int_a^b\int_a^b (F(y)-F(x))^2\,dx dy, \qquad (2.17)$$

where $R_G^*(f)$ is as given by (2.11).

If upper and lower bounds for the density function f are known, then we have the following result which was obtained from (2.17) in [7].

Theorem 11. *If f is supported on $[a,b]$ and there exist the constants $0<m, M<\infty$ such that*

$$m \le f(x) \le M \quad \textit{for a.e.} \quad x\in[a,b], \qquad (2.18)$$

then

$$\frac{1}{12}m^2(b-a)^3 \le \frac{1}{b-a}[b-E(f)][E(f)-a] - R_G(f) \le \frac{1}{12}M^2(b-a)^3. \quad (2.19)$$

3. Results from Steffensen's inequality

The following theorem is due to Steffensen [22] (see also [1] and [2]).

Theorem 12. *Let $h:[a,b]\to\mathbb{R}$ be a nondecreasing mapping on $[a,b]$ and $g:[a,b]\to\mathbb{R}$ be an integrable mapping on $[a,b]$ with*

$$-\infty < \phi \le g(x) \le \Phi < \infty \quad \textit{for all } x\in[a,b],$$

then

$$\Phi\int_a^{a+\lambda} h(x)\,dx + \phi\int_{a+\lambda}^b h(x)\,dx \le \int_a^b h(x)\,g(x)\,dx \qquad (3.1)$$
$$\le \phi\int_a^{b-\lambda} h(x)\,dx + \Phi\int_{b-\lambda}^b h(x)\,dx, \quad (3.2)$$

where

$$\lambda = \int_a^b G(x)\,dx, \quad G(x) = \frac{g(x)-\phi}{\Phi-\phi}, \quad \Phi\ne\phi. \qquad (3.3)$$

Remark 1. *We note that the result* (3.1) *may be rearranged to give Steffensen's better known result that*

$$\int_a^{a+\lambda} h(x)\,dx \le \int_a^b h(x)\,G(x)\,dx \le \int_{b-\lambda}^b h(x)\,dx, \tag{3.4}$$

where λ *is as given by* (3.3) *and* $0 \le G(x) \le 1$.

Equation (3.4) *has a very pleasant interpretation, as observed by Steffensen, that if we divide by* λ *then*

$$\frac{1}{\lambda}\int_a^{a+\lambda} h(x)\,dx \le \frac{\int_a^b G(x)\,h(x)\,dx}{\int_a^b G(x)\,dx} \le \frac{1}{\lambda}\int_{b-\lambda}^b h(x)\,dx. \tag{3.5}$$

Thus, the weighted integral mean of $h(x)$ *is bounded by the integral means over the end intervals of length* λ, *the total weight.*

Steffensen type inequalities have attracted considerable attention in the literature given the variety of applications and its generality. See for example [21] for a comprehensive survey and [20] wherein a number of generalisations have been provided.

Theorem 13. *Let* f *be supported on the interval* $[a,b]$ *and* $E(f)$ *exist. Then the Gini mean difference* $R_G(f)$ *satisfies*

$$\int_a^{a+\lambda} (a+\lambda-x)\,f(x)\,dx \le R_G(f) \le \lambda - \int_{b-\lambda}^b [x-(b-\lambda)]\,f(x)\,dx, \tag{3.6}$$

where $\lambda = E(f) - a$.

Proof. From the representation (2.10) for $R_G(f)$ we notice that $F(x)$ is non-decreasing and $0 \le 1 - F(x) \le 1$ for $x \in [a,b]$. Taking $h(x) = F(x)$ and $g(x) = 1 - F(x)$ in (3.1) we have that $\phi = 0$, $\Phi = 1$ and

$$\lambda = \int_a^b (1 - F(x))\,dx = E(f) - a. \tag{3.7}$$

For an interval $[c,d]$, we have on using integration by parts that

$$\int_c^d F(x)\,dx = dF(d) - cF(c) - \int_c^d xf(x)\,dx. \tag{3.8}$$

Result (3.7) is obtained on noticing that $F(b) = 1$ and $F(a) = 0$. Thus from (3.1) or (3.4) we have

$$\int_a^{a+\lambda} F(x)\,dx \le \int_a^b F(x)(1-F(x))\,dx \le \int_{b-\lambda}^b F(x)\,dx. \tag{3.9}$$

Now using (3.8), an integration by parts gives,

$$\int_a^{a+\lambda} F(x)\,dx = (a+\lambda)F(a+\lambda) - \int_a^{a+\lambda} xf(x)\,dx \tag{3.10}$$
$$= \int_a^{a+\lambda} (a+\lambda-x) f(x)\,dx$$

and similarly from (3.8),

$$\int_{b-\lambda}^{b} F(x)\,dx = \lambda - \int_{b-\lambda}^{b} (x-(b-\lambda)) f(x)\,dx. \tag{3.11}$$

Substitution of (3.10) and (3.11) into (3.9) upon noting (2.10), the stated result (3.6) holds with λ being given by (3.7). □

Remark 2. *We note that the result* (3.6) *may be compared with that of Gastwirth depicted by* (2.9). *The* (2.9) *result has the assumption that F is defined on* (a,∞) *and satisfies a DHR property giving* $\frac{\lambda}{2} \le R_G(f) \le \lambda$ *with* $\lambda = E(f) - a$.

We notice that the upper bound in (3.6) *is always less than* λ. *It is uncertain however as to whether the lower bound is greater or less than* $\frac{\lambda}{2}$.

Theorem 14. *Let* $f(x)$ *be a pdf on* $[a,b]$, $0 < \alpha \le xf(x) \le \beta$ *and* $\lambda = \frac{E(f)-\alpha(b-a)}{\beta-\alpha}$, *then the Gini mean difference* $R_G(f)$ *satisfies*

$$(\beta-\alpha)\int_a^{a+\lambda} (a+\lambda-x) f(x)\,dx \tag{3.12}$$
$$\le \frac{R_G(f)+E(f)}{2} - \alpha(b-E(f))$$
$$\le (\beta-\alpha)\left[\lambda - \int_{b-\lambda}^{b} (x-(b-\lambda)) f(x)\,dx\right].$$

Proof. (Sketch) From (2.2) we have the identity

$$\frac{R_G(f)+E(f)}{2} = \int_a^b xf(x)F(x)\,dx. \tag{3.13}$$

Associating $g(x)$ with $xf(x)$ and $h(x)$ with $F(x)$ in Theorem 12 and after some algebra gives the result. □

Theorem 15. *Let f be supported on the positive interval* $[a,b]$ *with* $0 \le a < b$ *and* $\phi \le f(x) \le \Phi$, $x \in [a,b]$ *and* $E(f)$ *exist. With* $\lambda = \frac{1-(b-a)\phi}{\Phi-\phi}$ *then the Gini mean* *difference* $R_G(f)$ *satisfies*

$$(\Phi-\phi)\int_a^{a+\lambda}\left[(a+\lambda)^2-x^2\right]f(x)\,dx \tag{3.14}$$
$$\leq R_G(f)+E(f)-\phi\int_a^b\left(b^2-x^2\right)f(x)\,dx$$
$$\leq(\Phi-\phi)\left\{\lambda(2b-\lambda)-\int_{b-\lambda}^b\left[x^2-(b-\lambda)^2\right]f(x)\,dx\right\}.$$

Proof. From (2.2) for f defined on the finite interval $[a,b]$ we have the identity

$$R_G(f)=2\int_a^b xf(x)F(x)\,dx-E(f). \tag{3.15}$$

In order to determine bounds for $R_G(f)$ consider $\int_a^b xf(x)F(x)\,dx$ from which we note that $xF(x)$ is nondecreasing. From Theorem 12 associating $g(x)$ with $f(x)$ and $h(x)$ with $xF(x)$, we have

$$(\Phi-\phi)\int_a^{a+\lambda}xF(x)\,dx\leq\int_a^b xf(x)F(x)\,dx-\phi\int_a^b xF(x)\,dx \tag{3.16}$$
$$\leq(\Phi-\phi)\int_{b-\lambda}^b xF(x)\,dx,$$

where

$$\lambda=\int_a^b\frac{f(x)-\phi}{\Phi-\phi}dx=\frac{1-(b-a)\phi}{\Phi-\phi}.$$

Now,

$$2\int_c^d xF(x)\,dx=d^2F(d)-c^2F(c)-\int_c^d x^2f(x)\,dx \tag{3.17}$$

so that with $F(a)=0$, $F(b)=1$ we have

$$\begin{cases} 2\int_a^b xF(x)\,dx = \int_a^b\left(b^2-x^2\right)f(x)\,dx, \\ 2\int_a^{a+\lambda}xF(x)\,dx=\int_a^{a+\lambda}\left[(a+\lambda)^2-x^2\right]f(x)\,dx, \\ 2\int_{b-\lambda}^b xF(x)\,dx=b^2-(b-\lambda)^2-\int_{b-\lambda}^b\left[x^2-(b-\lambda)^2\right]f(x)\,dx. \end{cases} \tag{3.18}$$

Substitution of (3.18) into (3.16) and noting (3.15) readily produces the result (3.14). □

4. Results with Karamata's inequality

In an interesting, but not well-known paper [19], Alexandru Lupaş generalised some results due to Karamata. These are presented and applications to bounding the Gini Mean Difference are demonstrated in the current section.

First some notation.

Let $-\infty < a < b < +\infty$ and $e_0(x) = 1$, $x \in [a,b]$. Further, let X be a real linear space with elements being real functions defined on $[a,b]$. By $\mathcal{F} : X \to \mathbb{R}$ we denote a positive linear functional normalised by $\mathcal{F}(e_0) = 1$. The following three results were obtained by Lupaş in [19].

Theorem 16. *Let $h, g \in X$ with*

$$m_1 \le h(x) \le M_1 \quad (M_1 \ne m_1), \qquad 0 < m_2 \le g(x) \le M_2 \quad x \in [a,b]. \tag{4.1}$$

If $D(h) = M_1 - \mathcal{F}(h)$, $d(h) = \mathcal{F}(h) - m_1$, then

$$\frac{m_1 M_2 D(h) + M_1 m_2 d(h)}{M_2 D(h) + m_2 d(h)} \le \frac{\mathcal{F}(hg)}{\mathcal{F}(g)} \le \frac{M_1 M_2 d(h) + m_1 m_2 D(h)}{M_2 d(h) + m_2 D(h)}. \tag{4.2}$$

The bounds in (4.2) are best possible.

Theorem 17. *Let h, g be elements from X which satisfy (4.1). If $\Delta(x) = M_1 - h(x)$, $\delta(x) = h(x) - m_1$, then*

$$|\mathcal{F}(h)\mathcal{F}(g) - \mathcal{F}(hg)| \le \frac{M_2 - m_2}{(M_1 - m_1)(M_2 + m_2)} [\mathcal{F}(\Delta)\mathcal{F}(\delta g) + \mathcal{F}(\delta)\mathcal{F}(\Delta g)]. \tag{4.3}$$

Theorem 18. *Let $h, g \in X$ with*

$$0 < m_1 \le h(x) \le M_1, \qquad 0 < m_2 \le g(x) \le M_2 \quad x \in [a,b]. \tag{4.4}$$

If $K = \frac{\sqrt{m_1 m_2} + \sqrt{M_1 M_2}}{\sqrt{m_1 M_2} + \sqrt{M_1 m_2}}$, then

$$\frac{1}{K^2} \le \frac{\mathcal{F}(hg)}{\mathcal{F}(h)\mathcal{F}(g)} \le K^2. \tag{4.5}$$

We note that Karamata established (4.2) and (4.5) in [16] and [17] for

$$\mathcal{F}(h) = \int_0^1 h(t)\,dt.$$

Further, h and g in Theorem 18 are assumed to be strictly positive and bounded whereas in Theorems 16 and 17, h is not allowed to be constant and the requirement for positivity is removed.

The following three theorems assume that the normalised positive linear functional $\mathcal{F}(\cdot)$ is given by

$$\mathcal{F}(h) = \frac{1}{b-a}\int_a^b h(x)\,dx. \tag{4.6}$$

Theorem 19. *Let $f(x)$ be a pdf on $[a,b]$ and $0 < \alpha \le x f(x) \le \beta$, then the Gini mean difference $R_G(f)$ satisfies*

$$\left(\frac{1 - pz}{1 + pz}\right) E(f) \le R_G(f) \le \left(\frac{p - z}{p + z}\right) E(f), \tag{4.7}$$

where $p = \frac{\beta}{\alpha}$ and $z = \frac{E(f) - a}{b - E(f)}$.

Proof. From (2.2) for f defined on the finite interval $[a,b]$, we have the identity (3.13). In order to obtain bounds for $R_G(f)$ consider $\int_a^b xf(x)F(x)\,dx$. Let $h(x)=F(x)$ and $g(x)=xf(x)$ in Theorem 16, then we have that $0\le F(x)\le 1$ and $0<\alpha\le xf(x)\le\beta$ from the postulates.

Now, from (4.6)

$$d(F)=\tfrac{1}{b-a}\int_a^b F(x)\,dx=\tfrac{b-E(f)}{b-a}\quad\text{and}\quad D(F)=1-d(F)=\tfrac{E(f)-a}{b-a}. \tag{4.8}$$

From (4.2) we have, on utilising (3.13),

$$\frac{\alpha d(F)}{\beta D(F)+\alpha d(F)}\le\frac{\int_a^b xf(x)F(x)\,dx}{E(f)}=\frac{R_G(f)+E(f)}{2E(f)}\le\frac{\beta d(F)}{\alpha D(F)+\beta d(F)},$$

giving

$$L(f)\le R_G(f)\le U(f), \tag{4.9}$$

where $L(f)=E(f)\left(\frac{\alpha d(F)-\beta D(F)}{\alpha d(F)+\beta D(F)}\right)$ and $U(f)=E(f)\left(\frac{\beta d(F)-\alpha D(F)}{\beta d(F)+\alpha D(F)}\right)$.

Simplification of (4.9) gives the stated result (4.7) where we have from (4.8) that $z=\frac{D(F)}{d(F)}$. □

Theorem 20. *Let $f(x)$ be a pdf on $[a,b]$ and $0<m\le f(x)\le M$, then the Gini mean difference $R_G(f)$ satisfies*

$$\frac{2b\xi}{b\rho-(\rho-1)\xi}-E(f)\le R_G(f)\le\frac{2b\rho\xi}{b+(\rho-1)\xi}-E(f), \tag{4.10}$$

where

$$\rho=\frac{M}{m},\quad \xi=\frac{b^2-\mathcal{M}_2}{2(b-a)}\quad\textit{and}\quad \mathcal{M}_2=\int_a^b x^2f(x)\,dx. \tag{4.11}$$

Proof. Following the proof of Theorem 19 so that in order to find bounds for $R_G(f)$ we consider $\int_a^b xf(x)F(x)\,dx$. Let $h(x)=xF(x)$ and $g(x)=f(x)$ in Theorem 16 then we have that $0\le xF(x)\le b$ and $0<m\le f(x)\le M$ from the postulates. Further, let the normalised positive linear functional $F(\cdot)$ be given by (4.6) so that

$$d(h)=\tfrac{1}{b-a}\int_a^b xF(x)\,dx=\tfrac{b^2-\mathcal{M}_2}{2(b-a)}\ \text{and}\ D(h)=b-d(h) \tag{4.12}$$

From (4.2) we have

$$\frac{bmd(h)}{MD(h)+md(h)}\le\int_a^b xf(x)F(x)\,dx\le\frac{bMd(h)}{Md(h)+mD(h)} \tag{4.13}$$

so that from (3.13) and (4.12),

$$L(h)\le R_G(f)\le U(h) \tag{4.14}$$

where

$$L(h)=\tfrac{2bmd(h)}{M(b-d(h))+md(h)}-E(f)\ \text{and}\ U(h)=\tfrac{2bMd(h)}{Md(h)+m(b-d(h))}-E(f). \tag{4.15}$$

Rearranging the denominator in (4.15), noting that $\xi = d(h)$ and division by m gives the required result (4.10)–(4.11). $\square$

Theorem 21. *Let $f(x)$ be a pdf on $[a,b]$ and $0 < m \le f(x) \le M$, then the Gini mean difference $R_G(f)$ satisfies*

$$\frac{E(f) + 2M\left[a\left(\frac{a+b}{2}\right) - bE(f)\right]}{2bM - 1} \le R_G(f) \le \frac{E(f) + 2M\left[b\left(\frac{a+b}{2}\right) - aE(f)\right]}{2aM - 1}. \tag{4.16}$$

Proof. Let $h(x) = f(x)F(x)$ and $g(x) = x$ in Theorem 16. Then we have from the postulates that $0 \le f(x)F(x) \le M$ and $0 < a \le x \le b$. Further, with the normalised positive linear functional $\mathcal{F}(\cdot)$ given by (4.6) produces

$$d(h) = \frac{1}{b-a}\int_a^b f(x)F(x)\,dx = \frac{1}{2(b-a)} \quad \text{and} \quad D(h) = M - d(h). \tag{4.17}$$

From (4.2) we have

$$\frac{aMd(h)}{b(M - d(h)) + ad(h)} \le \frac{\int_a^b xf(x)F(x)\,dx}{\int_a^b xdx} \le \frac{bMd(h)}{bd(h) + a(M - d(h))}$$

which from (3.13) and upon rearrangement gives

$$\frac{2aMd(h)\int_a^b xdx}{bM - (b-a)d(h)} - E(f) \le R_G(f) \le \frac{2bMd(h)\int_a^b xdx}{aM + (b-a)d(h)} - E(f).$$

That is, since from (4.17) $d(h)\int_a^b xdx = \frac{b+a}{4}$ we have

$$\frac{aM(b+a)}{2bM - 1} - E(f) \le R_G(f) \le \frac{bM(b+a)}{2aM - 1} - E(f). \tag{4.18}$$

Rearrangement of (4.18) readily produces the result (4.16). $\square$

Theorem 22. *Let $f(x)$ be a pdf on $[a,b]$ with $a > 0$ and $0 < m \le f(x) \le M$, $x \in [a,b]$. Then the Gini mean difference $R_G(f)$ satisfies*

$$\left(\frac{1-\rho\zeta}{1+\rho\zeta}\right)E(f) \le R_G(f) \le \left(\frac{\rho-\zeta}{\rho+\zeta}\right)E(f), \tag{4.19}$$

where $\rho = \frac{M}{m}$, $\zeta = \frac{\mathcal{M}_2 - a^2}{b^2 - \mathcal{M}_2}$ and $\mathcal{M}_2 = \int_a^b x^2 f(x)\,dx$, the second moment about zero.

Proof. Let the normalised positive linear functional $\mathcal{F}(\cdot)$ be given by

$$\mathcal{F}(h) = \frac{\int_a^b w(x)h(x)\,dx}{\int_a^b w(x)\,dx}. \tag{4.20}$$

Taking $w(x) = x$, $h(x) = F(x)$ and $g(x) = f(x)$ in Theorem 16, and that $0 \leq F(x) \leq 1$ and $0 < m \leq f(x) \leq M$, we then have that

$$\frac{md(F)}{md(F) + MD(F)} \leq \frac{\int_a^b f(x)\, xF(x)\, dx}{E(f)} \leq \frac{Md(F)}{Md(F) + mD(F)}. \tag{4.21}$$

Now, from (3.13) we have, upon simplification that

$$\left(\frac{md(F) - MD(F)}{md(F) + MD(F)}\right) \leq R_G(f) \leq \left(\frac{Md(F) - mD(F)}{md(F) + MD(F)}\right), \tag{4.22}$$

where

$$d(F) = \frac{\int_a^b xF(x)\, dx}{\int_a^b x dx} = \frac{b^2 - \mathcal{M}_2}{b^2 - a^2} \quad \text{and} \quad D(F) = 1 - d(F) = \frac{\mathcal{M}_2 - a^2}{b^2 - a^2}. \tag{4.23}$$

since

$$2\int_a^b xF(x)\, dx = x^2 F(x)\Big]_a^b - \int_a^b x^2 f(x)\, dx = b^2 - \mathcal{M}_2.$$

Simplifying (4.22) in a similar manner as in Theorem 16, we have the result (4.19). □

Remark 3. *The lower bounds in* (4.7) *and* (4.19) *are only useful when they are greater than* 0 *since* $R_G(f)$ *is known to be non-negative. This occurs for* $E(f) < \frac{a\beta + b\alpha}{\alpha + \beta}$ *and* $\mathcal{M}_2 < \frac{Ma^2 + mb^2}{M + m}$.

Acknowledgement

The author would like to thank the referee for his detailed comments on an earlier version of this paper.

References

[1] P. Cerone, *On an identity for the Chebychev functional and some ramifications*, J. Ineq. Pure and Appl. Math. **3**(1) Art. 4 (2002). `http://jipam.vu.edu.au/v3n1/`.

[2] P. Cerone, *On some generalisatons of Steffensen's inequality and related results*, J. Ineq. Pure and Appl. Math. **2**(3) Art. 28 (2001). `http://jipam.vu.edu.au/v2n3/`.

[3] P. Cerone and S.S. Dragomir, *A survey on bounds for the Gini Mean Difference*, Advances in Inequalities from Probability Theory and Statistics, N.S. Barnett and S.S. Dragomir (Eds.), Nova Science Publishers, (2008), 81–111.

[4] P. Cerone and S.S. Dragomir, *Bounds for the Gini mean difference via the Sonin identity*, Comp. Math. Modelling **50** (2005), 599–609.

[5] P. Cerone and S.S. Dragomir, *Bounds for the Gini mean difference via the Korkine identity*, J. Appl. Math. & Computing (Korea) **22**(3) (2006), 305–315.

[6] P. Cerone and S.S. Dragomir, *Bounds for the Gini mean difference of continuous distributions defined on finite intervals (I)*, Applied Mathematics Letters **20** (2007), 782–789.

[7] P. Cerone and S.S. Dragomir, *Bounds for the Gini mean difference of continuous distributions defined on finite intervals (II)*, Comput. Math. Appl. **52**(10-11) (2006), 1555–1562.

[8] P. Cerone and S.S. Dragomir, *A refinement of the Grüss inequality and applications*, Tamkang J. Math. **38**(1) (2007), 37–49. (*See also* RGMIA Res. Rep. Coll. **5**(2) (2002), Art. 14. [ONLINE: `http://rgmia.vu.edu.au/v5n2.html`]

[9] X.-L. Cheng and J. Sun, *A note on the perturbed trapezoid inequality*, J. Inequal. Pure & Appl. Math. **3**(2) (2002), Article 29, `http://jipam.vu.edu.au/article.php?sid=181`

[10] H.A. David, *Gini's mean difference rediscovered*, Biometrika **55** (1968), 573.

[11] J.L. Gastwirth, *The estimation of the Lorentz curve and Gini index*, Rev. Econom. Statist. **54** (1972), 305–316.

[12] C. Gini, *Variabilità e Metabilità, contributo allo studia della distribuzioni e relationi statistiche*, Studi Economica-Gicenitrici dell'Univ. di Coglani, **3** (1912), 1–158.

[13] G.M. Giorgi, *Bibliographic portrait of the Gini concentration ratio*, Metron **XLVIII** (1-4) (1990), 103–221.

[14] G.M. Giorgi, *Alcune considerazioni teoriche su di un vecchio ma per sempre attuale indice: il rapporto di concentrazione del Gini*, Metron **XLII**(3-4) (1984), 25–40.

[15] G.H. Hardy, J.E. Littlewood and G. Polya, *Inequalities*, Cambridge Univ. Press., 1934.

[16] J. Karamata, *O prvom stavu srednjih vrednosti odredenih integrala*, Glas Srpske Kraljevske Akademje **CLIV** (1933), 119–144.

[17] J. Karamata, *Sur certaines inégalités relatives aux quotients et à la différence de* $\int fg$ *et* $\int f \cdot \int g$, Acad. Serbe Sci. Publ. Inst. Math. **2** (1948), 131–145.

[18] M. Kendall and A. Stuart, *The Advanced Theory of Statistics*, Volume **1**, Distribution Theory, Fourth Edition, Charles Griffin & Comp. Ltd., London, 1977.

[19] A. Lupaş, *On two inequalities of Karamata*, Univ. Beograd. Publ. Elektrotechn. Fak. Ser. Mat. Fiz. **602–633** (1978), 119–123.

[20] P.R. Mercer, *Extensions of Steffensen's inequality*, J. Math. Anal. & Applics. **246** (2000), 325–329.

[21] D.S. Mitrinović, J.E. Pečarić and A.M. Fink, *Classical and New Inequalities in Analysis*, Kluwer Academic Publishers, Dordrecht/Boston/London, 1993.

[22] J.F. Steffensen, *On certain inequalities between mean values and their application to actuarial problems*, Skandinavisk Aktuarietidskrif (1918), 82–97.

Pietro Cerone
School of Computer Science and Mathematics
Victoria University, PO Box 14428
MCMC 8001 VIC, Australia
URL: `http://www.staff.vu.edu.au/RGMIA/cerone/`
e-mail: `pietro.cerone@vu.edu.au`

International Series of Numerical Mathematics, Vol. 157, 91–95

On Some Integral Inequalities

Bogdan Gavrea

Abstract. An extension of inequalities (1.2) and (1.3) ([1]) is given and an open problem raised in [1] is solved.

Mathematics Subject Classification (2000). 26D15.

Keywords. Inequality, increasing function.

1. Introduction

Let f be a continuous function defined on $[0,1]$. Assume also that f satisfies the following inequality:

$$\int_x^1 f(t)dt \geq \frac{1-x^2}{2}, \; \forall\, x \in [0,1]. \tag{1.1}$$

In [1] it is shown that for any positive function f which satisfies inequality (1.1), the following two inequalities hold:

$$\int_0^1 f^{\alpha+1}(x)dx \geq \int_0^1 x^\alpha f(x)dx \tag{1.2}$$

$$\int_0^1 f^{\alpha+1}(x)dx \geq \int_0^1 x f^\alpha(x)dx \tag{1.3}$$

for any positive real number α.

Now, let f be a continuous function on $[0,1]$ and g be a continuous and positive function such that:

$$\int_x^1 f(t)dt \geq \int_x^1 g(t)dt, \; \forall\, x \in [0,1]. \tag{1.4}$$

The aim of this note is to derive inequalities similar to (1.2) and (1.3) for functions f that verify inequality (1.4).

2. Auxiliary results

Lemma 2.1. *Let h be a continuous, increasing and positive function on $[0,1]$. If f satisfies inequality* (1.4) *then*

$$\int_u^1 h(x)f(x)dx \geq \int_u^1 h(x)g(x)dx, \ \forall\ u \in [0,1]. \tag{2.5}$$

Proof. Without loss of generality, we may assume that h is differentiable and its derivative is continuous on $[0,1]$.

Multiplying both sides of (1.4) by $h'(x)$ and than integrating on $[u,1]$ gives:

$$\int_u^1 \left[h'(x)\int_x^1 f(t)dt\right]dx \geq \int_u^1 \left[h'(x)\int_x^1 g(t)dt\right]dx. \tag{2.6}$$

By performing integration by parts in (2.6), we obtain:

$$-h(u)\int_u^1 f(t)dt + \int_u^1 h(x)f(x)dx \geq -h(u)\int_u^1 g(t)dt + \int_u^1 h(x)g(x)dx$$

or

$$\int_u^1 h(x)f(x)dx - \int_u^1 h(x)g(x)dx \geq h(u)\left[\int_u^1 f(t)dt - \int_u^1 g(t)dt\right]. \tag{2.7}$$

From (2.7) and (1.4) the conclusion of the Lemma follows. □

Lemma 2.2. *If f and g satisfy* (1.4) *with f positive and g increasing, then for any $\alpha \geq 1$ we have:*

$$\int_u^1 f^\alpha(x)dx \geq \int_u^1 g^\alpha(x)dx, \ \forall\ u \in [0,1]. \tag{2.8}$$

Proof. Let $\alpha > 1$ and let h be an increasing function on $[0,1]$. From Lemma 2.1 we have:

$$\int_u^1 h(x)f(x)dx \geq \int_u^1 h(x)g(x)dx. \tag{2.9}$$

By using Hölder's inequality we obtain:

$$\int_u^1 h(x)f(x)dx \leq \left(\int_u^1 [h(x)]^{\frac{\alpha}{\alpha-1}}\right)^{\frac{\alpha-1}{\alpha}} \left[\int_u^1 f^\alpha(x)dx\right]^{\frac{1}{\alpha}}. \tag{2.10}$$

From (2.10) we get:

$$\int_u^1 f^\alpha(x)dx \geq \frac{\left[\int_u^1 h(x)f(x)dx\right]^\alpha}{\left[\int_u^1 [h(x)]^{\frac{\alpha}{\alpha-1}}dx\right]^{\alpha-1}}. \tag{2.11}$$

Inequalities (2.9) and (2.11) give:

$$\int_u^1 f^\alpha(x)dx \geq \frac{\left[\int_u^1 h(x)g(x)dx\right]^\alpha}{\left[\int_u^1 [h(x)]^{\frac{\alpha}{\alpha-1}}dx\right]^{\alpha-1}} \tag{2.12}$$

Because g is an increasing function, it follows that

$$h = g^{\alpha-1}$$

is also an increasing function. Taking $h = g^{\alpha-1}$ in (2.12), we obtain the inequality (2.8). □

Remark 2.3. For $\alpha \in (0,1)$ the above result doesn't hold in general. To better see this, consider:

$$f(t) = \begin{cases} \frac{1}{2}t, & t \in \left[0, \frac{1}{3}\right) \\ \frac{3}{2}t - \frac{1}{3}, & t \in \left[\frac{1}{3}, 1\right] \end{cases}, \quad g(t) = t$$

Given f and g above we have:

$$\int_x^1 f(t)dt \geq \int_x^1 g(t)dt$$

and

$$\int_0^1 [f(t)]^{\frac{1}{2}}dt < \int_0^1 [g(t)]^{\frac{1}{2}}dt.$$

3. Main results

Theorem 3.1. *Assume f and g satisfy* (1.4) *with f positive and g increasing. Let $x \in [0,1]$ and $\alpha > 0$. Then*

$$\int_x^1 f^{\alpha+1}(t)dt \geq \int_x^1 g^\alpha(t)f(t)dx. \tag{3.13}$$

Proof. Using the Mean's Inequality, we obtain:

$$\frac{1}{\alpha+1}f^{\alpha+1}(t) + \frac{\alpha}{\alpha+1}g^{\alpha+1}(t) \geq f(t)g^\alpha(t). \tag{3.14}$$

Integrating both sides of (3.14) on $[x,1]$ gives:

$$\frac{1}{\alpha+1}\int_x^1 f^{\alpha+1}(t)dt + \frac{\alpha}{\alpha+1}\int_x^1 g^{\alpha+1}(t)dt \geq \int_x^1 f(t)g^\alpha(t)dt. \tag{3.15}$$

From Lemma 2.2, we obtain

$$\int_x^1 g^{\alpha+1}(t)dt \le \int_x^1 f^{\alpha+1}(t)dt \tag{3.16}$$

From (3.15) and (3.16) we get

$$\frac{1}{\alpha+1}\int_x^1 f^{\alpha+1}(t)dt + \frac{\alpha}{\alpha+1}\int_x^1 g^{\alpha+1}(t)dt \le \int_x^1 f^{\alpha+1}(t)dt. \tag{3.17}$$

Inequality (3.13) follows now from (3.15) and (3.17). □

Theorem 3.2. *Assume f and g satisfy* (1.4). *Then, for any $\alpha > 0$ and $x \in [0,1]$ we have:*

$$\int_x^1 f^{\alpha+1}(t)dt \ge \int_x^1 g(t)f^{\alpha}(t)dt.$$

Proof. Given that for nonnegative t, the function t^α is an increasing function, we have

$$(u^\alpha - v^\alpha)(u-v) \ge 0,\ \forall\ u,v \in [0,\infty). \tag{3.18}$$

Taking $u = f(t)$ and $v = g(t)$ gives

$$(f^\alpha(t) - g^\alpha(t))(f(t)-g(t)) \ge 0,\ \forall\ t \in [0,1]. \tag{3.19}$$

Integrating both sides of (3.19) on $[x,1]$, we obtain

$$\int_x^1 f^{\alpha+1}(t)dt - \int_x^1 f^\alpha(t)g(t)dt - \int_x^1 g^\alpha(t)f(t)dt + \int_x^1 g^{\alpha+1}(t)dt \ge 0 \tag{3.20}$$

or

$$\int_x^1 f^{\alpha+1}(t)dt \ge \int_x^1 f^\alpha(t)g(t)dt + \int_x^1 g^\alpha(t)f(t)dt - \int_x^1 g^{\alpha+1}dt. \tag{3.21}$$

Since g^α is an increasing function, by taking

$$h(t) = g^\alpha(t)$$

in Lemma 2.1, we obtain

$$\int_x^1 g^\alpha(t)f(t)dt \ge \int_x^1 g^{\alpha+1}(t)dt. \tag{3.22}$$

The conclusion of the theorem follows from (3.21) and (3.22). □

In [1], the following open problem is proposed: Let f be a continuous function on $[0,1]$ satisfying the inequality:

$$\int_x^1 f(t)dt \ge \int_x^1 tdt,\ \forall\ x \in [0,1].$$

Under what assumptions does the following inequality holds

$$\int_0^1 f^{\alpha+\beta}(x)dx \ge \int_0^1 x^\alpha f^\beta(x)dx?$$

In what follows, we assume that the functions f and g are continuous, f is positive, g is increasing and positive on $[0,1]$ and that inequality (1.4) holds.

Theorem 3.3. *If f is an increasing function, $\alpha \geq 1$, $\beta \geq 0$ and $x \in [0,1]$ then the inequality*

$$\int_x^1 f^{\alpha+\beta}(t)dt \geq \int_x^1 g^{\alpha}(t)f^{\beta}(t)dt$$

holds.

Proof. For $\alpha \geq 1$, from (2.8) we have:

$$\int_x^1 f^{\alpha}(t)dt \geq \int_x^1 g^{\alpha}(t)dt. \tag{3.23}$$

Because f^{β} is an increasing function, by taking $h = f^{\beta}$ in (2.5), it follows from (3.23) that

$$\int_x^1 f^{\beta}(t)f^{\alpha}(t)dt \geq \int_x^1 g^{\alpha}(t)f^{\beta}(t)dt.$$

The last inequality is equivalent to the desired result. □

Acknowledgements

We would like to thank the anonymous referee for important observations that led to the improvement of this work.

References

[1] Quôc Anh Ngô, Du Duc Thang, Tran Tat Dat, Dang Anh Tuan, *Notes on integral inequality*, J. Ineq. Pure Appl. Math. **7** (2006), Art. 120.

Bogdan Gavrea
Technical University of Cluj-Napoca
Department of Mathematics
Str. C. Daicoviciu 15
RO-3400 Cluj-Napoca, Romania
e-mail: `Bogdan.Gavrea@math.utcluj.ro`

International Series of Numerical Mathematics, Vol. 157, 97–107

A New Characterization of the Hardy and Its Limit Pólya-Knopp Inequality for Decreasing Functions

Maria Johansson

Abstract. In this paper we present and prove a new alternative weight characterization for the Hardy inequality for decreasing functions. We also give an alternative approach to the characterization of the Hardy inequality using a fairly new equivalence theorem. In fact, this result shows that there are infinitely many possibilities to characterize the considered Hardy inequality for decreasing functions. We also state the corresponding weight characterization for the Pólya-Knopp inequality for decreasing functions.

Mathematics Subject Classification (2000). Primary 26D10; Secondary 26D15, 26D07.

Keywords. Integral inequalities, weights, Hardy operator, equivalence theorem.

1. Introduction

The study of the Pólya-Knopp inequality

$$\left(\int_0^\infty \left(\exp\left(\frac{1}{x}\int_0^x \ln f(t)dt\right)\right)^q u(x)dx\right)^{\frac{1}{q}} \leq C\left(\int_0^\infty f^p(x)v(x)dx\right)^{\frac{1}{p}} \tag{1.1}$$

is closely connected to the study of the Hardy inequality

$$\left(\int_0^\infty \left(\frac{1}{x}\int_0^x f(t)\,dt\right)^q u(x)\,dx\right)^{\frac{1}{q}} \leq C\left(\int_0^\infty f^p(x)\,v(x)\,dx\right)^{\frac{1}{p}} \tag{1.2}$$

since the Pólya-Knopp inequality can be regarded as a limit inequality to the Hardy inequality. In [6] L.E. Persson and V.D. Stepanov proved a weight characterization for the inequality (1.1) by first characterizing the Hardy inequality (1.2) with a new weight criteria. Their result reads:

Theorem 1.1. *Let $1 < p \leq q < \infty$. Then the Hardy inequality* (1.2) *holds for $f \geq 0$ if and only if*

$$A_{PS} = A_{PS}(p,q,u,v) := \tag{1.3}$$
$$\sup_{t>0} \left(\int_0^t u(x) x^{-q} \left(\int_0^x v(y)^{1-p'} dy \right)^q dx \right)^{\frac{1}{q}} \left(\int_0^t v(x)^{1-p'} dx \right)^{-\frac{1}{p}} < \infty.$$

Moreover, the best constant C in (1.2) *can be estimated as follows:*

$$A_{PS} \leq C \leq p' A_{PS}..$$

Then by performing a limiting procedure they obtained the following limit result:

Theorem 1.2. *Let $0 < p \leq q < \infty$. Then the inequality* (1.1) *holds for all $f \geq 0$ if and only if*

$$D_{PS} = D_{PS}(p,q,w) := \sup_{t>0} t^{-\frac{1}{p}} \left(\int_0^t w(x) dx \right)^{\frac{1}{q}} < \infty, \tag{1.4}$$

where

$$w(x) = \exp\left(\frac{1}{x} \int_0^x \ln v(y) dy \right)^{-\frac{q}{p}} u(x). \tag{1.5}$$

Moreover, if C is the best possible constant in (1.1), *then*

$$D_{PS} \leq C \leq e^{\frac{1}{p}} D_{PS}.$$

As mentioned Persson and Stepanov (see Theorem 1.1) proved a new weight characterization for the Hardy inequality. Their motivation for doing that, was the fact that it was not possible to perform the mentioned limiting procedure by using the Muckenhoupt condition so they needed an equivalent characterization. Partly guided by this we also prove a new weight characterization for the Hardy inequality for decreasing functions, because it is not possible to use Sawyers conditions for the purpose of performing a limiting procedure.

More precisely, in Section 2 we present and prove our new alternative weight characterization of the Hardy inequality for decreasing functions (see Theorem 2.1). In Section 3 we use a fairly new equivalence theorem to give an alternative proof of Theorem 2.1 and Sawyer's result (see Remark 3.3). In fact, this result shows that there are infinitely many possibilities to characterize the considered Hardy inequality for decreasing functions. Finally, in Section 4 we state the corresponding Pólya-Knopp inequality for decreasing functions (see Remark 4.1).

In 1990 E. Sawyer [7] proved the following theorem by performing a general approach for general operators. In [9] V.D. Stepanov gave a direct proof, which also gave good estimates of the constants.

Theorem 1.3. *Let $1 < p \leq q < \infty$. Then the inequality* (1.2) *holds for all decreasing $f \geq 0$ if and only if*

$$A_0 = A_0(p,q,u,v) := \sup_{t>0} V(t)^{-\frac{1}{p}} \left(\int_0^t u(x) dx \right)^{\frac{1}{q}} < \infty, \tag{1.6}$$

and

$$A_1 = A_1(p,q,u,v) :=$$

$$\sup_{t>0}\left(\int_t^\infty u(x)x^{-q}dx\right)^{\frac{1}{q}}\left(\int_0^t x^{p'}V(x)^{-p'}v(x)dx\right)^{\frac{1}{p'}} < \infty,$$

where $p' = \frac{p}{p-1}$ *and*

$$V(t) = \int_0^t v(x)dx. \tag{1.7}$$

Moreover, if C *is the best possible constant in* (1.2), *then*

$$C \approx \max(A_0, A_1).$$

Here and in the sequel by decreasing we mean non-increasing.

2. A new weight characterization of the weighted Hardy inequality for decreasing functions

The main result of this section reads:

Theorem 2.1. *Let* $1 < p \le q < \infty$. *Then the Hardy inequality* (1.2) *holds for all decreasing* $f \ge 0$ *if and only if* (1.6) *and*

$$A_2 = A_2(p,q,u,v) := \tag{2.1}$$

$$\sup_{t>0}\left(\int_0^t\left(\int_0^x y^{p'}V(y)^{-p'}v(y)dy\right)^q u(x)x^{-q}dx\right)^{\frac{1}{q}}\left(\int_0^t x^{p'}V(x)^{-p'}v(x)dx\right)^{-\frac{1}{p}} < \infty$$

hold, where V *is defined by* (1.7). *Moreover, if* C *is the best possible constant in* (1.2), *then*

$$C \approx \max(A_0, A_2). \tag{2.2}$$

For the proof we need the following Lemma:

Lemma 2.2. *Let* $0 < p \le q < \infty$. *Then the inequality*

$$\left(\int_0^\infty f^q(x)u(x)dx\right)^{\frac{1}{q}} \le C\left(\int_0^\infty f^p(x)v(x)dx\right)^{\frac{1}{p}} \tag{2.3}$$

holds for all decreasing $f \ge 0$ *if and only if* (1.6) *holds. Moreover, the constant* $C = A_0$ *is the best possible, where* A_0 *is defined by* (1.6).

A proof of this Lemma can be found, e.g., in [9] (for more references see, e.g., the Ph.D. thesis [1] by S. Barza).

Proof. (Theorem 2.1). Since f is a decreasing function we can write $f(x) = \int_x^\infty h(y)dy$, for some $h(y) \ge 0$. First note that, by changing order of integration,

$$\frac{1}{x}\int_0^x f(t)dt = \frac{1}{x}\int_0^x\int_t^\infty h(y)dydt = \int_x^\infty h(y)dy + \frac{1}{x}\int_0^x yh(y)dy. \tag{2.4}$$

Consequently, by using (2.4) and Minkowski's inequality, we find that

$$\begin{aligned}
F &:= \left(\int_0^\infty \left(\frac{1}{x}\int_0^x f(t)dt\right)^q u(x)dx\right)^{\frac{1}{q}} \\
&= \left(\int_0^\infty \left(\int_x^\infty h(y)dy + \frac{1}{x}\int_0^x yh(y)dy\right)^q u(x)dx\right)^{\frac{1}{q}} \\
&\le \left(\int_0^\infty f^q(x)\,u(x)dx\right)^{\frac{1}{q}} + \left(\int_0^\infty \left(\frac{1}{x}\int_0^x yh(y)dy\right)^q u(x)dx\right)^{\frac{1}{q}} \\
&:= \left(\int_0^\infty f^q(x)u(x)dx\right)^{\frac{1}{q}} + F_1.
\end{aligned}$$

The first term on the right-hand side can be estimated by using Lemma 2.2 and, thus,

$$F \le A_0 \left(\int_0^\infty f^p(x)v(x)dx\right)^{\frac{1}{p}} + F_1. \tag{2.5}$$

Let us estimate the second term F_1. We have, setting $H = hV$, where V is defined by (1.7),

$$F_1^q = \int_0^\infty \left(\frac{1}{x}\int_0^x yh(y)dy\right)^q u(x)dx = \int_0^\infty \left(\frac{1}{x}\int_0^x \frac{yH(y)}{V(y)}dy\right)^q u(x)dx.$$

Moreover, by integrating by parts and setting $\int_0^y H(t)dt = G(y)$, we get that

$$\begin{aligned}
\int_0^x \frac{yH(y)}{V(y)}dy &= \frac{xG(x)}{V(x)} - \int_0^x \frac{G(y)}{V(y)}dy + \int_0^x \frac{yv(y)}{V^2(y)}G(y)dy \\
&\le \frac{xG(x)}{V(x)} + \int_0^x \frac{yv(y)}{V^2(y)}G(y)dy.
\end{aligned}$$

Hence, according to Minkowski's inequality,

$$\begin{aligned}
F_1 \le{}& \left(\int_0^\infty \left(\frac{G(x)}{V(x)}\right)^q u(x)dx\right)^{1/q} + \left(\int_0^\infty \left(\frac{1}{x}\int_0^x \frac{yv(y)}{V^2(y)}G(y)dy\right)^q u(x)dx\right)^{1/q} \\
:={}& F_2 + F_3.
\end{aligned} \tag{2.6}$$

Moreover, by integrating by parts and using an approximation argument, we find that

$$F_2^q = \int_0^\infty \left(\frac{G(x)}{V(x)}\right)^q u(x)dx \le q\int_0^\infty \frac{G(x)^q v(x)}{V(x)^{q+1}}\left(\int_0^x u(t)dt\right)dx.$$

Hence, by applying condition (1.6), we have that

$$F_2^q \le qA_0^q \int_0^\infty \frac{G(x)^q v(x)}{V(x)^{q+1-q/p}}dx. \tag{2.7}$$

Moreover,

$$G(y) = \int_0^y h(x) \int_0^x v(t)dtdx = \int_0^y v(t) \left(\int_t^y h(x)dx \right) dt \tag{2.8}$$
$$\leq \int_0^y v(t) \left(\int_t^\infty h(x)dx \right) dt \leq \int_0^y f(t)v(t)dt.$$

Consequently, by inserting (2.8) into (2.7), we get that

$$F_2^q \leq qA_0^q \int_0^\infty \left(\int_0^x f(t)v(t)dt \right)^q \frac{v(x)}{V(x)^{q+1-q/p}} dx.$$

If we apply Theorem 1.1 to the function $f(t)\, v(t)$ instead of $f(t)$ and the weights $V(x)^{\frac{q}{p}-q-1} v(x)\, x^q$ instead of $u(x)$ and $v(x)^{1-p}$ instead of $v(x)$ we find that the condition (1.3) becomes

$$A_{PS}^q = \sup_{t>0} \left(\int_0^t V(x)^{q/p-1-q} v(x) \left(\int_0^x v(y)^{(1-p)(1-p')} dy \right)^q dx \right)$$
$$\times \left(\int_0^t v(x)^{(1-p)(1-p')} dx \right)^{-\frac{q}{p}}$$
$$= \sup_{t>0} \left(\int_0^t V(x)^{q/p-1} v(x))dx \right) \left(\int_0^t v(x)dx \right)^{-\frac{q}{p}} = \frac{p}{q},$$

and, thus,

$$F_2^q \leq p\,(p')^q A_0^q \left(\int_0^\infty f^p(x)v(x)dx \right)^{\frac{q}{p}}. \tag{2.9}$$

Denoting

$$\frac{yv(y)}{V^2(y)} G(y) = \Phi(y)$$

and

$$x^{p'} V(x)^{-p'} v(x) = \Psi(x)^{1-p'} \tag{2.10}$$

and, again applying Theorem 1.1, (2.1) and (2.8), we find that

$$F_3^q = \int_0^\infty \left(\frac{1}{x} \int_0^x \Phi(y)dy \right)^q u(x)dx \leq (p')^q A_2^q \left(\int_0^\infty \Phi^p(x)\Psi(x)dx \right)^{\frac{q}{p}}$$
$$= (p')^q A_2^q \left(\int_0^\infty G^p(x)v(x)V(x)^{-p}dx \right)^{\frac{q}{p}}$$
$$\leq (p')^q A_2^q \left(\int_0^\infty \left(\int_0^x f(y)v(y)dy \right)^p v(x)V(x)^{-p}dx \right)^{\frac{q}{p}}$$
$$\leq (p')^{2q} A_2^q \left(\int_0^\infty f(x)^p v(x)dx \right)^{\frac{q}{p}}. \tag{2.11}$$

Here in the last step we apply Theorem 1.1 with $q = p$, $f(x)$ replaced by $f(x)v(x)$, $v(x)$ replaced by $v^{1-p}(x)$ and $u(x)$ replaced by $v(x)V^{-p}(x)x^p$ so that the condition (1.3) $A_{PS} = 1$. By combining the estimates (2.5), (2.6), (2.9) and (2.11) we find that (1.2) holds with a constant C satisfying

$$C \leq \left(1 + p^{1/q}p'\right) A_0 + (p')^2 A_2,$$

i.e., the upper estimate in (2.2) holds.

For the necessity condition, assume that the inequality (1.2) holds for all decreasing $f \geq 0$. Consider, for fixed $y > 0$, the decreasing test function

$$f_y(s) = \left(\int_s^y t^{p'}V(t)^{-p'-1}v(t)dt\right)^{\frac{1}{p}} \chi_{[0,y]}(s).$$

Applying this function to the right-hand side of (1.2) and changing the order of integration we obtain that

$$\begin{aligned}
\left(\int_0^\infty f_y(s)^p v(s)ds\right)^{1/p} &= \left(\int_0^y \left(\int_s^y t^{p'}V(t)^{-p'-1}v(t)dt\right) v(s)ds\right)^{\frac{1}{p}} \\
&= \left(\int_0^y t^{p'}V(t)^{-p'-1}\left(\int_0^t v(s)ds\right) dt\right)^{\frac{1}{p}} \\
&= \left(\int_0^y t^{p'}V(t)^{-p'}v(t)dt\right)^{\frac{1}{p}}. \qquad (2.12)
\end{aligned}$$

For the left-hand side of (1.2) we have that

$$\begin{aligned}
&\left(\int_0^\infty \left(\frac{1}{x}\int_0^x f(s)ds\right)^q u(x)dx\right)^{\frac{1}{q}} \\
&\geq \left(\int_0^y \left(\int_0^x \left(\int_s^y t^{p'}V(t)^{-p'-1}v(t)dt\right)^{\frac{1}{p}} ds\right)^q u(x)x^{-q}dx\right)^{\frac{1}{q}}. \qquad (2.13)
\end{aligned}$$

For the inner integral it yields that

$$\begin{aligned}
&\int_0^x \left(\int_s^y t^{p'}V(t)^{-p'-1}v(t)dt\right)^{\frac{1}{p}} ds \\
&\geq \int_0^x s^{\frac{p'}{p}} \left(\int_s^y V(t)^{-p'-1}v(t)dt\right)^{\frac{1}{p}} ds \\
&\geq \frac{1}{p}\int_0^x s^{\frac{p'}{p}} \left(\int_s^x \left(\int_t^y V(z)^{-p'-1}v(z)dz\right)^{-\frac{1}{p'}} V(t)^{-p'-1}v(t)dt\right) ds := I
\end{aligned}$$

Now, by changing the order of integration, we find that

$$
\begin{aligned}
I &= \frac{1}{p}\int_0^x V(t)^{-p'-1}v(t)\left(\int_0^t s^{\frac{p'}{p}}ds\right)\left(\int_t^y V(z)^{-p'-1}v(z)dz\right)^{-\frac{1}{p'}}dt \\
&= \frac{1}{pp'}\int_0^x V(t)^{-p'-1}v(t)t^{p'}\left(\int_t^y V(z)^{-p'-1}v(z)dz\right)^{-\frac{1}{p'}}dt \\
&\geq \frac{1}{p\,(p')^{\frac{1}{p}}}\int_0^x V(t)^{-p'}v(t)t^{p'}\,dt.
\end{aligned} \tag{2.14}
$$

Therefore, by combining (2.12) and (2.13) with (2.14), we obtain that

$$
\left(\int_0^y t^{p'}V(t)^{-p'}v(t)dt\right)^{-\frac{1}{p}}\left(\int_0^y\left(\int_0^x t^{p'}V^{-p'}(t)v(t)dt\right)^q u(x)x^{-q}dx\right)^{\frac{1}{q}} \leq p\,(p')^{\frac{1}{p}}\,C
$$

Hence $C \geq \frac{A_2}{p(p')^{\frac{1}{p}}}$. Moreover, since $f(x) \leq \frac{1}{x}\int_0^x f(t)dt$ when f is decreasing, it follows that $A_0 \leq C$, by choosing $f_t\,(x) = \chi_{[0,t]}\,(x)$ and taking supremum. Hence also the lower estimate of (2.2) holds and the proof is complete. □

Remark 2.3. The crucial part of the proof above is to use the results in [6], [8] and the technique in [9] combined with finding explicitly the corresponding test functions. For the estimates $C \approx \max(A_0, A_2)$ we have found that

$$
\max\left(A_0, A_2\frac{1}{p\,(p')^{1/p}}\right) \leq C \leq \left(1 + p^{1/q}p'\right)A_0 + (p')^2 A_2 \tag{2.15}
$$

but we strongly believe that these estimates can be improved.

3. Another approach via an equivalence theorem

In [7] E. Sawyer showed that to characterize the inequality (1.2) for decreasing functions f is equivalent to characterize the following inequalities:

$$
\left(\int_0^\infty\left(\int_0^x f(t)dt\right)^{p'} V(x)^{-p'}v(x)dx\right)^{\frac{1}{p'}} \leq C\left(\int_0^\infty f^{q'}(x)u(x)^{1-q'}dx\right)^{\frac{1}{q'}}, \tag{3.1}
$$

and

$$
\left(\int_0^\infty\left(\int_x^\infty \frac{f(t)}{t}dt\right)^{p'} x^{p'}V(x)^{-p'}v(x)dx\right)^{\frac{1}{p'}} \leq C\left(\int_0^\infty f^{q'}(x)u(x)^{1-q'}dx\right)^{\frac{1}{q'}}, \tag{3.2}
$$

where V is defined by (1.7). Then, by using the Hardy inequality and the so-called Muckenhoupt condition to characterize the inequalities (3.1) and (3.2), E. Sawyer received the before mentioned result in Theorem 1.3.

Recently the weighted Hardy inequality has been characterized with some new conditions (see, e.g., [4] and [10]). More generally, nowadays we know that the Hardy inequality (1.2) for $1 < p \le q < \infty$ holds for all $f \ge 0$ if and only if just one of the following (infinite many equivalent) conditions is satisfied (see [2]):

Theorem 3.1. *Let* $1 < p \le q < \infty, 0 < s < \infty$ *and define*

$$\mathcal{A}_1(s) := \sup_{t>0} \left(\int_t^\infty u(x) \left(\int_0^x v(y)^{1-p'} dy \right)^{q\left(\frac{1}{p'}-s\right)} dx \right)^{\frac{1}{q}} \left(\int_0^t v(x)^{1-p'} dx \right)^s ,$$

$$\mathcal{A}_2(s) := \sup_{t>0} \left(\int_0^t u(x) \left(\int_0^x v(y)^{1-p'} dy \right)^{q\left(\frac{1}{p'}+s\right)} dx \right)^{\frac{1}{q}} \left(\int_0^t v(x)^{1-p'} dx \right)^{-s} ,$$

$$\mathcal{A}_3(s) := \sup_{t>0} \left(\int_0^t v(x)^{1-p'} \left(\int_x^\infty u(y) dy \right)^{p'\left(\frac{1}{q}-s\right)} dx \right)^{\frac{1}{p'}} \left(\int_t^\infty u(x) dx \right)^s ,$$

$$\mathcal{A}_4(s) := \sup_{t>0} \left(\int_t^\infty v(x)^{1-p'} \left(\int_x^\infty u(y) dy \right)^{p'\left(\frac{1}{q}+s\right)} dx \right)^{\frac{1}{p'}} \left(\int_t^\infty u(x) dx \right)^{-s} . \tag{3.3}$$

Then the Hardy inequality

$$\left(\int_0^\infty \left(\int_0^x f(t)\, dt \right)^q u(x)\, dx \right)^{\frac{1}{q}} \le C \left(\int_0^\infty f^p(x)\, v(x)\, dx \right)^{\frac{1}{p}} \tag{3.4}$$

holds for all measurable functions $f \ge 0$ *if and only if any of the quantities* $\mathcal{A}_i(s)$ *is finite. Moreover, for the best constant* C *in* (3.4) *we have* $C \approx \mathcal{A}_i(s), i = 1, 2, 3, 4.$

This gives us the possibility to characterize the weighted Hardy inequality for decreasing functions with some new conditions, by just using the technique of E. Sawyer described above. We can therefore formulate the following generalization of our Theorem 2.1:

Theorem 3.2. *Let* $1 < p \le q < \infty$. *Then the inequality* (1.2) *holds for all decreasing* $f \ge 0$ *if and only if for any* s, r, $0 < s, r < \infty$, *one of the quantities*

$$\mathcal{B}_1(s) := \sup_{t>0} \left(\int_t^\infty V(x)^{-p'} v(x) \left(\int_0^x u(y) dy \right)^{p'\left(\frac{1}{q}-s\right)} dx \right)^{\frac{1}{p'}} \left(\int_0^t u(x) dx \right)^s ,$$

$$\mathcal{B}_2(s) := \sup_{t>0} \left(\int_0^t V(x)^{-p'} v(x) \left(\int_0^x u(y) dy \right)^{p'\left(\frac{1}{q}+s\right)} dx \right)^{\frac{1}{p'}} \left(\int_0^t u(x) dx \right)^{-s} ,$$

$$\mathcal{B}_3(s) := \sup_{t>0} \left(\int_0^t u(x) \left(\int_x^\infty V(y)^{-p'} v(y)\, dy \right)^{q\left(\frac{1}{p'}-s\right)} dx \right)^{\frac{1}{q}} \times \left(\int_t^\infty V(x)^{-p'} v(x)\, dx \right)^s ,$$

$$\mathcal{B}_4(s) := \sup_{t>0} \left(\int_t^\infty u(x) \left(\int_x^\infty V(y)^{-p'} v(y)\, dy \right)^{q\left(\frac{1}{p'}+s\right)} dx \right)^{\frac{1}{q}} \times \left(\int_t^\infty V(x)^{-p'} v(x)\, dx \right)^{-s}$$

and any of the quantities

$$\mathcal{E}_1(r) := \sup_{t>0} \left(\int_t^\infty u(x)\, x^{-q} \left(\int_0^x \Psi(y)^{1-p'} dy \right)^{q\left(\frac{1}{p'}-r\right)} dx \right)^{\frac{1}{q}} \left(\int_0^t \Psi(x)^{1-p'} dx \right)^r,$$

$$\mathcal{E}_2(r) := \sup_{t>0} \left(\int_0^t u(x)\, x^{-q} \left(\int_0^x \Psi(y)^{1-p'} dy \right)^{q\left(\frac{1}{p'}+r\right)} dx \right)^{\frac{1}{q}} \times \left(\int_0^t \Psi(x)^{1-p'} dx \right)^{-r},$$

$$\mathcal{E}_3(r) := \sup_{t>0} \left(\int_0^t \Psi(x)^{1-p'} \left(\int_x^\infty u(y) y^{-q} dy \right)^{p'\left(\frac{1}{q}-r\right)} dx \right)^{\frac{1}{p'}} \left(\int_t^\infty u(x) x^{-q} dx \right)^r,$$

$$\mathcal{E}_4(r) := \sup_{t>0} \left(\int_t^\infty \Psi(x)^{1-p'} \left(\int_x^\infty u(y) y^{-q} dy \right)^{p'\left(\frac{1}{q}+r\right)} dx \right)^{\frac{1}{p'}} \times \left(\int_t^\infty u(x) x^{-q} dx \right)^{-r}$$

are finite, where V is defined by (1.7) *and Ψ is defined by* (2.10).

Remark 3.3. By using $\mathcal{B}_1$ with $s = \frac{1}{q}$ and $V(\infty) = \infty$ (this is no real restriction see, e.g., [7, p. 148] and [5, p. 316]) and integrating, and also using $\mathcal{E}_1$ with $r = \frac{1}{p'}$ we get the result due to Sawyer (Theorem 1.3). Furthermore, if we again use $\mathcal{B}_1$ as above with $s = \frac{1}{q}$ and also $\mathcal{E}_2$ with $r = \frac{1}{p}$ we get our result (Theorem 2.1).

Proof. (Theorem 3.2). We apply the above mentioned technique by E. Sawyer [7]. First we give weight characterizations for the inequality (3.1). If we use Theorem 3.1 with $p, q, u(x), v(x)$ replaced by q', p',$V(x)^{-p'} v(x)$ and $u(x)^{1-q'}$, respectively, then we obtain the first bunch of conditions in Theorem 3.2.

Now, we characterize the inequality (3.2) and first we note that, by duality (see, e.g., [5, p. 314]), (3.2) is equivalent to the Hardy inequality

$$\left(\int_0^\infty \left(\int_0^x f(t) dt \right)^q u(x) x^{-q} dx \right)^{\frac{1}{q}} \le C \left(\int_0^\infty f^p(x) \Psi(x) dx \right)^{\frac{1}{p}},$$

where $\Psi(x)$ is defined by (2.10) so we can again use Theorem 3.1 now with $s, u(x)$ and $v(x)$ replaced by r, $u(x)x^{-q}$and $\Psi(x)$ respectively, and we get the second bunch of conditions in Theorem 3.2. The proof is complete. □

Remark 3.4. By using Theorem 3.2 and arguing as in the proof of Theorem 4.1 we can get a number of other characterizations of the Pólya-Knopp inequality (1.1) besides that stated in Theorem 4.1.

4. Concluding remarks

In this Section we will give a weight characterization of the Pólya-Knopp inequality (1.1) for decreasing functions for the case when $v(x)$ is decreasing. In fact, the following Pólya-Knopp type inequality may be regarded as a limit result of our Theorem 2.1:

Remark 4.1. Let $0 < p \leq q < \infty$. Let $u(x)$ and $v(x)$ be weight functions and assume that $v(x)$ is decreasing. Then the inequality (1.1) holds for all decreasing $f \geq 0$ if and only if

$$D_{PS} = D_{PS}(p,q,w) := \sup_{t>0} t^{-\frac{1}{p}} \left(\int_0^t w(x)dx \right)^{\frac{1}{q}} < \infty, \tag{4.1}$$

holds. Moreover, if C is the best possible constant in (1.1), then

$$D_{PS} \leq C \leq \inf_{r>1} k(r) D_{PS}, \tag{4.2}$$

where

$$k(r) = \left(1 + r^{\frac{p}{qr}} r' + (r')^2\right)^{\frac{r}{p}},$$

The details of the proof can be found in [3].

Remark 4.2. We have obtained the same condition for characterizing the Pólya-Knopp inequality for decreasing functions as for all positive functions in [6] but our upper estimate of the best constant C in (4.2) is different.

Acknowledgement: The author express her deep gratitude to Professor Lars-Erik Persson for fruitful discussions.

References

[1] S. Barza, *On Some Integral Inequalities Weighted Multidimensional Integral Inequalities and Applications*, Ph.D. Thesis, Department of Mathematics, Luleå University of Technology, 199).

[2] A. Gogatishvili, A. Kufner, L.-E. Persson and A. Wedestig, *On Some Integral Inequalities An equivalence theorem for integral conditions related to Hardy's inequality*, Real Anal. Exchange **29**(2) (2003/04), 867–880.

[3] M. Johansson, *Carleman type inequalities and Hardy type inequalities for monotone functions*, Ph.D. Thesis, Department of Mathematics, Luleå University of Technology, 2007.

[4] A. Kufner, L.-E. Persson and A. Wedestig, *A study of some constants characterizing the weighted Hardy inequality*, Orlicz centenary volume, Banach Center Publ, Polish Acad. Sci., Warsaw **64** (2004) 135–146.

[5] A. Kufner and L.-E. Persson, *Weighted inequalities of Hardy type*, World Scientific Publishing Co, Singapore/ New Jersey / London / Hong Kong, 2003.

[6] L.-E. Persson and V.D. Stepanov, *Weighted integral inequalities with the geometric mean operator*, J. Inequal. Appl. **7**(5) (2002), 727–746.

[7] E. Sawyer, *Boundedness of classical operators on classical Lorentz spaces*, Studia Math. **96**(2) (1990), 145–158.

[8] V.D. Stepanov, *Integral operators on the cone of monotone functions*, J. London Math. Soc. **48**(2) (1993), 465–487.

[9] V.D. Stepanov, *The weighted Hardy's inequality for nonincreasing functions*, Trans. Amer. Math. Soc. **338**(1) (1993), 173–186.

[10] A. Wedestig, *Some new Hardy type inequalities and their limiting inequalities*, J. Inequal. Pure Appl. Math. **4**(3) (2003), Article 61.

Maria Johansson
Department of Mathematics
Luleå University of Technology
97187 Luleå, SWEDEN
e-mail: maria.l.johansson@ltu.se

International Series of Numerical Mathematics, Vol. 157, 109–120

Euler-Grüss Type Inequalities Involving Measures

Ambroz Čivljak, Ljuban Dedić and Marko Matić

Abstract. An inequality of Grüss type for a real Borel measure μ is proved. Some Euler-Grüss type inequalities are given, by using general Euler identities involving μ-harmonic sequences of functions with respect to a real Borel measure μ.

Mathematics Subject Classification (2000). 26D15, 26D20.

Keywords. Grüss type inequalities, Euler identities, harmonic sequences, Borel measures.

1. Introduction

For $a, b \in \mathbb{R}$, $a < b$, let $C[a,b]$ be the Banach space of all continuous functions $f : [a,b] \to \mathbb{R}$ with the max norm, and $M[a,b]$ the Banach space of all real Borel measures on $[a,b]$ with the total variation norm. For $\mu \in M[a,b]$ define function $\check{\mu}_n : [a,b] \to \mathbb{R}$, $n \geq 1$, by

$$\check{\mu}_n(t) = \tfrac{1}{(n-1)!} \int_{[a,t]} (t-s)^{n-1} \mathrm{d}\mu(s). \tag{1.1}$$

The function $\check{\mu}_n$ is differentiable, $\check{\mu}_n'(t) = \check{\mu}_{n-1}(t)$ and $\check{\mu}_n(a) = 0$, for every $n \geq 2$, while for $n = 1$

$$\check{\mu}_1(t) = \int_{[a,t]} \mathrm{d}\mu(s) = \mu([a,t]), \tag{1.2}$$

which means that $\check{\mu}_1(t)$ is equal to the distribution function of μ. From (1.1) and (1.2) using the Fubini theorem we easily get the following two formulas

$$\check{\mu}_n(t) = \tfrac{1}{(n-2)!} \int_a^t (t-s)^{n-2} \check{\mu}_1(s) \mathrm{d}s, \quad n \geq 2$$

and

$$\int_a^t \check{\mu}_n(s) \mathrm{d}s = \check{\mu}_{n+1}(t), \quad n \geq 1. \tag{1.3}$$

Also, $g(s) = (t-s)^{n-1}$ is nonincreasing on $[a,t]$ so that from (1.1) we get the estimate

$$|\check{\mu}_n(t)| \leq \tfrac{1}{(n-1)!} (t-a)^{n-1} \|\mu\|, \; t \in [a,b], \; n \geq 1 \tag{1.4}$$

where $\|\mu\|$ denotes the total variation of μ. Note also that in the case when $\mu \geq 0$ every $\check{\mu}_n(\cdot)$ is nondecreasing on $[a,b]$.

A sequence of functions $P_n : [a,b] \to \mathbb{R}$, $n \geq 1$, is called μ-**harmonic sequence** of functions on $[a,b]$ if

$$P_n'(t) = P_{n-1}(t),\ n \geq 2;\quad P_1(t) = c + \check{\mu}_1(t),\quad t \in [a,b],$$

for some $c \in \mathbb{R}$.

The sequence $(\check{\mu}_n, n \geq 1)$ is an example of μ-harmonic sequence of functions on $[a,b]$.

Assume that $(P_n(t), n \geq 1)$ is a μ-harmonic sequence of functions on $[a,b]$. Define $P_n^*(t)$, for $n \geq 1$, to be a periodic function of period 1, related to $P_n(t)$ as

$$P_n^*(t) = [P_n(a + (b-a)t)]/\,(b-a)^n\,,\ 0 \leq t < 1;\quad P_n^*(t+1) = P_n^*(t),\ t \in \mathbb{R}.$$

Thus, for $n \geq 2$, $P_n^*(t)$ is continuous on $\mathbb{R}\backslash\mathbb{Z}$ and has a jump of

$$\alpha_n = [P_n(a) - P_n(b)]/\,(b-a)^n \tag{1.5}$$

at every $k \in \mathbb{Z}$, whenever $\alpha_n \neq 0$. Also note that $P_n^*(t)$ is differentiable on $\mathbb{R}\backslash\mathbb{Z}$ for $n \geq 2$ and

$$P_n^{*\prime}(t) = P_{n-1}^*(t),\ n \geq 2,\ \ t \in \mathbb{R}\backslash\mathbb{Z}.$$

Let $f : [a,b] \to \mathbb{R}$ be such that $f^{(n-1)}$ is a continuous function of bounded variation on $[a,b]$ for some $n \geq 1$. In the recent paper [1] the following two identities have been proved:

$$\mu([a,b])f(x) = \int_{[a,b]} f_x(t)\mathrm{d}\mu(t) + S_n(x) + R_n^1(x), \tag{1.6}$$

and for $n \geq 2$

$$\begin{aligned}\mu([a,b])f(x) = {} & \int_{[a,b]} f_x(t)\mathrm{d}\mu(t) + S_{n-1}(x) \\ & + [P_n(a) - P_n(b)]\, f^{(n-1)}(x) + R_n^2(x),\end{aligned} \tag{1.7}$$

where

$$S_m(x) = \sum_{k=1}^m P_k(x)\,[f^{(k-1)}(b) - f^{(k-1)}(a)] + \sum_{k=2}^m [P_k(a) - P_k(b)]\, f^{(k-1)}(x),$$

for $1 \leq m \leq n$, with convention $S_1(x) = P_1(x)\,[f(b) - f(a)]$, and

$$f_x(t) = \begin{cases} f(a+x-t), & a \leq t \leq x \\ f(b+x-t), & x < t \leq b \end{cases}, \tag{1.8}$$

while for every $x \in [a,b]$

$$R_n^1(x) = -(b-a)^n \int_{[a,b]} P_n^*\left(\tfrac{x-t}{b-a}\right)\mathrm{d}f^{(n-1)}(t) \tag{1.9}$$

and

$$R_n^2(x) = -(b-a)^n \int_{[a,b]} [P_n^*\left(\tfrac{x-t}{b-a}\right) - \tfrac{1}{(b-a)^n} P_n(x)]\mathrm{d}f^{(n-1)}(t). \tag{1.10}$$

Identities (1.6) and (1.7) are called the **general Euler harmonic identities**. They are generalizations of similar identities considered in [2] and [3].

Recently the following theorem was proved [5, Theorem 4]:

Theorem A *Let $f, g : [a, b] \to \mathbb{R}$ be integrable functions such that*

$$\gamma \leq f(t) \leq \Gamma, \quad a.e. \text{ and } \int_a^b g(t)\mathrm{d}t = 0,$$

for some $\gamma, \Gamma \in \mathbb{R}$, Then

$$\left| \int_a^b f(t)g(t)\mathrm{d}t \right| \leq \tfrac{1}{2}(\Gamma - \gamma) \int_a^b |g(t)| \,\mathrm{d}t,$$

with equality if and only if either

$$f(t) = \Gamma, \ t \in I_+ \ and \ f(t) = \gamma, \ t \in I_-, \quad a.e. \ on \ I_+ \cup I_-$$

or

$$f(t) = \gamma, \ t \in I_+ \ and \ f(t) = \Gamma, \ t \in I_-, \quad a.e. \ on \ I_+ \cup I_-,$$

where $I_+ = \{t \in [a, b]; g(t) > 0\}$, $I_- = \{t \in [a, b]; g(t) < 0\}$.

Remark 1. *The assumption $\int_a^b g(t)\mathrm{d}t = 0$ is not essential since $g(t)$ can be replaced with $\tilde{g}(t) = g(t) - \frac{1}{b-a}\int_a^b g(s)\mathrm{d}s$. Such result was proved by Matić* [4, Theorem 3].

The aim of this paper is to give generalizations of Theorem A (see Theorem 1 and Remark 4), and applying them to formulae (1.6) and (1.7) to prove some general Euler-Grüss type inequalities. Our Theorem 1 has many applications. For some of them see recent papers [6] and [7].

2. Some inequalities of Grüss type

Let $X \subset \mathbb{R}^m$ be a Borel set in $\mathbb{R}^m$, $m \geq 1$, and let $M(X)$ denotes the Banach space of all real Borel measures on X with the total variation norm. For $\mu \in M(X)$ let $\mu = \mu_+ - \mu_-$ be the Jordan-Hahn decomposition of μ, where μ_+ and μ_- are orthogonal and positive measures. Then we have $|\mu| = \mu_+ + \mu_-$ and

$$\|\mu\| = |\mu|(X) = \|\mu_+\| + \|\mu_-\| = \mu_+(X) + \mu_-(X).$$

Measure $\mu \in M(X)$ is called **balanced** if $\mu(X) = 0$. This is equivalent to

$$\|\mu_+\| = \|\mu_-\| = \tfrac{1}{2}\|\mu\|.$$

Theorem 1. *For balanced measure $\mu \in M(X)$ let $f \in L_\infty(X, \mu)$ be such that*

$$\gamma \leq f(t) \leq \Gamma, \quad t \in X, \ \mu - a.e., \tag{2.1}$$

for some $\gamma, \Gamma \in \mathbb{R}$. Then

$$\left| \int f(t)\mathrm{d}\mu(t) \right| \leq \tfrac{1}{2}(\Gamma - \gamma)\|\mu\|, \tag{2.2}$$

with the equality if and only if either

$$f(t) = \Gamma, \ t \in I_+ \ and \ f(t) = \gamma, \ t \in I_-, \quad \mu - a.e. \tag{2.3}$$

or

$$f(t) = \gamma, \ t \in I_+ \ and \ f(t) = \Gamma, \ t \in I_-, \quad \mu - a.e., \tag{2.4}$$

where I_+ and I_- are disjoint Borel sets satisfying

$$\mu_+(I_+) = \|\mu_+\|, \quad \mu_-(I_-) = \|\mu_-\|, \quad \mu_+(I_-) = \mu_-(I_+) = 0.$$

Proof. Integrating the relation (2.1) with respect to μ_+ and μ_- we get

$$\tfrac{1}{2}\gamma \|\mu\| \leq \textstyle\int f(t)\mathrm{d}\mu_+(t) \leq \tfrac{1}{2}\Gamma \|\mu\| \tag{2.5}$$

and

$$-\tfrac{1}{2}\Gamma \|\mu\| \leq -\textstyle\int f(t)\mathrm{d}\mu_-(t) \leq -\tfrac{1}{2}\gamma \|\mu\| . \tag{2.6}$$

Adding these relations together we get our inequality.

The equality case occurs in (2.2) if and only if we have the equality either in the both right-hand sides of (2.5) and (2.6), or in the both left-hand sides of (2.5) and (2.6). The former case is equivalent to (2.3), while the later case to (2.4). □

Remark 2. *Let f and g be from Theorem* A *and let $\mu \in M([a,b])$ be defined by* $\mathrm{d}\mu(t) = g(t)\mathrm{d}t$. *Then μ_+ and μ_- are given by* $\mathrm{d}\mu_+(t) = g_+(t)\mathrm{d}t$, $\mathrm{d}\mu_-(t) = g_-(t)\mathrm{d}t$, *where*

$$g_+(t) = \tfrac{1}{2}\left[|g(t)| + g(t)\right], \; g_-(t) = \tfrac{1}{2}\left[|g(t)| - g(t)\right],$$

and we have

$$\mu([a,b]) = \textstyle\int_a^b g(t)\mathrm{d}t = 0,$$

which means that μ is balanced. Now we see that Theorem 1 *reduces to Theorem* A *since* $\|\mu\| = \int_a^b |g(t)|\,\mathrm{d}t$.

Remark 3. *The inequality* (2.2) *is obviously sharp. Namely for the function f defined as $f(t) = \Gamma\chi_{I_+}(t) + \gamma\chi_{I_-}(t)$, $t \in X$, we have equality in* (2.2). *Clearly, the same is true for the function $f(t) = \gamma\chi_{I_+}(t) + \Gamma\chi_{I_-}(t)$, $t \in X$.*

Corollary 1. *Let $(c_k, k \geq 1)$ be a sequence in $\mathbb{R}$ such that $\sum_{k\geq 1} |c_k| < \infty$ and $\sum_{k\geq 1} c_k = 0$. Then for every bounded sequence $(d_k, k \geq 1)$ in $\mathbb{R}$ we have*

$$\left|\textstyle\sum_{k\geq 1} c_k d_k\right| \leq \tfrac{1}{2}(\Gamma - \gamma)\textstyle\sum_{k\geq 1} |c_k| ,$$

where $\Gamma = \sup\{d_k : c_k \neq 0\}$ and $\gamma = \inf\{d_k : c_k \neq 0\}$. The equality occurs if and only if either

$$d_k = \Gamma, \; k \in I_+ \text{ and } d_k = \gamma, \; k \in I_-,$$

or

$$d_k = \gamma, \; k \in I_+ \text{ and } d_k = \Gamma, \; k \in I_-,$$

where $I_+ = \{k : c_k > 0\}$ and $I_- = \{k : c_k < 0\}$.

Proof. Choose any sequence $(x_k, k \geq 1)$ of distinct points $x_k \in \mathbb{R}$ and set $X = \{x_k : k \geq 1\}$. Then apply the theorem above for the measure $\mu = \sum_{k\geq 1} c_k \delta_{x_k}$, where δ_y is the Dirac measure at $y \in \mathbb{R}$, i.e., the measure defined by $\int f(t)\mathrm{d}\delta_y(t) = f(y)$, and for the function $f : X \to \mathbb{R}$ defined as $f(x_k) = d_k$, $k \geq 1$. In this case $\|\mu\| = \sum_{k\geq 1} |c_k| < \infty$ and $\mu(X) = \sum_{k\geq 1} c_k = 0$, which means that μ is balanced, while $\int f(t)\mathrm{d}\mu(t) = \sum_{k\geq 1} c_k f(x_k) = \sum_{k\geq 1} c_k d_k$. □

Corollary 2. *For $\mu \in M(X)$ let $f : X \to \mathbb{R}$ be a Borel function such that $\gamma \leq f(t) \leq \Gamma$, $t \in X$. Then for every $x \in X$ we have*

$$\left|\mu(X)f(x) - \int f(t)\mathrm{d}\mu(t)\right| \leq \tfrac{1}{2}(\Gamma - \gamma)\left[|\mu(X)| + \|\mu\|\right].$$

Proof. For $x \in X$ define measure ν_x by $\nu_x = \mu(X)\delta_x - \mu$. Then $\nu_x(X) = \mu(X) - \mu(X) = 0$, and $\|\nu_x\| = \|\mu(X)\delta_x - \mu\| \leq |\mu(X)| + \|\mu\|$,while $\int f(t)\mathrm{d}\nu_x(t) = \mu(X)f(x) - \int f(t)\mathrm{d}\mu(t)$. Apply now the theorem above. □

Corollary 3. *Let $\mu, \nu \in M(X)$ be probability measures and let $f : X \to \mathbb{R}$ be a Borel function such that $\gamma \leq f(t) \leq \Gamma$, $t \in X$, μ and ν-a.e. Then we have*

$$\left|\int f(t)\mathrm{d}\mu(t) - \int f(t)\mathrm{d}\nu(t)\right| \leq \Gamma - \gamma.$$

Proof. Apply the theorem above for $\mu - \nu$ and note that $(\mu - \nu)(X) = 1 - 1 = 0$ and $\|\mu - \nu\| \leq \|\mu\| + \|\nu\| = 2$. □

Corollary 4. *For a probability measure $\mu \in M(X)$ and a Borel function $f : X \to \mathbb{R}$ such that $\gamma \leq f(t) \leq \Gamma$, $t \in X$, and for every $x \in X$ we have*

$$\left|f(x) - \int f(t)\mathrm{d}\mu(t)\right| \leq \Gamma - \gamma.$$

Proof. Apply Corollary 3 for μ and $\nu = \delta_x$. □

3. Some Euler-Grüss type inequalities

Throughout this section we use the same notations as in the previous one for the special case $X = [a, b] \subset \mathbb{R}$. Hence μ denotes a real Borel measure on $[a, b]$. Also, whenever $f : [a, b] \to \mathbb{R}$ is such that $f^{(k)}$ exists and is bounded, for some $k \geq 1$, we assume that there are some real constants γ_k and Γ_k such that

$$\gamma_k \leq f^{(k)}(t) \leq \Gamma_k, \ t \in [a, b]. \tag{3.1}$$

Theorem 2. *Let $f : [a, b] \to \mathbb{R}$ be such that $f^{(n-1)}$ is a continuous function of bounded variation, for some $n \geq 2$. Then for every μ-harmonic sequence $(P_k, k \geq 1)$ we have*

$$\left|R_n^2(x)\right| \leq \tfrac{1}{2}(\Gamma_{n-1} - \gamma_{n-1})[|P_n(b) - P_n(a)| + \int_a^b |P_{n-1}(t)|\,\mathrm{d}t], \tag{3.2}$$

where $R_n^2(x)$ is given by (1.10).

Proof. We shall rewrite $R_n^2(x)$ in more suitable form. Integration by parts yields

$$\begin{aligned} R_n^2(x) = &-(b-a)^n[P_n^*(\tfrac{x-t}{b-a}) - \tfrac{1}{(b-a)^n}P_n(x)]f^{(n-1)}(t)\,\Big|_a^b \\ &+ (b-a)^n\int_{[a,b]} f^{(n-1)}(t)\mathrm{d}P_n^*(\tfrac{x-t}{b-a}). \end{aligned} \tag{3.3}$$

For $a \leq x < b$ we have $P_n^*(\frac{x-b}{b-a}) = P_n^*(\frac{x-a}{b-a}) = \frac{1}{(b-a)^n}P_n(x)$ so that from (3.3) we get

$$R_n^2(x) = (b-a)^n\int_{[a,b]} f^{(n-1)}(t)\mathrm{d}P_n^*(\tfrac{x-t}{b-a}). \tag{3.4}$$

For $a < x < b$ the function $\varphi_n(x;t) = P_n^*(\frac{x-t}{b-a})$ is differentiable on $[a,b] \setminus \{x\}$ and its derivative is equal to $\frac{-1}{b-a}P_{n-1}^*(\frac{x-t}{b-a})$. Further, it has a jump of $\varphi_n(x;x+0) - \varphi_n(x;x-0) = -\alpha_n$ at x, where α_n is given by (1.5), so that from (3.4) we get

$$R_n^2(x) = -(b-a)^{n-1}\int_a^b f^{(n-1)}(t)P_{n-1}^*(\tfrac{x-t}{b-a})\mathrm{d}t - [P_n(a) - P_n(b)]\, f^{(n-1)}(x). \quad (3.5)$$

We claim that this formula holds for all $x \in [a,b]$. For $x = a$ the function $\varphi_n(a;t) = P_n^*(\frac{a-t}{b-a})$ is differentiable on (a,b) and its derivative is equal to $\frac{-1}{b-a}P_n^*(\frac{a-t}{b-a})$. Further, it has a jump of $\varphi_n(a;a+0) - \varphi_n(a;a) = -\alpha_n$ at a, while $\varphi_n(a;b) - \varphi_n(a;b-0) = 0$, so that from (3.4) we get formula (3.5) with $x = a$. It remains to see that formula (3.5) holds for $x = b$. In this case we have $P_n^*(\frac{b-b}{b-a}) = P_n^*(\frac{b-a}{b-a}) = \frac{1}{(b-a)^n}P_n(a)$ so that from (3.3) we get

$$R_n^2(b) = [P_n(b) - P_n(a)]\,[f^{(n-1)}(b) - f^{(n-1)}(a)] + (b-a)^n \int_{[a,b]} f^{(n-1)}(t)\mathrm{d}P_n^*(\tfrac{b-t}{b-a}).$$

Now, the function $\varphi_n(b;t) = P_n^*(\frac{b-t}{b-a})$ is identically equal to $P_n^*(\frac{a-t}{b-a})$ so that from the formula above we easily get formula (3.5) for $x = b$.

Define measures ν_n and ξ_n by $\mathrm{d}\xi_n(t) = -(b-a)^{n-1}P_{n-1}^*(\frac{x-t}{b-a})\mathrm{d}t$ and $\nu_n = \xi_n - [P_n(a) - P_n(b)]\,\delta_x$, and note that by (3.5) we have

$$R_n^2(x) = \int_{[a,b]} f^{(n-1)}(t)\mathrm{d}\nu_n(t).$$

For all $k \geq 1$ we have

$$P_k^*(\tfrac{x-t}{b-a}) = \frac{1}{(b-a)^k} \times \begin{cases} P_k(a+x-t), & \text{for } a \leq t \leq x \\ P_k(b+x-t), & \text{for } x < t \leq b \end{cases} \quad (3.6)$$

so that

$$\begin{aligned} \nu_n([a,b]) &= -(b-a)^{n-1}\int_a^b P_{n-1}^*(\tfrac{x-t}{b-a})\mathrm{d}t - [P_n(a) - P_n(b)] \\ &= -\int_a^b P_{n-1}(t)\,\mathrm{d}t - [P_n(a) - P_n(b)] \\ &= P_n(a) - P_n(b) - [P_n(a) - P_n(b)] = 0, \end{aligned}$$

which means that ν_n is balanced. Further

$$\begin{aligned} \|\nu_n\| &= (b-a)^{n-1}\int_a^b |P_{n-1}^*(\tfrac{x-t}{b-a})|\mathrm{d}t + |P_n(a) - P_n(b)| \\ &= \int_a^b |P_{n-1}(t)|\,\mathrm{d}t + |P_n(a) - P_n(b)|\,. \end{aligned}$$

Now we can apply Theorem 1 to obtain (3.2). □

Corollary 5. *Let f and μ be as in Theorem 2. Then we have*

$$\begin{aligned} &\left|\mu([a,b])f(x) - \int_{[a,b]} f_x(t)\mathrm{d}\mu(t) - \check{S}_{n-1}(x) + \check{\mu}_n(b)\, f^{(n-1)}(x)\right| \\ &\leq \tfrac{1}{2}(\Gamma_{n-1} - \gamma_{n-1})[|\check{\mu}_n(b)| + \int_a^b |\check{\mu}_{n-1}(t)|\,\mathrm{d}t] \\ &\leq \tfrac{1}{(n-1)!}(b-a)^{n-1}\,[\Gamma_{n-1} - \gamma_{n-1}]\,\|\mu\|\,, \end{aligned}$$

where

$$\check{S}_m(x) = \sum_{k=1}^m \check{\mu}_k(x)\,[f^{(k-1)}(b) - f^{(k-1)}(a)] - \sum_{k=2}^m \check{\mu}_k(b)\, f^{(k-1)}(x). \quad (3.7)$$

Proof. The first inequality follows by Theorem 2 applied to the μ-harmonic sequence $(\check{\mu}_k, k \geq 1)$ defined by (1.1) and (1.2). In that case $S_{n-1}(x)$ becomes $\check{S}_{n-1}(x)$ since $\check{\mu}_k(a) = 0$, $k \geq 2$. The second inequality follows from inequality (1.4). □

Corollary 6. *Let f and μ be as in Theorem 2. If $\mu \geq 0$, then we have*

$$\begin{aligned}&\left|\mu([a,b])f(x) - \int_{[a,b]} f_x(t)\mathrm{d}\mu(t) - \check{S}_{n-1}(x) + \check{\mu}_n(b)\, f^{(n-1)}(x)\right|\\ &\leq [\Gamma_{n-1} - \gamma_{n-1}]\, \check{\mu}_n(b)\\ &\leq \tfrac{1}{(n-1)!}(b-a)^{n-1}[\Gamma_{n-1} - \gamma_{n-1}]\, \|\mu\|\,.\end{aligned}$$

Proof. Apply Corollary 5. Since $\mu \geq 0$, we have $\check{\mu}_k \geq 0$, for $k \geq 1$ and $\int_a^b \check{\mu}_{n-1}(t)\, \mathrm{d}t = \check{\mu}_n(b)$, by (1.3). □

Corollary 7. *Let f be as in Theorem 2. Then for every $x \in [a,b]$*

$$\begin{aligned}&\left|(b-a)f(x) - \int_a^b f(t)\mathrm{d}t - U_{n-1}(x) + \tfrac{1}{n!}(b-a)^n f^{(n-1)}(x)\right|\\ &\leq \tfrac{1}{n!}(b-a)^n [\Gamma_{n-1} - \gamma_{n-1}]\,,\end{aligned}$$

where

$$U_m(x) = \sum_{k=1}^m \tfrac{1}{k!}(x-a)^k [f^{(k-1)}(b) - f^{(k-1)}(a)] - \sum_{k=2}^m \tfrac{1}{k!}(b-a)^k f^{(k-1)}(x).$$

Proof. Apply Corollary 6 for the case when μ is the Lebesgue measure on $[a,b]$. For any $x \in [a,b]$ we then have $\int_{[a,b]} f_x(t)\mathrm{d}\mu(t) = \int_a^b f(t)\mathrm{d}t$. Also we have $\check{\mu}_k(t) = \frac{1}{k!}(t-a)^k$, $k \geq 1$, so that $\check{S}_{n-1}(x)$ becomes $U_{n-1}(x)$. □

Corollary 8. *Let f be as in Theorem 2. Then for every $x, y \in [a,b]$, $y \leq x$ we have*

$$\begin{aligned}&\left|f(x) - f(a+x-y) - T_{n-1}(x,y) + \tfrac{1}{(n-1)!}(b-y)^{n-1} f^{(n-1)}(x)\right|\\ &\leq \tfrac{1}{(n-1)!}(b-y)^{n-1}[\Gamma_{n-1} - \gamma_{n-1}]\,,\end{aligned}$$

where $T_{n-1}(x,y)$ is equal to

$$\sum_{k=1}^{n-1} \tfrac{1}{(k-1)!}(x-y)^{k-1}[f^{(k-1)}(b) - f^{(k-1)}(a)] - \sum_{k=2}^{n-1} \tfrac{1}{(k-1)!}(b-y)^{k-1} f^{(k-1)}(x).$$

Proof. Apply Corollary 6 for $\mu = \delta_y$, $a \leq y \leq x$. Then $\|\mu\| = 1$. Also $\check{S}_{n-1}(x)$ becomes $T_{n-1}(x,y)$, since $\check{\mu}_n(t) = \frac{1}{(n-1)!}(t-y)^{n-1}$, $y \leq t \leq b$, and $\check{\mu}_n(t) = 0$, $a \leq t < y$. □

Corollary 9. *If f is such that f' is a continuous function of bounded variation, then for every $\mu \in M[a,b]$ and every real constant c the absolute value of*

$$\mu([a,b])f(x) - \int_{[a,b]} f_x(t)\mathrm{d}\mu(t) - [c + \check{\mu}_1(x)]\,[f(b) - f(a)] + [c(b-a) + \check{\mu}_2(b)]\, f'(x)$$

is less than or equal to

$$\tfrac{1}{2}(\Gamma_1 - \gamma_1)[|c(b-a) + \check{\mu}_2(b)| + \int_a^b |c + \check{\mu}_1(t)|\, \mathrm{d}t].$$

Proof. For any constant $c \in \mathbb{R}$, $P_1(t) = c + \check{\mu}_1(t)$ and $P_2(t) = c(t-a) + \check{\mu}_2(t)$ are two beginning terms of a μ-harmonic sequence of functions on $[a,b]$. So we can apply Theorem 2 for $n = 2$ and use the fact that $S_1(x) = [c + \check{\mu}_1(x)]\,[f(b) - f(a)]$ and $P_2(a) - P_2(b) = -c(b-a) - \check{\mu}_2(b)$. □

Corollary 10. *Under assumptions of Corollary 9 if $\mu \geq 0$ and $c \geq 0$, then the absolute value of*

$$\mu([a,b])f(x) - \int_{[a,b]} f_x(t)\mathrm{d}\mu(t) - [c + \check{\mu}_1(x)]\,[f(b) - f(a)] + [c(b-a) + \check{\mu}_2(b)]\,f'(x)$$

is less than or equal to

$$(\Gamma_1 - \gamma_1)\,[c(b-a) + \check{\mu}_2(b)] \leq (\Gamma_1 - \gamma_1)\,(b-a)\,(c + \|\mu\|)\,.$$

Proof. Apply Corollary 9 and note that in this case

$$|c(b-a) + \check{\mu}_2(b)| + \int_a^b |c + \check{\mu}_1(t)|\,\mathrm{d}t = 2\,[c(b-a) + \check{\mu}_2(b)]$$
$$\text{and} \qquad \check{\mu}_2(b) \leq (b-a)\,\|\mu\|\,.$$
□

Theorem 3. *Let $f : [a,b] \to \mathbb{R}$ be such that $f^{(n-1)}$ is absolutely continuous for some $n \geq 1$, and let*

$$\gamma_n \leq f^{(n)} \leq \Gamma_n, \quad \text{a.e.}$$

for some real constants γ_n and Γ_n. If $(P_k, k \geq 1)$ is a μ-harmonic sequence such that

$$P_{n+1}(a) = P_{n+1}(b)$$

for that particular n, then for every $x \in [a,b]$ we have

$$\left|\mu([a,b])f(x) - \int_{[a,b]} f_x(t)\mathrm{d}\mu(t) - S_n(x)\right| \leq \tfrac{1}{2}(\Gamma_n - \gamma_n)\int_a^b |P_n(t)|\,\mathrm{d}t, \tag{3.8}$$

where $f_x(t)$ is defined by (1.8).

Proof. Since $f^{(n-1)}$ is absolutely continuous its derivative $f^{(n)}$ exists *a.e.* and $R_n^1(x)$ from (1.6), which is defined by (1.9), can be rewritten as

$$R_n^1(x) = -(b-a)^n \int_a^b P_n^*\left(\tfrac{x-t}{b-a}\right) f^{(n)}(t)\mathrm{d}t = \int_a^b f^{(n)}(t)\mathrm{d}\nu_n(t),$$

where measure ν_n is defined by $\mathrm{d}\nu_n(t) = -(b-a)^n P_n^*\left(\frac{x-t}{b-a}\right)\mathrm{d}t$. Using (3.6) we get

$$\nu_n([a,b]) = -(b-a)^n \int_a^b P_n^*\left(\tfrac{x-t}{b-a}\right)\mathrm{d}t = -\int_a^b P_n(t)\,\mathrm{d}t = P_{n+1}(a) - P_{n+1}(b) = 0,$$

which means that ν_n is balanced. Further,

$$\|\nu_n\| = (b-a)^n \int_a^b \left|P_n^*\left(\tfrac{x-t}{b-a}\right)\right|\mathrm{d}t = \int_a^b |P_n(t)|\,\mathrm{d}t.$$

Now (3.8) follows immediately by Theorem 1, since the left-hand side of (3.8) is equal to $|R_n^1(x)|$. □

Corollary 11. *Let $f : [a,b] \to \mathbb{R}$ be an absolutely continuous function and let*

$$\gamma_1 \le f' \le \Gamma_1, \quad a.e.$$

for some real constants γ_1 and Γ_1. If $\mu \in M[a,b]$ and $c \in \mathbb{R}$ are such that

$$c(b-a) + \check{\mu}_2(b) = 0,$$

then for every $x \in [a,b]$ we have

$$\left| \mu([a,b]) f(x) - \int_{[a,b]} f_x(t) \mathrm{d}\mu(t) - [c + \check{\mu}_1(x)] \left[f(b) - f(a) \right] \right| \le \frac{1}{2}(\Gamma_1 - \gamma_1) \int_a^b |c + \check{\mu}_1(t)| \mathrm{d}t.$$

Proof. Note that $P_1(t) = c + \check{\mu}_1(t)$ and $P_2(t) = c(t-a) + \check{\mu}_2(t)$ are two beginning terms of a μ-harmonic sequence of functions on $[a,b]$. Also note that the condition $P_2(a) = P_2(b)$ reduces to $c(b-a) + \check{\mu}_2(b) = 0$ and then apply Theorem 3. $\square$

Measure $\mu \in M[a,b]$ is called k-**balanced** if $\check{\mu}_k(b) = 0$. We see that 1-balanced measure is the same as balanced measure. We also define the kth moment of μ as

$$m_k(\mu) = \textstyle\int_{[a,b]} t^k \mathrm{d}\mu(t), \ k \ge 0.$$

Theorem 4. *For any $\mu \in M[a,b]$ the following assertions hold:*

1) *For any $n \ge 1$ we have*

$$\check{\mu}_n(b) = \textstyle\sum_{k=0}^{n-1} \frac{(-1)^k}{(n-1)!} \binom{n-1}{k} b^{n-1-k} m_k(\mu).$$

2) *For any $n \ge 0$ we have*

$$m_n(\mu) = \textstyle\sum_{k=0}^{n} (-1)^k k! \binom{n}{k} b^{n-k} \check{\mu}_{k+1}(b).$$

3) *μ is k-balanced for every $k \in \{1,\dots,n\}$ if and only if $m_k(\mu) = 0$ for every $k \in \{0,\dots,n-1\}$.*
4) *μ is uniquely determined by the sequence $(\check{\mu}_k(b), k \ge 1)$.*

Proof. 1) By definition of $\check{\mu}_n$ we have $\check{\mu}_n(b) = \frac{1}{(n-1)!} \int_{[a,b]} (b-s)^{n-1} \mathrm{d}\mu(s)$.The stated identity follows from the binomial formula applied to $(b-s)^{n-1}$.
2) For every real α, by a simple calculation we have

$$\textstyle\sum_{k \ge 0} \alpha^k \check{\mu}_{k+1}(b) = \int_{[a,b]} \exp(\alpha(b-s)) \mathrm{d}\mu(s)$$

and

$$\textstyle\sum_{k \ge 0} \alpha^k \check{\mu}_{k+1}(b) \exp(-\alpha b) = \int_{[a,b]} \exp(-\alpha s) \mathrm{d}\mu(s).$$

Expand both sides of this identity in Taylor series in variable α and then equate the coefficients to get the formula.
3) Follows immediately from 1) and 2).
4) Every compactly supported real Borel measure μ in $\mathbb{R}$ is uniquely determined by its moments, and therefore $\mu \in M[a,b]$ is uniquely determined by $(\check{\mu}_k(b), k \ge 1)$, because of 1) and 2). $\square$

Corollary 12. *Let $f : [a,b] \to \mathbb{R}$, γ_n and Γ_n be as in Theorem 3. Then for every $(n+1)$-balanced measure $\mu \in M[a,b]$ and for every $x \in [a,b]$ we have*

$$\left|\mu([a,b])f(x) - \int_{[a,b]} f_x(t)\mathrm{d}\mu(t) - \check{S}_n(x)\right| \le \tfrac{1}{2}(\Gamma_n - \gamma_n)\int_a^b |\check{\mu}_n(t)|\,\mathrm{d}t$$
$$\le \tfrac{1}{2}(\Gamma_n - \gamma_n)\tfrac{1}{n!}(b-a)^n \|\mu\|$$

where $\check{S}_n(x)$ is defined by (3.7).

Proof. To obtain the first inequality apply Theorem 3 for μ-harmonic sequence $(\check{\mu}_k, k \ge 1)$ and note that the condition $P_{n+1}(a) = P_{n+1}(b)$ reduces to $\check{\mu}_{n+1}(b) = 0$, which means that μ is $(n+1)$-balanced. The second inequality follows by (1.4). □

Corollary 13. *Let $f : [a,b] \to \mathbb{R}$, γ_n and Γ_n be as in Theorem 3. Then for every $\mu \in M[a,b]$, such that all k-moments of μ are zero, for $k = 0,\dots,n$, and for any $x \in [a,b]$ we have*

$$\left|\int_{[a,b]} f_x(t)\mathrm{d}\mu(t) - \sum_{k=1}^{n} \check{\mu}_k(x)\left[f^{(k-1)}(b) - f^{(k-1)}(a)\right]\right| \le \frac{1}{2}(\Gamma_n - \gamma_n)\int_a^b |\check{\mu}_n(t)|\,\mathrm{d}t$$
$$\le \frac{1}{2}(\Gamma_n - \gamma_n)\frac{1}{n!}(b-a)^n \|\mu\|.$$

Proof. By Theorem 4 the condition $m_k(\mu) = 0$, $k = 0,\dots,n$, is equivalent to $\check{\mu}_k(b) = 0$, $k = 1,\dots,n+1$. Apply now Corollary 12 and note that in this case $\mu([a,b]) = 0$ and $\check{S}_n(x) = \sum_{k=1}^{n} \check{\mu}_k(x)\left[f^{(k-1)}(b) - f^{(k-1)}(a)\right]$. □

Corollary 14. *Let $f : [a,b] \to \mathbb{R}$, γ_n and Γ_n be as in Theorem 3. Then for every $\mu \in M[a,b]$, such that all k-moments of μ are zero, for $k = 0,\dots,n$, we have*

$$\left|\int_{[a,b]} f(t)\mathrm{d}\mu(t)\right| \le \tfrac{1}{2}(\Gamma_n - \gamma_n)\int_a^b |\check{\mu}_n(t)|\,\mathrm{d}t \le \tfrac{1}{2(n!)}(\Gamma_n - \gamma_n)(b-a)^n \|\mu\|.$$

Proof. Put $x = b$ in Corollary 13. Then $\check{S}_n(b) = 0$, and we can replace $f_b(t) = f(a+b-t)$ by $f(t)$ since the constants Γ_n and γ_n are the same for both $f_x(t)$ and $f(t)$. □

Remark 4. *The inequality of Corollary 14*

$$\left|\int_{[a,b]} f(t)\mathrm{d}\mu(t)\right| \le \tfrac{1}{2(n!)}(\Gamma_n - \gamma_n)(b-a)^n \|\mu\|$$

can be regarded as an nth order generalization of inequality (2.2) of Theorem 1.

Corollary 15. *Let $f : [a,b] \to \mathbb{R}$, γ_1 and Γ_1 be as in Corollary 11. Then for every 2-balanced measure μ and for any $x \in [a,b]$ we have*

$$\left|\mu([a,b])f(x) - \int_{[a,b]} f_x(t)\mathrm{d}\mu(t) - \check{\mu}_1(x)\left[f(b) - f(a)\right]\right| \le \tfrac{1}{2}(\Gamma_1 - \gamma_1)\int_a^b |\check{\mu}_1(t)|\,\mathrm{d}t$$
$$\le \tfrac{1}{2}(\Gamma_1 - \gamma_1)(b-a)\|\mu\|.$$

Proof. Put $n = 1$ in Corollary 12. □

Corollary 16. *Let $f:[a,b]\to\mathbb{R}$, γ_1 and Γ_1 be as in Corollary 11. Then for every $\mu\in M[a,b]$ such that*

$$\int_{[a,b]}\mathrm{d}\mu(t)=\int_{[a,b]}t\mathrm{d}\mu(t)=0,$$

and for any $x\in[a,b]$ we have

$$\begin{aligned}\left|\int_{[a,b]}f_x(t)\mathrm{d}\mu(t)-\check{\mu}_1(x)\left[f(b)-f(a)\right]\right| &\le \tfrac{1}{2}(\Gamma_1-\gamma_1)\int_a^b\left|\check{\mu}_1(t)\right|\mathrm{d}t\\ &\le \tfrac{1}{2}(\Gamma_1-\gamma_1)(b-a)\left\|\mu\right\|\end{aligned}$$

and

$$\left|\int_{[a,b]}f(t)\mathrm{d}\mu(t)\right|\le\tfrac{1}{2}(\Gamma_1-\gamma_1)\int_a^b\left|\check{\mu}_1(t)\right|\mathrm{d}t\le\tfrac{1}{2}(\Gamma_1-\gamma_1)(b-a)\left\|\mu\right\|.$$

Proof. Put $n=1$ in Corollaries 13 and 14. □

Corollary 17. *Let $\{x_k : k\ge1\}$ be a subset of $[a,b]$ of different points and let $(c_k, k\ge1)$ be a sequence in $\mathbb{R}$ such that*

$$\sum_{k\ge1}|c_k|<\infty,\ \sum_{k\ge1}c_k(b-x_k)=0.$$

Then for every $x\in[a,b]$ and $f:[a,b]\to\mathbb{R}$ such that $\gamma_1\le f'(x_k)\le\Gamma_1$, $k\ge1$ we have

$$\begin{aligned}&\left|f(x)\sum_{k\ge1}c_k-\sum_{k\ge1}c_kf_x(x_k)-\sum_{k\ge1}c_k\chi_{[a,x]}(x_k)\left[f(b)-f(a)\right]\right|\\ &\le\tfrac{1}{2}(\Gamma_1-\gamma_1)(b-a)\sum_{k\ge1}|c_k|,\end{aligned}$$

where $\chi_{[a,x]}$ is the indicator function of $[a,x]$ and $f_x(t)$ is defined by (1.8).

Proof. Apply Corollary 15 for discrete measure $\mu=\sum_{k\ge1}c_k\delta_{x_k}$. In this case $\|\mu\|=\sum_{k\ge1}|c_k|$, $\check{\mu}_1=\sum_{k\ge1}c_k\chi_{[x_k,b]}$, $\check{\mu}_2(b)=\sum_{k\ge1}c_k(b-x_k)=0$ and $\int_{[a,b]}f_x(t)\mathrm{d}\mu(t)=\sum_{k\ge1}c_kf_x(x_k)$. □

Corollary 18. *Let $\{x_k; k\ge1\}$ be a subset of $[a,b]$ of different points and let $(c_k, k\ge1)$ be a sequence in $\mathbb{R}$ such that*

$$\sum_{k\ge1}|c_k|<\infty,\ \sum_{k\ge1}c_k=\sum_{k\ge1}c_kx_k=0.$$

Then for every $f:[a,b]\to\mathbb{R}$, such that $\gamma_1\le f'(x_k)\le\Gamma_1$, $k\ge1$ we have

$$\left|\sum_{k\ge1}c_kf(x_k)\right|\le\tfrac{1}{2}(\Gamma_1-\gamma_1)(b-a)\sum_{k\ge1}|c_k|.$$

Proof. Apply Corollary 16 for discrete measure $\mu=\sum_{k\ge1}c_k\delta_{x_k}$. In this case we have $\|\mu\|=\sum_{k\ge1}|c_k|$, $\check{\mu}_1(b)=\sum_{k\ge1}c_k=0$ and $\check{\mu}_2(b)=\sum_{k\ge1}c_k(b-x_k)=0$, while $\int_{[a,b]}f(t)\mathrm{d}\mu(t)=\sum_{k\ge1}c_kf(x_k)$. □

References

[1] A. Čivljak, Lj. Dedić and M. Matić, *Euler harmonic identities for measures*, Nonlinear Funct. Anal. Appl. (accepted)

[2] Lj. Dedić, M. Matić, J. Pečarić and A. Aglić Aljinović, *On weighted Euler harmonic identities with applications*, Math. Inequal. Appl. **8**(2) (2005), 237–257.

[3] Lj. Dedić, M. Matić and J. Pečarić, A. Vukelić, *On generalizations of Ostrowski inequality via Euler harmonic identities*, J. Inequal. Appl. **7**(6) (2002), 787–805.

[4] M. Matić, *Improvement of some estimations related to the remainder in generalized Taylor's formula*, Math. Inequal. Appl. **5**(4) (2002), 637–648.

[5] M. Matić, *Improvement of some inequalities of Euler-Grüss type*, Comput. Math. Applic. **46** (2003), 1325–1336.

[6] A. Čivljak, Lj. Dedić and M. Matić, *On an integration-by-parts formula for measures*, JIPAM, Volume 8, Issue 4, Article 93 (2007).

[7] A. Čivljak, Lj. Dedić and M. Matić, *On Ostrowski and Euler-Grüss type inequalities involving measures*JMI, Volume 1, Number 1 (2007), 65–81.

Ambroz Čivljak
American College of Management and Technology
Rochester Institute of Technology
Don Frana Bulica 6
20000 Dubrovnik, Croatia
e-mail: acivljak@acmt.hr

Ljuban Dedić and Marko Matić
Department of Mathematics
Faculty of Natural Sciences, Mathematics and Education
University of Split
Teslina 12
21000 Split, Croatia
e-mail: ljuban@pmfst.hr
e-mail: mmatic@pmfst.hr

International Series of Numerical Mathematics, Vol. 157, 121–132

The ρ-quasiconcave Functions and Weighted Inequalities

William Desmond Evans, Amiran Gogatishvili and Bohumír Opic

Abstract. We present some facts from a general theory of ρ-quasiconcave functions defined on the interval $I = (a, b) \subseteq \mathbb{R}$ and show how to use them to characterize the validity of weighted inequalities involving ρ-quasiconcave operators.

Mathematics Subject Classification (2000). 26D10, 46E30, 26A99.

Keywords. ρ-quasiconcave function, representation of ρ-quasiconcave functions, weighted Lebesgue space, ρ-fundamental function, ρ-quasiconcave operator, discretization of weighted quasi-norms, weighted inequalities.

1. Introduction and basic definitions

Throughout the paper we assume that $I := (a, b) \subseteq \mathbb{R}$ and that ρ is a positive, continuous and strictly increasing function on the interval I. Such a function ρ is called *admissible* on I – notation $\rho \in Ads(I)$. A non-negative function h is said to be ρ-*quasiconcave* on I – notation $h \in Q_\rho(I)$ – if h is non-decreasing on I and h/ρ is non-increasing on I.

Note that when $I = (0, +\infty)$ and the function ρ is the identity map on I, then the class $Q_\rho(I)$ coincides with the well-known class $Q((0, +\infty))$ of all quasiconcave functions on the interval $(0, +\infty)$. In the literature the class $Q((0, +\infty))$ and its subclass $Q^0((0, +\infty))$ have been especially investigated and used (cf., for example, [2, Chapter 3], [10], [11], [9]); the set $Q^0((0, +\infty))$ consists of those $h \in Q((0, +\infty))$ which are such that $h(t)$, $t/h(t) \to 0$ as $t \to 0_+$, and $h(t)$, $t/h(t) \to +\infty$ as $t \to +\infty$. Another subclass of $Q_\rho((0, +\infty))$ with $\rho(t) = t^k$, $t > 0$, can be found in [8]. The subclass $Q^0_\rho((0, +\infty))$ of those $h \in Q_\rho((0, +\infty))$ which are such that $h(t)$, $(\rho/h)(t) \to 0$ as $t \to 0_+$, and $h(t)$, $(\rho/h)(t)) \to +\infty$ as $t \to +\infty$ was used in [7] to characterize the associate space to $\Gamma^p(v)$. A particular case of $Q_\rho(I)$ with $I =$

The research was supported by grant no. 201/05/2033 of the Grant Agency of the Czech Republic, by the INTAS grant no. 05-1000008-8157 and by the Institutional Research Plan no. AV0Z10190503 of the Academy of Sciences of the Czech Republic.

$(1, +\infty)$ and $\rho(t) = t^\lambda$, $t \in I$, $\lambda > 0$, was considered, e.g., in [13], where relations between summability of functions and their Fourier series were investigated.

By $\mathcal{B}^+(I)$ we denote the collection of all non-negative Borel measures on the interval I. Let $\mu \in \mathcal{B}^+(I)$. The symbol $\mathcal{M}(I, \mu)$ stands for the set of all μ-measurable functions on I, while $\mathcal{M}^+(I, \mu)$ is used to denote the collection of all $f \in \mathcal{M}(I, \mu)$ which are non-negative on I. The family of all *weight functions* on I is given by

$$\mathcal{W}(I, \mu) = \{w \in \mathcal{M}(I, \mu);\ w > 0\ \mu\text{-a.e. on } I\}.$$

If the measure μ is the Lebesgue measure on I, then we omit the symbol μ in the notation and, for example, we write simply $\mathcal{M}(I)$ instead of $\mathcal{M}(I, \mu)$.

For $p \in (0, +\infty]$ and $w \in \mathcal{M}^+(I, \mu)$, we define the functional $\|\cdot\|_{p,w,I,\mu}$ on $\mathcal{M}(I, \mu)$ by

$$\|f\|_{p,w,I,\mu} = \begin{cases} (\int_I |fw|^p \, d\mu)^{1/p} & \text{if } p < +\infty \\ \operatorname*{ess\,sup}_I |fw| & \text{if } p = +\infty. \end{cases}$$

If, in addition, $w \in \mathcal{W}(I, \mu)$, then the *weighted Lebesgue space* $L^p(w, I, \mu)$ is given by

$$L^p(w, I, \mu) = \{f \in \mathcal{M}(I, \mu);\ \|f\|_{p,w,I,\mu} < +\infty\}$$

and it is equipped with the quasi-norm $\|\cdot\|_{p,w,I,\mu}$.

When $w \equiv 1$ on I, we write simply $L^p(I, \mu)$ and $\|\cdot\|_{p,I,\mu}$ instead of $L^p(w, I, \mu)$ and $\|\cdot\|_{p,w,I,\mu}$, respectively. Furthermore, if μ is the Lebesgue measure, then we use symbols $L^p(I)$, $\|\cdot\|_{p,I}$, $L^p(w, I)$ and $\|\cdot\|_{p,w,I}$ instead of $L^p(I, \mu)$, $\|\cdot\|_{p,I,\mu}$, $L^p(w, I, \mu)$ and $\|\cdot\|_{p,w,I,\mu}$, respectively.

Finally, if $p \in (0, +\infty]$, $\mathcal{Z} \subseteq \mathbb{Z}$, $\mathcal{Z} \neq \emptyset$, then the discrete analogue of $L^p(I)$ is denoted by $\ell^p = \ell^p(\mathcal{Z})$. Sometimes we write $\|a_k\|_{\ell^p(\mathcal{Z})}$ instead of $\|\{a_k\}\|_{\ell^p(\mathcal{Z})}$.

The function φ associated to the space $L^p(w, I, \mu)$ and defined by

$$\varphi(x) := \|\min\{\rho(\cdot), \rho(x)\}\|_{p,w,I,\mu}, \quad x \in I,$$

is called the *ρ-fundamental function* of the space $L^p(w, I, \mu)$. This function is an important example of a ρ-quasiconcave function on the interval I (cf. [5]) and it plays a crucial role in what follows.

The operator T, whose domain $\mathcal{D}(T)$ is a subset of all non-negative functions on I, is called ρ-quasiconcave provided that $Tf \in Q_\rho(I)$ for all $f \in \mathcal{D}(T)$. For example, the following operators are ρ-quasiconcave (cf. [5]):

$$(Tg)(x) := \sup_{a<t\le x} \rho(t) \int_t^b g(s)\,ds, \quad g \in \mathcal{M}^+(I), \quad \rho \in Ads(I), \tag{1.1}$$

$$(Tg)(x) := \|\rho(t)\|g\|_{\infty,(t,b)}\|_{\infty,(a,x)}, \quad g \in \mathcal{M}^+(I), \quad \rho \in Ads(I), \tag{1.2}$$

$$(Tg)(x) := \int_a^x u(t)\Big(\int_t^b g(s)\,ds\Big)\,dt, \quad g \in \mathcal{M}^+(I), \tag{1.3}$$

$$(Tg)(x) := \int_a^x u(t)\|g\|_{\infty,(t,b)}\,dt, \quad g \in \mathcal{M}^+(I), \tag{1.4}$$

where (in the case of (1.3) and (1.4))

$$\rho(t) := \int_a^x u(t)\,dt, \quad \text{with} \quad u \in \mathcal{W}(I), \quad \text{satisfies} \quad \rho \in Ads(I). \tag{1.5}$$

Throughout the paper we write $A \lesssim B$ (or $A \gtrsim B$) if $A \le cB$ (or $cA \ge B$) for some positive constant c independent of appropriate quantities involved in the expressions A and B, and $A \approx B$ (and say that A *is equivalent to* B) if $A \lesssim B$ and $A \gtrsim B$.

In Section 2 we present a representation of ρ-quasiconcave functions on I by means of non-negative Borel measures on I. This generalizes the well-known result about the representation of quasiconcave functions on $(0, +\infty)$ (cf., for example, [3, page 117]). Furthermore, we assign to any $h \in Q_\rho(I)$ a sequence $\{x_k\}$ contained in the closure of the interval I and we use this sequence to decompose the interval I into a system $\{I_k\}$ of disjoint subintervals I_k with the property that, for all $x, y \in I_k$, either $h(x) \approx h(y)$ or $(h/\rho)(x) \approx (h/\rho)(y)$. Such a decomposition corresponding to the ρ-fundamental function of the weighted Lebesgue space $L^q(w, I, \mu)$ is applied in Section 3 to discretize $L^q(w, I, \mu)$-quasinorms involving ρ-quasiconcave functions on I and ρ-quasiconcave operators.

In Section 4 we make use of our results to characterize the validity of weighted inequalities involving the ρ-quasiconcave operators T given by (1.1) and (1.3). Our method consists in a discretization of the inequalities in question. We solve them locally (which represents an easier task) to obtain a discrete characterization of the original problem. Finally, we apply an antidiscretization to convert the discrete characterization to a continuous one.

For related weighted inequalities involving the operator $\overline{T}$ given by

$$(\overline{T}g)(x) := \sup_{x \le t < +\infty} \omega(t) \int_0^t g(s)\,ds, \quad g \in \mathcal{M}^+((0,+\infty)), \text{ with } \omega \in \mathcal{W}((0,+\infty)),$$

we refer to [6], where quite different methods were used to prove the boundedness of $\overline{T} : L^p(v, I) \mapsto L^q(w, I)$. Since the operator

$$g \mapsto \widetilde{\rho}(x) \sup_{x<t<b} \frac{1}{\widetilde{\rho}(t)} \int_a^t g(s)\,ds, \quad g \in \mathcal{M}^+(I), \quad \text{with} \quad \widetilde{\rho} \in Ads(I),$$

is $\widetilde{\rho}$-quasiconcave (cf. [5]), the methods of this paper can be used to treat the boundedness of the operator $\overline{T} : L^p(v, I) \mapsto L^q(w, I)$ provided that $1/\omega \in Ads((0, +\infty))$.

In Section 4 (see Remark 4.6) we also explain how our results on the boundedness of the operator $T : L^p(v, I) \mapsto L^q(w, I)$, with T given by (1.3), can bc used to solve similar problems for the Stieltjes transform treated in [1], [14] and [15].

Note that the complete proofs of all our results, together with other ones, can be found in [5].

In the paper we use the abbreviation LHS($*$) (RHS($*$)) for the left- (right-) hand side of the relation ($*$). We also adopt the following convention.

Convention 1.1.

(i) Throughout the paper we put $1/(+\infty) = 0$, $1/0 = +\infty$, $(+\infty)/(+\infty) = 0$, $0/0 = 0$, $0 \cdot (\pm\infty) = 0$, $(+\infty)^\alpha = +\infty$ and $\alpha^0 = 1$ if $\alpha \in (0, +\infty)$.

(ii) If $I = (a,b) \subseteq \mathbb{R}$ and g is a monotone function on I, then by $g(a)$ and $g(b)$ we mean the limits $\lim_{x\to a_+} g(x)$ and $\lim_{x\to b_-} g(x)$, respectively.

2. The ρ-quasiconcave functions

We have noted that the ρ-fundamental function of the space $L^p(w, I, \mu)$ is ρ-quasiconcave on the interval I. Consequently, the function

$$h(x) = \int_I \min\{\rho(t), \rho(x)\}\, d\mu(t), \ \ x \in I,$$

belongs to $Q_\rho(I)$. The following theorem implies that any function $h \in Q_\rho(I)$ can be represented in a similar form.

Theorem 2.1. *Let $I = (a,b) \subseteq \mathbb{R}$, $\rho \in Ads(I)$ and let $h \in Q_\rho(I)$. Then there is a non-negative Borel measure μ on I such that, for all $x \in I$,*

$$h(x) \le \alpha + \beta\rho(x) + \int_I \min\{\rho(t), \rho(x)\}\, d\mu(t) \le 4h(x), \tag{2.1}$$

where

$$\alpha = \lim_{x\to a_+} h(x) \quad \textit{and} \quad \beta = \lim_{x\to b_-} \frac{h(x)}{\rho(x)}.$$

By (2.1),

$$h(x) \approx \alpha + \beta\rho(x) + \int_I \min\{\rho(t), \rho(x)\}\, d\mu(t) \quad \text{for all } x \in I,$$

which is the desired representation of the function $h \in Q_\rho(I)$.

If $h \in Q_\rho(I)$ and $p \in (0, +\infty)$, then it is clear that

$$h^p \in Q_{\rho^p}(I) \quad \text{and} \quad \left(\frac{\rho}{h}\right)^p \in Q_{\rho^p}(I).$$

In this connection a natural question arises: What is the relationship between representations of h and h^p or between representations of h and $(\rho/h)^p$? The following two theorems solve this problem provided that the non-negative Borel measure μ satisfies

$$d\mu(t) = w(t)\, dt, \quad \text{where} \quad w \in \mathcal{W}(I).$$

Theorem 2.2. *Let $I = (a,b) \subseteq \mathbb{R}$, $\rho \in Ads(I)$, $h \in Q_\rho(I)$, $\alpha, \beta \in [0, +\infty]$, $w \in \mathcal{W}(I)$, $p \in (0, +\infty)$ and let*

$$h(x) \approx \alpha + \beta\rho(x) + \int_I \min\{\rho(x), \rho(t)\} w(t)\, dt \quad \textit{for all } x \in I.$$

Then, for all $x \in I$,

$$h^p(x) \approx \alpha^p + \beta^p \rho^p(x) + \int_I \min\{\rho^p(x), \rho^p(t)\} \Big(\frac{H}{\rho}\Big)^{p-1}(t)\, w(t)\, dt,$$

where

$$H(x) = \int_I \min\{\rho(x), \rho(t)\} w(t)\, dt, \quad x \in I.$$

Theorem 2.3. *Let* $I = (a,b) \subseteq \mathbb{R}$, $\rho \in Ads(I)$, $h \in Q_\rho(I)$, $0 \not\equiv h \not\equiv +\infty$, $\alpha, \beta \in [0,+\infty)$, $w \in \mathcal{W}(I)$, $p \in (0,+\infty)$ *and let*

$$h(x) \approx \alpha + \beta\rho(x) + \int_I \min\{\rho(x), \rho(t)\} w(t)\, dt \quad \textit{for all} \quad x \in I.$$

Then

$$\Big(\frac{\rho}{h}\Big)^p(x) \approx \alpha_1^p + \beta_1^p \rho^p(x) + V(x) \quad \textit{for all} \quad x \in I,$$

where

$$V(x) = \int_I \min\{\rho^p(x), \rho^p(t)\} h^{-p-2}(t) \Big(\alpha + \int_a^t \rho(s) w(s)\, ds\Big) \times \Big(\beta + \int_t^b w(s)\, ds\Big)\, d\rho(t),$$

$$\alpha_1 = \lim_{t \to a_+} \Big(\frac{\rho}{h}\Big)(t) \quad \textit{and} \quad \beta_1 = \lim_{t \to b_-} \Big(\frac{1}{h}\Big)(t).$$

Another result, which is also needed in the antidiscretization process, reads as follows.

Lemma 2.4. *Let* $I = (a,b) \subseteq \mathbb{R}$, $\rho \in Ads(I)$ *and* $h \in Q_\rho(I)$. *Then*

$$\Big(\frac{\rho}{h}\Big)(x) = \| \min\{\rho(\cdot), \rho(x)\} \|_{\infty, 1/h, I} \quad \textit{for all } x \in I.$$

To any function $h \in Q_\rho(I)$, $0 \not\equiv h \not\equiv +\infty$, we can assign an increasing sequence $\{t_j\}_{j=J_-}^{J_+} \subset I := (a,b)$ which will be used to define a convenient decomposition of the interval I.

Definition 2.5. Let $I = (a,b) \subseteq \mathbb{R}$, $\rho \in Ads(I)$ and $h \in Q_\rho(I)$. A strictly increasing sequence $\{t_j\}_{j=J_-}^{J_+} \subset I$, where $-\infty \le J_- \le 0 \le J_+ \le +\infty$, is said to be a ρ-*discretizing sequence* of h if:

(i) There is $\alpha \in (1,+\infty)$ such that the inequalities

$$\alpha h(t_{j-1}) \le h(t_j) \quad \text{and} \quad \alpha\Big(\frac{\rho}{h}\Big)(t_{j-1}) \le \Big(\frac{\rho}{h}\Big)(t_j) \tag{2.2}$$

hold for all j satisfying $J_- < j \le J_+$.

(ii) There is a positive constant C such that for any $t \in I$ there exists an integer $k = k(t) \in [J_-, J_+]$ satisfying

$$C^{-1} h(t_k) \le h(t) \le C h(t_k)$$

or

$$C^{-1}\Big(\frac{\rho}{h}\Big)(t_k) \le \Big(\frac{\rho}{h}\Big)(t) \le C\Big(\frac{\rho}{h}\Big)(t_k).$$

The set of all ρ-discretizing sequences of the function $h \in Q_\rho(I)$ is denoted by $DS(h,\rho,I)$. The symbol $DS(h,\rho,I,\alpha)$ stands for the subset of those elements of $DS(h,\rho,I)$ which satisfy (2.2) with a given $\alpha \in (1,+\infty)$. One can show (cf. [5]) that this set is nonempty for any $\alpha \in (1,+\infty)$.

If $h \in Q_\rho(I)$, then $\frac{\rho}{h} \in Q_\rho(I)$ and also $h^p \in Q_{\rho^p}(I)$ for any $p \in (0,+\infty)$. Moreover, it is easy to verify that the following implications hold:

$$\{t_j\}_{j=J_-}^{J_+} \in DS(h,\rho,I,\alpha) \Rightarrow \{t_j\}_{j=J_-}^{J_+} \in DS\Big(\frac{\rho}{h},\rho,I,\alpha\Big),$$

$$\{t_j\}_{j=J_-}^{J_+} \in DS(h,\rho,I,\alpha) \Rightarrow \{t_j\}_{j=J_-}^{J_+} \in DS(h^p,\rho^p,I,\alpha^p),\ p \in (0,+\infty).$$

In the next lemma we mention some more properties of ρ-discretizing sequences of $h \in Q_\rho(I)$.

Lemma 2.6. *Let $I=(a,b)\subseteq \mathbb{R}$, $h \in Q_\rho(I)$, $0 \not\equiv h \not\equiv +\infty$, $\alpha \in (1,+\infty)$ and let $\{t_j\}_{j=J_-}^{J_+} \in DS(h,\rho,I,\alpha)$.*

(i) *Then the inequalities*

$$\alpha\rho(t_{j-1}) \le \rho(t_j)$$

hold for all j satisfying $J_- < j \le J_+$.

(ii) *If $J_+ = +\infty$, then*

$$\lim_{j\to+\infty} t_j = b.$$

(iii) *$J_+ = +\infty$ if and only if*

$$\lim_{t\to b_-} h(t) = +\infty \quad \textit{and} \quad \lim_{t\to b_-}\Big(\frac{\rho}{h}\Big)(t) = +\infty.$$

(iv) *If $J_- = -\infty$, then*

$$\lim_{j\to-\infty} t_j = a.$$

(v) *$J_- = -\infty$ if and only if*

$$\lim_{t\to a_+} h(t) = 0 \quad \textit{and} \quad \lim_{t\to a_+}\Big(\frac{\rho}{h}\Big)(t) = 0.$$

Now, let $h \in Q_\rho(I)$, $I=(a,b)$. We are going to assign to any ρ-discretizing sequence $\{t_j\}_{j=J_-}^{J_+}$ of h another strictly increasing sequence, say $\{x_k\}_{k=K_-}^{K_+}$, which we shall call the *ρ-covering sequence* of h and which is defined as follows:

(i) If $J_+ < +\infty$, we put $K_+ = J_+ + 1$ and $x_{K_+} = b$.
(ii) If $J_- > -\infty$, we put $K_- = J_- - 1$ and $x_{K_-} = a$.
(iii) If $J_+ = +\infty$, we put $K_+ = J_+$.
(iv) If $J_- = -\infty$, we put $K_- = J_-$.
(v) For all $k \in [J_-, J_+]$ we put $x_k = t_k$.

We use the symbol $CS(h,\rho,I)$ to denote the set of all ρ-covering sequences of $h \in Q_\rho(I)$. Note that the rule which was used to assign to a ρ-discretizing sequence of $h \in Q_\rho(I)$ the corresponding ρ-covering sequence defines in fact a one to one mapping between the sets $DS(h,\rho,I)$ and $CS(h,\rho,I)$. If $\alpha \in (1,+\infty)$, then we denote by $CS(h,\rho,I,\alpha)$ the subset of $CS(h,\rho,I)$ consisting of all ρ-covering sequences of $h \in Q_\rho(I)$ which correspond to elements of $DS(h,\rho,I,\alpha)$. To clarify our terminology, note the following. If $\{x_k\}_{k=K_-}^{K_+} \in CS(h,\rho,I)$, then the system of intervals $\{[x_{k-1},x_k)\}_{k=K_-}^{K_+}$ forms a covering of the interval I.

We close this section with the following theorem, which is the desired result, mentioned in the Introduction, on a decomposition of the interval I for a given ρ-quasiconcave function h on I.

Theorem 2.7. *Let $I=(a,b) \subseteq \mathbb{R}$, $h \in Q_\rho(I)$ and let $\{x_k\}_{k=K_-}^{K_+} \in CS(h,\rho,I)$. Put $\mathcal{Z} = \{k \in \mathbb{Z};\ K_- < k \le K_+\}$. Then there is a decomposition*

$$\mathcal{Z} = \mathcal{Z}_1 \cup \mathcal{Z}_2, \quad \mathcal{Z}_1 \cap \mathcal{Z}_2 = \emptyset,$$

such that

$$h(x) \approx h(x_k) \quad \text{for all } x \in [x_{k-1},x_k] \text{ and every } k \in \mathcal{Z}_1$$

and

$$\Big(\frac{\rho}{h}\Big)(x) \approx \Big(\frac{\rho}{h}\Big)(x_k) \quad \text{for all } x \in [x_{k-1},x_k] \text{ and every } k \in \mathcal{Z}_2.$$

3. Discretization of weighted quasi-norms

Our aim in this section is to discretize weighted quasi-norms of ρ-quasiconcave functions. We start with some notation. Assume that $I=(a,b) \subseteq \mathbb{R}$, $\rho \in Ads(I)$, $\varphi \in Q_\rho(I)$ and $\{x_k\}_{k=K_-}^{K_+} \in CS(\varphi,\rho,I)$. Then we put

$$\mathcal{K}_-^+ = \{k \in \mathbb{Z};\ K_- \le k \le K_+\} \quad \text{and} \quad \mathcal{K}^+ = \{k \in \mathbb{Z};\ K_- < k \le K_+\}.$$

Lemma 3.1. *Let $I=(a,b) \subseteq \mathbb{R}$, $\rho \in Ads(I)$, $p \in (0,+\infty]$, $\mu \in \mathcal{B}^+(I)$ and $w \in \mathcal{W}(I,\mu)$. Put*

$$\varphi(x) := \|\min\{\rho(\cdot),\rho(x)\}\|_{p,w,I,\mu}, \quad x \in I,$$

and assume that

$$\varphi(\bar{x}) < +\infty \quad \text{for some} \quad \bar{x} \in I.$$

Let $\{x_k\}_{k=K_-}^{K_+} \in CS(\varphi,\rho,I,\alpha)$ with $\alpha > 2^{1/p}$. Then, for all $f \in Q_\rho(I)$,

$$\|f\|_{p,w,I,\mu} \approx \Big\|\Big(\frac{f\varphi}{\rho}\Big)(x_k)\Big\|_{\ell^p(\mathcal{K}_-^+)}, \tag{3.1}$$

where

$$\Big(\frac{f\varphi}{\rho}\Big)(x_k) := \lim_{x\to a_+}\Big(\frac{f\varphi}{\rho}\Big)(x) \quad \text{if} \quad x_k = a \tag{3.2}$$

and

$$\Big(\frac{f\varphi}{\rho}\Big)(x_k) := \lim_{x\to b_-}\Big(\frac{f\varphi}{\rho}\Big)(x) \quad \text{if} \quad x_k = b. \tag{3.3}$$

If one has more information about the ρ-quasiconcave function f, then one can prove even more than (3.1). For example, this is the case when f belongs to the range of some ρ-quasiconcave operator T. In our next assertion we consider the case when the operator T is given by (1.1). (*From now on we still assume* (3.2) *and* (3.3) *when f is any ρ-quasiconcave function on I and* $\{x_k\}_{k=K_-}^{K_+} \in CS(\varphi,\rho,I)$.)

Theorem 3.2. *Let* $I=(a,b)\subseteq\mathbb{R}$, $\rho\in Ads(I)$, $q\in(0,+\infty]$, $\mu\in\mathcal{B}^+(I)$ *and let* $w\in\mathcal{W}(I,\mu)$. *Put*

$$\varphi(x) := \|\min\{\rho(\cdot),\rho(x)\}\|_{q,w,I,\mu}, \quad x\in I,$$

and suppose that

$$\varphi(\bar{x}) < +\infty \quad \text{for some} \quad \bar{x}\in I.$$

Let $\{x_k\}_{k=K_-}^{K_+} \in CS(\varphi,\rho,I,\alpha)$ *with* $\alpha>2^{1/q}$. *Then, for all* $g\in\mathcal{M}^+(I)$,

$$\begin{aligned}\Big\|\sup_{a<t\le x}\rho(t)\int_t^b g(s)\,ds\Big\|_{q,w,I,\mu} &\approx \Big\|\frac{\varphi(x_k)}{\rho(x_k)}\sup_{a<t\le x_k}\rho(t)\int_t^b g(s)\,ds\Big\|_{\ell^q(\mathcal{K}_-^+)}\\ &\approx \Big\|\sup_{x_{k-1}<t\le x_k}\varphi(t)\int_t^{x_k} g(s)\,ds\Big\|_{\ell^q(\mathcal{K}^+)}.\end{aligned}$$

Our next result is a modification of Theorem 3.2 and concerns the case when the operator T from (1.1) is replaced by the operator T defined by (1.2). In this context note that, on changing the order of the essential suprema, we obtain, for all $g\in\mathcal{M}^+(I)$,

$$(Tg)(x) = \|\rho(t)\|g\|_{\infty,(t,b)}\|_{\infty,(a,x)} = \|\min\{\rho(\cdot),\rho(x)\}\|_{\infty,g,I}.$$

Theorem 3.3. *Let* $I=(a,b)\subseteq\mathbb{R}$, $\rho\in Ads(I)$, $q\in(0,+\infty]$, $\mu\in\mathcal{B}^+(I)$ *and let* $w\in\mathcal{W}(I,\mu)$. *Put*

$$\varphi(x) := \|\min\{\rho(\cdot),\rho(x)\}\|_{q,w,I,\mu}, \quad x\in I,$$

and suppose that

$$\varphi(\bar{x}) < +\infty \text{ for some } \bar{x}\in I.$$

Let $\{x_k\}_{k=K_-}^{K_+} \in CS(\varphi,\rho,I,\alpha)$ *with* $\alpha>2^{1/q}$. *Then, for all* $g\in\mathcal{M}^+(I)$,

$$\|\,\|\min\{\rho(\cdot),\rho(x)\}\|_{\infty,g,I}\|_{q,w,I,\mu} \approx \|\,\|\varphi\|_{\infty,g,(x_{k-1},x_k)}\|_{\ell^q(\mathcal{K}^+)}.$$

We close this section with one more result involving the operator T given by (1.3). In this context we mention that, by Fubini's theorem,

$$(Tg)(x) = \int_a^x u(t)\Big(\int_t^b g(s)\,ds\Big)\,dt = \|\min\{\rho(\cdot),\rho(x)\}\|_{1,g,I} \tag{3.4}$$

for all $g\in\mathcal{M}^+(I)$, with ρ from (1.5).

Theorem 3.4. *Let $I = (a,b) \subseteq \mathbb{R}$, $\rho \in Ads(I)$, $q \in (0,+\infty]$, $\mu \in \mathcal{B}^+(I)$ and let $w \in \mathcal{W}(I,\mu)$. Put*

$$\varphi(x) := \|\min\{\rho(\cdot),\rho(x)\}\|_{q,w,I,\mu}, \quad x \in I,$$

and assume that

$$\varphi(\bar{x}) < +\infty \quad \text{for some} \quad \bar{x} \in I.$$

Let $\{x_k\}_{k=K_-}^{K_+} \in CS(\varphi,\rho,I,\alpha)$ with $\alpha > 2^{1/q}$. Then, for all $g \in \mathcal{M}^+(I)$,

$$\|\,\|\min\{\rho(\cdot),\rho(x)\}\|_{1,g,I}\|_{q,w,I,\mu} \approx \left\| \int_{(x_{k-1},x_k]} \varphi(t)g(t)\,dt \right\|_{\ell^q(\mathcal{K}^+)}.$$

4. Weighted inequalities

In this section we are going to apply our previous results to characterize the validity of weighted inequalities involving ρ-quasiconcave operators. Our method consists of several steps. Firstly, we discretize both sides of the inequality in question. Secondly, we solve our problem locally (which represents an easier task) to get a discrete characterization of the original inequality. Finally, we apply an antidiscretization to convert the discrete characterization to a continuous one.

We define p' by $1/p+1/p' = 1$ if $p \in [1,+\infty]$. Moreover, we put $p^* = p/(1-p)$ when $p \in (0,1]$.

The first two theorems concern the operator (1.1).

Theorem 4.1. *Let $I = (a,b) \subseteq \mathbb{R}$, $\rho \in Ads(I)$, $w,v \in \mathcal{W}(I)$ and let $1 \le p \le q \le +\infty$. Then*

$$\left\| \sup_{a<t\le x} \rho(t) \int_t^b g(s)\,ds \right\|_{q,w,I} \lesssim \|g\|_{p,v,I} \quad \text{for all } g \in \mathcal{M}^+(I) \tag{4.1}$$

if and only if

$$\left\| \left\| \min\left\{\frac{\rho(\cdot)}{\rho(x)},1\right\} \right\|_{q,w,I} \sup_{a<t\le x} \rho(t)\|v^{-1}\|_{p',(t,b)} \right\|_{\infty,I} < +\infty. \tag{4.2}$$

The idea of the proof.

Step 1. Let the function $\varphi \in Q_\rho(I)$ be given by

$$\varphi(x) := \|\min\{\rho(\cdot),\rho(x)\}\|_{q,w,I}, \quad x \in I,$$

and assume (for simplicity) that $\varphi(x) < +\infty$ for all $x \in I$. Let $\{x_k\}_{k=K_-}^{K_+} \in CS(\varphi,\rho,I,\alpha)$ with $\alpha > 2^{1/q}$. Then, by Theorem 3.2,

$$\text{LHS(4.1)} \approx \left\| \sup_{t\in I_k} \varphi(t) \int_t^{x_k} g(s)\,ds \right\|_{\ell^q(\mathcal{K}^+)},$$

where $I_k = (x_{k-1}, x_k]$. Since RHS(4.1)$= \| \, \|gv\|_{p,I_k} \|_{\ell^p(\mathcal{K}^+)}$, inequality (4.1) can be rewritten as

$$\Big\| \sup_{t \in I_k} \varphi(t) \int_t^{x_k} g(s)\, ds \Big\|_{\ell^q(\mathcal{K}^+)} \lesssim \| \, \|gv\|_{p,I_k} \|_{\ell^p(\mathcal{K}^+)} \quad \text{for all } g \in \mathcal{M}^+(I). \tag{4.3}$$

Step 2. We solve (4.3) locally. That is, for any $k \in \mathcal{K}^+$, we solve the Hardy-type inequality

$$\sup_{t \in I_k} \varphi(t) \int_t^{x_k} h(s)\, ds \le A_k \|hv\|_{p,I_k}, \quad h \in \mathcal{M}^+(I_k), \tag{4.4}$$

where A_k is the best possible constant in (4.4), i.e.,

$$A_k := \sup_{h \in \mathcal{M}^+(I_k)} \Big(\sup_{t \in I_k} \varphi(t) \int_t^{x_k} h(s)\, ds \Big) / \|hv\|_{p,I_k};$$

here we use the convention that $0/0 = 0$ and $(+\infty)/(+\infty) = 0$ – cf. Convention 1.1 (i). It is well known (cf. [12]) that if $1 \le p \le +\infty$, then

$$A_k \approx \sup_{t \in I_k} \|\varphi\|_{\infty,(x_{k-1},t)} \|v^{-1}\|_{p',(t,x_k)} = \sup_{t \in I_k} \varphi(t) \|v^{-1}\|_{p',(t,x_k)};$$

the last equality holds since φ is continuous and non-decreasing on I.

Step 3. It is not hard to show that inequality (4.3) (and so (4.1) as well) is equivalent to

$$\|\{A_k\}\|_{\ell^\infty(\mathcal{K}^+)} < +\infty. \tag{4.5}$$

Step 4. We apply results of Section 3 to convert the discrete characterization (4.5) to a continuous one. □

The following result is an analogue of Theorem 4.1 and concerns the case when $1 \le p \le +\infty$, $0 < q < +\infty$ and $q < p$.

Theorem 4.2. *Let $I = (a,b) \subseteq \mathbb{R}$, $\rho \in Ads(I)$, $w, v \in \mathcal{W}(I)$, $1 \le p \le +\infty$, $0 < q < +\infty$ and $q < p$. Put $1/r := 1/q - 1/p$. Then* (4.1) *holds if and only if*

$$\Big\| \Big\| \min\Big\{ \frac{\rho(\cdot)}{\rho(x)}, 1 \Big\} \Big\|_{q,w,I}^{1-\frac{q}{r}} w^{\frac{q}{r}}(x) \sup_{a < t \le x} \rho(t) \|v^{-1}\|_{p',(t,b)} \Big\|_{r,I} < +\infty.$$

The next theorem is an analogue of Theorem 4.1 for the operator (1.3).

Theorem 4.3. *Let $I = (a,b) \subseteq \mathbb{R}$, $w, v, u \in \mathcal{W}(I)$ and let $1 \le p \le q \le +\infty$. Put $\rho(x) := \int_a^x u(t)\, dt$, $x \in I$, and assume that $\rho \in Ads(I)$. Then*

$$\Big\| \int_a^x u(t) \Big(\int_t^b g(s)\, ds \Big) dt \Big\|_{q,w,I} \lesssim \|g\|_{p,v,I} \quad \textit{for all } g \in \mathcal{M}^+(I) \tag{4.6}$$

if and only if

$$\Big\| \Big\| \min\Big\{ \frac{\rho(\cdot)}{\rho(x)}, 1 \Big\} \Big\|_{q,w,I} \Big\| \min\{\rho(\cdot), \rho(x)\} \Big\|_{p',v^{-1},I} \Big\|_{\infty,I} < +\infty.$$

We continue with an analogue of Theorem 4.2.

Theorem 4.4. *Let $I = (a,b) \subseteq \mathbb{R}$, $w, v, u \in \mathcal{W}(I)$, $1 \le p \le +\infty$, $0 < q < +\infty$, $q < p$ and $1/r := 1/q - 1/p$. Put $\rho(x) := \int_a^x u(t)\,dt$, $x \in I$, and assume that $\rho \in Ads(I)$. Then* (4.6) *holds if and only if*

$$\Big\| \Big\| \min\Big\{\frac{\rho(\cdot)}{\rho(x)}, 1\Big\}\Big\|_{q,w,I}^{1-\frac{q}{r}} w^{\frac{q}{r}}(x) \Big\| \min\{\rho(\cdot), \rho(x)\}\Big\|_{p',v^{-1},I} \Big\|_{r,I} < +\infty.$$

Now, we investigate the reverse inequality to (4.6).

Theorem 4.5. *Let $I = (a,b) \subseteq \mathbb{R}$, $w, v, u \in \mathcal{W}(I)$ and let $0 < q \le p \le 1$. Put $\rho(x) := \int_a^x u(t)\,dt$, $x \in I$, and assume that $\rho \in Ads(I)$. Then*

$$\|g\|_{p,v,I} \lesssim \Big\| \int_a^x u(t)\Big(\int_t^b g(s)\,ds\Big)\,dt \Big\|_{q,w,I} \quad \text{for all } g \in \mathcal{M}^+(I)$$

if and only if

$$\Big\| \Big\| \min\Big\{1, \frac{\rho(x)}{\rho(\cdot)}\Big\}\Big\|_{p^*,v,I} \|\min\{\rho(\cdot), \rho(x)\}\|_{q,w,I}^{-1} \Big\|_{\infty,I} < +\infty.$$

Remark 4.6. Since

$$\begin{aligned}\|\min\{\rho(\cdot),\rho(x)\}\|_{1,g,I} &= \Big\| \frac{1}{\max\{\frac{1}{\rho(\cdot)}, \frac{1}{\rho(x)}\}} \Big\|_{1,g,I} \approx \Big\| \frac{1}{\frac{1}{\rho(\cdot)} + \frac{1}{\rho(x)}} \Big\|_{1,g,I} \\ &= \Big\| \frac{\rho(x)\rho(\cdot)}{\rho(x)+\rho(\cdot)} \Big\|_{1,g,I} = \rho(x) \Big\| \frac{h(\cdot)}{\rho(x)+\rho(\cdot)} \Big\|_{1,I}, \end{aligned} \tag{4.7}$$

where $h := \rho g$, we see from (3.4) and (4.7) that Theorems 4.3–4.5 can be used to characterize the validity of weighted inequalities involving the operator S defined by

$$(Sh)(x) = \int_a^b \frac{h(t)\,dt}{\rho(x)+\rho(t)}, \quad h \in \mathcal{M}^+(I).$$

We call this operator the *generalized Stieltjes transform*; the usual *Stieltjes transform* is obtained on putting $(a,b) = (0,+\infty)$ and $\rho(x) \equiv x$.

In the case that $(a,b) = (0,+\infty)$ and $\rho(x) \equiv x^\lambda$, $\lambda > 0$, the boundedness of the operator S between weighted L^p and L^q spaces was investigated in [1] (when $1 \le p \le q \le +\infty$) and in [14], [15] (when $1 \le q < p \le +\infty$).

References

[1] K. Andersen, *Weighted inequalities for the Stieltjes transform and Hilbert's double series*, Proc. Royal. Soc. Edinburgh Sect. A **86** (1980), 75–84.

[2] Yu.A. Brudnyĭ and N.Ya. Krugljak, *Interpolation Functors and Interpolation Spaces* **1**, North-Holland, Amsterdam, 1991.

[3] J. Bergh and J. Löfström, *Interpolation Spaces. An Introduction*, Springer-Verlag, Berlin, 1976.

[4] W.D. Evans, A. Gogatishvili and B. Opic, *The reverse Hardy inequality with measures*, Math. Inequal. Appl. **11** (2008), 43–74.

[5] W.D. Evans, A. Gogatishvili and B. Opic, *Weighted inequalities involving ρ-quasi-concave operators*, Preprint, Prague, 2003.

[6] A. Gogatishvili, B. Opic and L. Pick, *Weighted inequalities for Hardy-type operators involving suprema* Collect. Math. **57** (2006), 227–255.

[7] A. Gogatishvili and L. Pick, *Discretization and anti-discretization of rearrangement invariant norms*, Publ. Mat. **47** (2003), 311–358.

[8] M.L. Gol'dman, *On integral inequalities on a cone of functions with monotonicity properties*, Soviet Math. Dokl. **44** (1992), 581–587.

[9] M.L. Gol'dman, H.P. Heinig and V.D. Stepanov, *On the principle of duality in Lorentz spaces*, Can. J. Math. **48** (1996), 959–979.

[10] S. Janson, *Minimal and maximal method of interpolation*, J. Funct. Anal. **44** (1981), 50–73.

[11] P. Nilsson, *Interpolation of Banach lattices*, Studia Math. **82** (1985), 135–154.

[12] B. Opic and A. Kufner, *Hardy-Type Inequalities*, Pitman Res. Notes in Math. Ser. 219, Longman Sci. & Tech., Harlow (1990).

[13] L.E. Persson, *Relations between summability of functions and their Fourier series*, Acta Math. Acad. Sci. Hung. **27** (1976), 267–280.

[14] G. Sinnamon, *A note on the Stieltjes transform*, Proc. Royal. Soc. Edinburgh Sect. A **110** (1988), 73–78.

[15] G. Sinnamon, *Operators on Lebesgue spaces with general measures*, Doctoral thesis, McMaster University, 1987.

William Desmond Evans
School of Mathematics
Cardiff University
Cardiff CF24 4AG, Wales, UK
e-mail: evanswd@cardiff.ac.uk

Amiran Gogatishvili
Institute of Mathematics
Academy of Sciences of the Czech Republic
Žitná 25, 115 67 Praha 1, Czech Republic
e-mail: gogatish@math.cas.cz

Bohumír Opic
Institute of Mathematics
Academy of Sciences of the Czech Republic
Žitná 25, 115 67 Praha 1, Czech Republic
or
Department of Mathematics and Didactics of Mathematics
Pedagogical Faculty
Technical University of Liberec
Hálkova 6, 46117 Liberec, Czech Republic
e-mail: opic@math.cas.cz

Part III

Inequalities for Operators

PHB '07

International Series of Numerical Mathematics, Vol. 157, 135–146

Inequalities for the Norm and Numerical Radius of Composite Operators in Hilbert Spaces

Silvestru S. Dragomir

Abstract. Some new inequalities for the norm and the numerical radius of composite operators generated by a pair of operators are given.

Mathematics Subject Classification (2000). 47A12.

Keywords. Numerical range, numerical radius, bounded linear operators, Hilbert spaces.

1. Introduction

Let $(H; \langle\cdot,\cdot\rangle)$ be a complex Hilbert space. The *numerical range* of an operator T is the subset of the complex numbers $\mathbb{C}$ given by $W(T) = \{\langle Tx, x\rangle,\ x \in H,\ \|x\| = 1\}$, see for instance [6, p. 1]. It is well known that (see [6]):

(i) The numerical range of an operator is convex;
(ii) The spectrum of an operator is contained in the closure of its numerical range;
(iii) T is self-adjoint if and only if $W(T)$ is real.

The *numerical radius* $w(T)$ of an operator T on H is defined by $w(T) := \sup\{|\lambda|, \lambda \in W(T)\} = \sup\{|\langle Tx, x\rangle|, \|x\| = 1\}$, [6, p. 8]. It is well known that $w(\cdot)$ is a norm on the Banach algebra $B(H)$ of all bounded linear operators acting on H and the following inequality holds true:

$$w(T) \le \|T\| \le 2w(T). \tag{1.1}$$

We recall some classical results involving the numerical radius of two linear operators A, B.

The following general result for the product of two operators holds [6, p. 37]:

Theorem 1. *If A, B are two bounded linear operators on the Hilbert space $(H, \langle\cdot,\cdot\rangle)$, then $w(AB) \le 4w(A)\,w(B)$.*

In the case that $AB = BA$, then $w(AB) \le 2w(A)\,w(B)$.

The following results are also well known [6, p. 38].

Theorem 2. *If A is a unitary operator that commutes with another operator B, then*

$$w(AB) \le w(B). \tag{1.2}$$

If A is an isometry and $AB = BA$, then (1.2) *also holds true.*

We say that A and B *double commute* if $AB = BA$ and $AB^* = B^*A$. The following result holds [6, p. 38].

Theorem 3 (Double commute). *If the operators A and B double commute, then $w(AB) \le w(B)\|A\|$.*

As a consequence of the above, we have [6, p. 39]:

Corollary 1. *Let A be a normal operator commuting with B. Then $w(AB) \le w(A)w(B)$.*

For other results and historical comments on the above see [6, pp. 39–41]. For more results on the numerical radius, see [7].

The main aim of this paper is to establish some new inequalities for composite operators generated by a pair of operators (A, B) under various assumptions. Namely, in one side, several inequalities involving the norm $\left\|\frac{A^*A+B^*B}{2}\right\|$ and the numerical radius $w(B^*A)$ are established. On the other side, upper bounds for the nonnegative quantities $\|A\|\|B\| - w(B^*A)$ and $\|A\|^2\|B\|^2 - w^2(B^*A)$ under special conditions for the operators involved are also given. These results provide various generalizations for some inequalities recently obtained by the author in [1]–[3].

2. General inequalities

The following result may be stated:

Theorem 4. *Let $A, B : H \to H$ be two bounded linear operators on the Hilbert space $(H, \langle\cdot,\cdot\rangle)$. If $r > 0$ and*

$$\|A - B\| \le r, \tag{2.1}$$

then

$$\left\|\frac{A^*A + B^*B}{2}\right\| \le w(B^*A) + \frac{1}{2}r^2. \tag{2.2}$$

Proof. For any $x \in H$, $\|x\| = 1$, we have from (2.1) that

$$\|Ax\|^2 + \|Bx\|^2 \le 2\operatorname{Re}\langle Ax, Bx\rangle + r^2. \tag{2.3}$$

However

$$\|Ax\|^2 + \|Bx\|^2 = \langle (A^*A + B^*B)x, x\rangle$$

and by (2.3) we obtain

$$\langle (A^*A + B^*B)x, x\rangle \le 2|\langle (B^*A)x, x\rangle| + r^2 \tag{2.4}$$

for any $x \in H$, $\|x\| = 1$.

Taking the supremum over $x \in H$, $\|x\| = 1$ in (2.4) we get

$$w(A^*A + B^*B) \le 2w(B^*A) + r^2 \tag{2.5}$$

and since the operator $A^*A + B^*B$ is self-adjoint, hence $w(A^*A + B^*B) = \|A^*A + B^*B\|$ and by (2.5) we deduce the desired inequality (2.2). □

Remark 1. *We observe that, from the proof of the above theorem, we have the inequalities*

$$0 \le \left\| \frac{A^*A + B^*B}{2} \right\| - w(B^*A) \le \frac{1}{2} \|A - B\|^2, \tag{2.6}$$

provided that A, B are bounded linear operators in H.

The second inequality in (2.6) *is obvious while the first inequality follows by the fact that*

$$\langle (A^*A + B^*B) x, x \rangle = \|Ax\|^2 + \|Bx\|^2 \ge 2 |\langle (B^*A) x, x \rangle|$$

for any $x \in H$.

The inequality (2.2) is obviously a reach source of particular inequalities of interest.

Indeed, if we assume, for $\lambda \in \mathbb{C}$ and a bounded linear operator T, that we have $\|T - \lambda T^*\| \le r$, for a given positive number r, then by (2.6) we deduce the inequality

$$0 \le \left\| \frac{T^*T + |\lambda|^2 TT^*}{2} \right\| - |\lambda| w(T^2) \le \frac{1}{2} r^2. \tag{2.7}$$

Now, if we assume that for $\lambda \in \mathbb{C}$ and a bounded linear operator V we have that $\|V - \lambda I\| \le r$, where I is the identity operator on H, then by (2.2) we deduce the inequality

$$0 \le \left\| \frac{V^*V + |\lambda|^2 I}{2} \right\| - |\lambda| w(V) \le \frac{1}{2} r^2.$$

As a dual approach, the following result may be noted as well:

Theorem 5. *Let $A, B : H \to H$ be two bounded linear operators on the Hilbert space H. Then*

$$\left\| \frac{A + B}{2} \right\|^2 \le \frac{1}{2} \left[\left\| \frac{A^*A + B^*B}{2} \right\| + w(B^*A) \right]. \tag{2.8}$$

Proof. We obviously have

$$\begin{aligned} \|Ax + Bx\|^2 &= \|Ax\|^2 + 2 \operatorname{Re} \langle Ax, Bx \rangle + \|Bx\|^2 \\ &\le \langle (A^*A + B^*B) x, x \rangle + 2 |\langle (B^*A) x, x \rangle| \end{aligned}$$

for any $x \in H$.

Taking the supremum over $x \in H$, $\|x\| = 1$, we get

$$\|A + B\|^2 \le \|A^*A + B^*B\| + 2w(B^*A),$$

from where we get the desired inequality (2.8). □

Remark 2. *The inequality* (2.8) *can generate some interesting particular results such as the following inequality*

$$\left\|\frac{T+T^*}{2}\right\|^2 \le \frac{1}{2}\left[\left\|\frac{T^*T+TT^*}{2}\right\| + w\left(T^2\right)\right], \tag{2.9}$$

holding for each bounded linear operator $T : H \to H$.

The following result may be stated as well.

Theorem 6. *Let* $A, B : H \to H$ *be two bounded linear operators on the Hilbert space* H *and* $p \ge 2$. *Then*

$$\left\|\frac{A^*A+B^*B}{2}\right\|^{\frac{p}{2}} \le \frac{1}{4}\left[\|A-B\|^p + \|A+B\|^p\right]. \tag{2.10}$$

Proof. We use the following inequality for vectors in inner product spaces obtained by Dragomir and Sándor in [4]:

$$2\left(\|a\|^p + \|b\|^p\right) \le \|a+b\|^p + \|a-b\|^p \tag{2.11}$$

for any $a, b \in H$ and $p \ge 2$.

Utilising (2.11) we may write

$$2\left(\|Ax\|^p + \|Bx\|^p\right) \le \|Ax+Bx\|^p + \|Ax-Bx\|^p \tag{2.12}$$

for any $x \in H$.

Now, observe that $\|Ax\|^p + \|Bx\|^p = \left(\|Ax\|^2\right)^{\frac{p}{2}} + \left(\|Bx\|^2\right)^{\frac{p}{2}}$ and by the elementary inequality $\frac{\alpha^q+\beta^q}{2} \ge \left(\frac{\alpha+\beta}{2}\right)^q$, $\alpha, \beta \ge 0$ and $q \ge 1$ we have

$$\left(\|Ax\|^2\right)^{\frac{p}{2}} + \left(\|Bx\|^2\right)^{\frac{p}{2}} \ge 2^{1-\frac{p}{2}}\left[\langle (A^*A+B^*B)\,x, x\rangle\right]^{\frac{p}{2}}. \tag{2.13}$$

Combining (2.12) with (2.13) we get

$$\frac{1}{4}\left[\|Ax-Bx\|^p + \|Ax+Bx\|^p\right] \ge \left|\left\langle\left(\frac{A^*A+B^*B}{2}\right)x, x\right\rangle\right|^{\frac{p}{2}} \tag{2.14}$$

for any $x \in H$, $\|x\| = 1$. Taking the supremum over $x \in H$, $\|x\| = 1$, and taking into account that $w\left(\frac{A^*A+B^*B}{2}\right) = \left\|\frac{A^*A+B^*B}{2}\right\|$, we deduce the desired result (2.10). □

Remark 3. *If* $p = 2$, *then we have the inequality:* $\left\|\frac{A^*A+B^*B}{2}\right\| \le \left\|\frac{A-B}{2}\right\|^2 + \left\|\frac{A+B}{2}\right\|^2$, *for any* A, B *bounded linear operators. This result can be obtained directly on utilising the parallelogram identity as well. We also should observe that for* $A = T$ *and* $B = T^*$, T *a normal operator, the inequality* (2.10) *becomes* $\|T\|^p \le \frac{1}{4}\left[\|T-T^*\|^p + \|T+T^*\|^p\right]$, *where* $p \ge 2$.

The following result may be stated as well.

Theorem 7. *Let $A, B : H \to H$ be two bounded linear operators on the Hilbert space H and $r \geq 1$. If $A^*A \geq B^*B$ in the operator order or, equivalently, $\|Ax\| \geq \|Bx\|$ for any $x \in H$, then:*

$$\left\| \frac{A^*A + B^*B}{2} \right\|^r \leq \|A\|^{r-1} \|B\|^{r-1} w(B^*A) + \frac{1}{2} r^2 \|A\|^{2r-2} \|A - B\|^2 . \quad (2.15)$$

Proof. We use the following inequality for vectors in inner product spaces due to Goldstein, Ryff and Clarke [5]:

$$\|a\|^{2r} + \|b\|^{2r} \leq 2 \|a\|^{r-1} \|b\|^{r-1} \operatorname{Re} \langle a, b \rangle + r^2 \|a\|^{2r-2} \|a - b\|^2 , \quad (2.16)$$

where $r \geq 1$, $a, b \in H$ and $\|a\| \geq \|b\|$.

Utilising (2.16) we can state that:

$$\|Ax\|^{2r} + \|Bx\|^{2r} \leq 2 \|Ax\|^{r-1} \|Bx\|^{r-1} |\langle Ax, Bx \rangle| + r^2 \|Ax\|^{2r-2} \|Ax - Bx\|^2 , \quad (2.17)$$

for any $x \in H$. As in the proof of Theorem 6, we also have

$$2^{1-r} [\langle (A^*A + B^*B) x, x \rangle]^r \leq \|Ax\|^{2r} + \|Bx\|^{2r} , \quad (2.18)$$

for any $x \in H$. Therefore, by (2.17) and (2.18) we deduce

$$\left[\left\langle \left(\frac{A^*A + B^*B}{2} \right) x, x \right\rangle \right]^r \leq \|Ax\|^{r-1} \|Bx\|^{r-1} |\langle Ax, Bx \rangle| + \frac{1}{2} r^2 \|Ax\|^{2r-2} \|Ax - Bx\|^2 \quad (2.19)$$

for any $x \in H$.

Taking the supremum in (2.19) we obtain the desired result (2.15). □

Remark 4. *If we choose in (2.15) $A = V$ and $B = V^*$, then, on taking into account that $w(V^2) \leq \|V\|^2$, we get the inequality*

$$\left\| \frac{V^*V + VV^*}{2} \right\|^r \leq \|V\|^{2r-2} \left[\|V\|^2 + \frac{1}{2} r^2 \|V - V^*\|^2 \right], \quad (2.20)$$

holding for any operator V and any $r \geq 1$.

3. Further inequalities for an invertible operator

In this section we assume that $B : H \to H$ is an invertible bounded linear operator and let $B^{-1} : H \to H$ be its inverse. Then, obviously,

$$\|Bx\| \geq \frac{1}{\|B^{-1}\|} \|x\| \quad \text{for any} \quad x \in H, \quad (3.1)$$

where $\|B^{-1}\|$ denotes the norm of the inverse B^{-1}.

Theorem 8. *Let $A, B : H \to H$ be two bounded linear operators on H and B is invertible such that, for a given $r > 0$,*

$$\|A - B\| \leq r. \tag{3.2}$$

Then:

$$\|A\| \leq \left\|B^{-1}\right\| \left[w\left(B^{*}A\right) + \frac{1}{2}r^{2}\right]. \tag{3.3}$$

Proof. The condition (3.2) is obviously equivalent to:

$$\|Ax\|^{2} + \|Bx\|^{2} \leq 2 \operatorname{Re}\left\langle \left(B^{*}A\right)x, x\right\rangle + r^{2} \tag{3.4}$$

for any $x \in H$, $\|x\| = 1$. Since, by (3.1), $\|Bx\|^{2} \geq \frac{1}{\|B^{-1}\|^{2}}\|x\|^{2}$, $x \in H$ and $\operatorname{Re}\left\langle \left(B^{*}A\right)x, x\right\rangle \leq \left|\left\langle \left(B^{*}A\right)x, x\right\rangle\right|$, hence by (3.4) we get

$$\|Ax\|^{2} + \frac{\|x\|^{2}}{\|B^{-1}\|^{2}} \leq 2\left|\left\langle \left(B^{*}A\right)x, x\right\rangle\right| + r^{2} \tag{3.5}$$

for any $x \in H$, $\|x\| = 1$. Taking the supremum over $x \in H$, $\|x\| = 1$ in (3.5), we have

$$\|A\|^{2} + \frac{1}{\|B^{-1}\|^{2}} \leq 2w\left(B^{*}A\right) + r^{2}. \tag{3.6}$$

By the elementary inequality

$$\frac{2\|A\|}{\|B^{-1}\|} \leq \|A\|^{2} + \frac{1}{\|B^{-1}\|^{2}} \tag{3.7}$$

and by (3.6) we then deduce the desired result (3.3). □

Remark 5. *If we choose above $B = \lambda I$, $\lambda \neq 0$, then we get the inequality*

$$(0 \leq)\, \|A\| - w(A) \leq \frac{1}{2|\lambda|}r^{2}, \tag{3.8}$$

provided $\|A - \lambda I\| \leq r$. This result has been obtained in the earlier paper [1].

Also, if we assume that $B = \lambda A^{*}$, A is invertible, then we obtain

$$\|A\| \leq \left\|A^{-1}\right\| \left[w\left(A^{2}\right) + \frac{1}{2|\lambda|}r^{2}\right], \tag{3.9}$$

provided $\|A - \lambda A^{*}\| \leq r$, $\lambda \neq 0$.

The following result may be stated as well:

Theorem 9. *Let $A, B : H \to H$ be two bounded linear operators on H. If B is invertible and for $r > 0$,*

$$\|A - B\| \leq r, \tag{3.10}$$

then

$$(0 \leq)\, \|A\| \|B\| - w\left(B^{*}A\right) \leq \frac{1}{2}r^{2} + \frac{\|B\|^{2}\left\|B^{-1}\right\|^{2} - 1}{2\left\|B^{-1}\right\|^{2}}. \tag{3.11}$$

Proof. The condition (3.10) is obviously equivalent to

$$\|Ax\|^2 + \|Bx\|^2 \le 2\,\mathrm{Re}\,\langle Ax, Bx\rangle + r^2$$

for any $x \in H$, with $\|x\| = 1$, which is clearly equivalent to

$$\|Ax\|^2 + \|B\|^2 \le 2\,\mathrm{Re}\,\langle B^*Ax, x\rangle + r^2 + \|B\|^2 - \|Bx\|^2. \tag{3.12}$$

Since

$$\mathrm{Re}\,\langle B^*Ax, x\rangle \le |\langle B^*Ax, x\rangle|, \quad \|Bx\|^2 \ge \frac{1}{\|B^{-1}\|^2}\|x\|^2$$

and $\|Ax\|^2 + \|B\|^2 \ge 2\|B\|\|Ax\|$ for any $x \in H$, hence by (3.12) we get

$$2\|B\|\|Ax\| \le 2|\langle B^*Ax, x\rangle| + r^2 + \frac{\|B\|^2\|B^{-1}\|^2 - 1}{\|B^{-1}\|^2} \tag{3.13}$$

for any $x \in H$, $\|x\| = 1$. Taking the supremum over $x \in H$, $\|x\| = 1$ we deduce the desired result (3.11). □

Remark 6. *If we choose in Theorem 9, $B = \lambda A^*$, $\lambda \neq 0$, A is invertible, then we get the inequality:*

$$(0 \le)\,\|A\|^2 - w\left(A^2\right) \le \frac{1}{2|\lambda|}r^2 + |\lambda| \cdot \frac{\|A\|^2\|A^{-1}\|^2 - 1}{2\|A^{-1}\|^2} \tag{3.14}$$

provided $\|A - \lambda A^*\| \le r$.

The following result may be stated as well.

Theorem 10. *Let $A, B : H \to H$ be two bounded linear operators on H. If B is invertible and for $r > 0$ we have*

$$\|A - B\| \le r < \|B\|, \tag{3.15}$$

then

$$\|A\| \le \frac{1}{\sqrt{\|B\|^2 - r^2}}\left(w(B^*A) + \frac{\|B\|^2\|B^{-1}\|^2 - 1}{2\|B^{-1}\|^2}\right). \tag{3.16}$$

Proof. The first part of condition (3.15) is obviously equivalent to

$$\|Ax\|^2 + \|Bx\|^2 \le 2\,\mathrm{Re}\,\langle Ax, Bx\rangle + r^2$$

for any $x \in H$, with $\|x\| = 1$, which is clearly equivalent to

$$\|Ax\|^2 + \|B\|^2 - r^2 \le 2\,\mathrm{Re}\,\langle B^*Ax, x\rangle + \|B\|^2 - \|Bx\|^2. \tag{3.17}$$

Since

$$\mathrm{Re}\,\langle B^*Ax, x\rangle \le |\langle B^*Ax, x\rangle|, \|Bx\|^2 \ge \frac{1}{\|B^{-1}\|^2}\|x\|^2$$

and, by the second part of (3.15), $\|Ax\|^2 + \|B\|^2 - r^2 \geq 2\sqrt{\|B\|^2 - r^2}\,\|Ax\|$, for any $x \in H$, hence by (3.17) we get

$$2\|Ax\|\sqrt{\|B\|^2 - r^2} \leq 2|\langle B^*Ax, x\rangle| + \frac{\|B\|^2\|B^{-1}\|^2 - 1}{\|B^{-1}\|^2} \tag{3.18}$$

for any $x \in H$, $\|x\| = 1$. Taking the supremum over $x \in H$, $\|x\| = 1$ in (3.18), we deduce the desired inequality (3.16). □

Remark 7. *The above Theorem 10 has some particular cases of interest. For instance, if we choose $B = \lambda I$, with $|\lambda| > r$, then (3.15) is obviously fulfilled and by (3.16) we get*

$$\|A\| \leq \frac{w(A)}{\sqrt{1 - \left(\frac{r}{|\lambda|}\right)^2}}, \tag{3.19}$$

provided $\|A - \lambda I\| \leq r$. This result has been obtained in the earlier paper [1].

On the other hand, if in the above we choose $B = \lambda A^*$ with $\|A\| > \frac{r}{|\lambda|}$ $(\lambda \neq 0)$, and A is invertible, then by (3.16) we get

$$\|A\| \leq \frac{1}{\sqrt{\|A\|^2 - \left(\frac{r}{|\lambda|}\right)^2}}\left[w\left(A^2\right) + |\lambda| \cdot \frac{\|A\|^2\|A^{-1}\|^2 - 1}{2\|A^{-1}\|^2}\right], \tag{3.20}$$

provided $\|A - \lambda A^*\| \leq r$.

The following result may be stated as well.

Theorem 11. *Let A, B and r be as in Theorem 8. Moreover, if*

$$\|B^{-1}\| < \frac{1}{r}, \tag{3.21}$$

then

$$\|A\| \leq \frac{\|B^{-1}\|}{\sqrt{1 - r^2\|B^{-1}\|^2}} w(B^*A). \tag{3.22}$$

Proof. Observe that, by (3.6) we have

$$\|A\|^2 + \frac{1 - r^2\|B^{-1}\|^2}{\|B^{-1}\|^2} \leq 2w(B^*A). \tag{3.23}$$

Utilising the elementary inequality

$$2\frac{\|A\|}{\|B^{-1}\|}\sqrt{1 - r^2\|B^{-1}\|^2} \leq \|A\|^2 + \frac{1 - r^2\|B^{-1}\|^2}{\|B^{-1}\|^2}, \tag{3.24}$$

which can be stated since (3.21) is assumed to be true, hence by (3.23) and (3.24) we deduce the desired result (3.22). □

Remark 8. *If we assume that* $B = \lambda A^*$ *with* $\lambda \neq 0$ *and* A *an invertible operator, then, by applying Theorem* 11, *we get the inequality:*

$$\|A\| \leq \frac{\|A^{-1}\| \, w(A^2) \, |\lambda|}{\sqrt{|\lambda|^2 - r^2 \|A^{-1}\|^2}}, \tag{3.25}$$

provided $\|A - \lambda A^*\| \leq r$ *and* $\|A^{-1}\| < \frac{|\lambda|}{r}$.

The following result may be stated as well.

Theorem 12. *Let* $A, B : H \to H$ *be two bounded linear operators. If* $r > 0$ *and* B *is invertible with the property that* $\|A - B\| \leq r$ *and*

$$\frac{1}{\sqrt{r^2+1}} \leq \|B^{-1}\| < \frac{1}{r}, \tag{3.26}$$

then

$$\|A\|^2 \leq w^2(B^*A) + 2w(B^*A) \cdot \frac{\|B^{-1}\| - \sqrt{1 - r^2 \|B^{-1}\|^2}}{\|B^{-1}\|}. \tag{3.27}$$

Proof. Let $x \in H$, $\|x\| = 1$. Then by (3.5) we have

$$\|Ax\|^2 + \frac{1}{\|B^{-1}\|^2} \leq 2|\langle B^*Ax, x\rangle| + r^2, \tag{3.28}$$

and since $\frac{1}{\|B^{-1}\|^2} - r^2 > 0$, we can conclude that $|\langle B^*Ax, x\rangle| > 0$ for any $x \in H$, $\|x\| = 1$.

Dividing in (3.28) with $|\langle B^*Ax, x\rangle| > 0$, we obtain

$$\frac{\|Ax\|^2}{|\langle B^*Ax, x\rangle|} \leq 2 + \frac{r^2}{|\langle B^*Ax, x\rangle|} - \frac{1}{\|B^{-1}\|^2 |\langle B^*Ax, x\rangle|}. \tag{3.29}$$

Subtracting $|\langle B^*Ax, x\rangle|$ from both sides of (3.29), we get

$$\begin{aligned} &\frac{\|Ax\|^2}{|\langle B^*Ax, x\rangle|} - |\langle B^*Ax, x\rangle| \\ &\leq 2 - |\langle B^*Ax, x\rangle| - \frac{1 - r^2\|B^{-1}\|^2}{|\langle B^*Ax, x\rangle| \, \|B^{-1}\|^2} \\ &= 2 - \frac{2\sqrt{1 - r^2\|B^{-1}\|^2}}{\|B^{-1}\|} - \left(\sqrt{|\langle B^*Ax, x\rangle|} - \frac{\sqrt{1 - r^2\|B^{-1}\|^2}}{\|B^{-1}\| \sqrt{|\langle B^*Ax, x\rangle|}}\right)^2 \\ &\leq 2\left(\frac{\|B^{-1}\| - \sqrt{1 - r^2\|B^{-1}\|^2}}{\|B^{-1}\|}\right), \end{aligned} \tag{3.30}$$

which gives:

$$\|Ax\|^2 \le |\langle B^*Ax, x\rangle|^2 + 2\,|\langle B^*Ax, x\rangle| \, \frac{\|B^{-1}\| - \sqrt{1 - r^2 \|B^{-1}\|^2}}{\|B^{-1}\|}. \tag{3.31}$$

We also remark that, by (3.26) the quantity

$$\|B^{-1}\| - \sqrt{1 - r^2 \|B^{-1}\|^2} \ge 0,$$

hence, on taking the supremum in (3.31) over $x \in H$, $\|x\| = 1$, we deduce the desired inequality. □

Remark 9. *It is interesting to remark that if we assume $\lambda \in \mathbb{C}$ with $0 < r \le |\lambda| \le \sqrt{r^2+1}$ and $\|A - \lambda I\| \le r$, then by (3.2) we can state the following inequality:*

$$\|A\|^2 \le |\lambda|^2 w^2(A) + 2\,|\lambda| \left(1 - \sqrt{|\lambda|^2 - r^2}\right) w(A). \tag{3.32}$$

Also, if $\|A - A^*\| \le r$, A is invertible and $\frac{1}{\sqrt{r^2+1}} \le \|A^{-1}\| < \frac{1}{r}$, then, by (3.27) we also have

$$\|A\|^2 \le w^2(A^2) + 2w(A^2) \cdot \frac{\|A^{-1}\| - \sqrt{1 - r^2 \|A^{-1}\|^2}}{\|A^{-1}\|}. \tag{3.33}$$

One can also prove the following result.

Theorem 13. *Let $A, B : H \to H$ be two bounded linear operators. If $r > 0$ and B is invertible with the property that $\|A - B\| \le r$ and $\|B^{-1}\| < \frac{1}{r}$, then*

$$\begin{aligned} (0 \le)\, &\|A\|^2 \|B\|^2 - w^2(B^*A) \\ &\le 2w(B^*A) \cdot \frac{\|B\|}{\|B^{-1}\|} \left(\|B\|\,\|B^{-1}\| - \sqrt{1 - r^2 \|B^{-1}\|^2}\right). \end{aligned} \tag{3.34}$$

Proof. We subtract the quantity $\frac{|\langle B^*Ax, x\rangle|}{\|B\|^2}$ from both sides of (3.29) to obtain

$$\begin{aligned} 0 &\le \frac{\|Ax\|^2}{|\langle B^*Ax, x\rangle|} - \frac{|\langle B^*Ax, x\rangle|}{\|B\|^2} \\ &\le 2 - 2 \cdot \frac{\sqrt{1 - r^2 \|B^{-1}\|^2}}{\|B\|\,\|B^{-1}\|} - \left(\frac{\sqrt{|\langle B^*Ax, x\rangle|}}{\|B\|} - \frac{\sqrt{1 - r^2 \|B^{-1}\|^2}}{\sqrt{|\langle B^*Ax, x\rangle|}\,\|B^{-1}\|}\right)^2 \\ &\le 2 \cdot \frac{\left(\|B\|\,\|B^{-1}\| - \sqrt{1 - r^2 \|B^{-1}\|^2}\right)}{\|B\|\,\|B^{-1}\|}, \end{aligned} \tag{3.35}$$

which is equivalent with

$$\begin{aligned}(0 \leq) \|Ax\|^2 \|B\|^2 - |\langle B^* Ax, x\rangle|^2 \\ \leq 2\frac{\|B\|}{\|B^{-1}\|} |\langle B^* Ax, x\rangle| \left(\|B\| \|B^{-1}\| - \sqrt{1 - r^2 \|B^{-1}\|^2} \right)\end{aligned} \tag{3.36}$$

for any $x \in H$, $\|x\| = 1$.

The inequality (3.36) also shows that $\|B\| \|B^{-1}\| \geq \sqrt{1 - r^2 \|B^{-1}\|^2}$ and then, by (3.36), we get

$$\begin{aligned}\|Ax\|^2 \|B\|^2 \leq |\langle B^* Ax, x\rangle|^2 \\ + 2\frac{\|B\|}{\|B^{-1}\|} |\langle B^* Ax, x\rangle| \left(\|B\| \|B^{-1}\| - \sqrt{1 - r^2 \|B^{-1}\|^2} \right)\end{aligned} \tag{3.37}$$

for any $x \in X$, $\|x\| = 1$. Taking the supremum in (3.37) we deduce the desired inequality (3.34). □

Remark 10. *The above Theorem* 13 *has some particular instances of interest as follows. If, for instance, we choose $B = \lambda I$ with $|\lambda| \geq r > 0$ and $\|A - \lambda I\| \leq r$, then by* (3.34) *we obtain the inequality*

$$(0 \leq) \|A\|^2 - w^2(A) \leq 2|\lambda| w(A) \left(1 - \sqrt{1 - \frac{r^2}{|\lambda|^2}} \right). \tag{3.38}$$

Also, if A is invertible, $\|A - \lambda A^*\| \leq r$ and $\|A^{-1}\| \leq \frac{|\lambda|}{r}$, then by (3.34) *we can state:*

$$\begin{aligned}(0 \leq) \|A\|^4 - w^2(A^2) \\ \leq 2|\lambda| w(A^2) \cdot \frac{\|A\|}{\|A^{-1}\|} \left(\|A\| \|A^{-1}\| - \sqrt{1 - \frac{r^2}{|\lambda|^2} \|A^{-1}\|^2} \right).\end{aligned} \tag{3.39}$$

Acknowledgement

The author would like to thank the anonymous referee for valuable comments that have been implemented in the final version of this paper.

References

[1] S.S. Dragomir, *Reverse inequalities for the numerical radius of linear operators in Hilbert spaces*, Bull. Austral. Math. Soc. **73** (2006), 255–262.

[2] S.S. Dragomir, *Some inequalities for the Euclidean operator radius of two operators in Hilbert spaces*, Linear Algebra Appl. **419** (2006), 256–264.

[3] S.S. Dragomir, *Inequalities for the norm and the numerical radius of linear operators in Hilbert spaces*, Demonstratio Math. **40**(2007), 411–417.

[4] S.S. Dragomir and J. Sándor, *Some inequalities in prehilbertian spaces*, Studia Univ. "Babeş-Bolyai" – Mathematica **32**(1) (1987), 71–78.

[5] A. Goldstein, J.V. Ryff and L.E. Clarke, *Problem 5473*, Amer. Math. Monthly **75**(3) (1968), 309.

[6] K.E. Gustafson and D.K.M. Rao, *Numerical Range*, Springer-Verlag, New York, Inc., 1997.

[7] P.R. Halmos, *Introduction to Hilbert Space and the Theory of Spectral Multiplicity*, Chelsea Pub. Comp, New York, N.Y., 1972.

Silvestru S. Dragomir
School of Computer Science and Mathematics
Victoria University, PO Box 14428
Melbourne City, Victoria 8001, Australia
URL: http://rgmia.vu.edu.au/dragomir
e-mail: sever.dragomir@vu.edu.au

International Series of Numerical Mathematics, Vol. 157, 147–154

Norm Inequalities for Commutators of Normal Operators

Fuad Kittaneh

Abstract. Let S, T, and X be bounded linear operators on a Hilbert space. It is shown that if S and T are normal with the Cartesian decompositions $S = A+iC$ and $T = B+iD$ such that $a_1 \leq A \leq a_2$, $b_1 \leq B \leq b_2$, $c_1 \leq C \leq c_2$, and $d_1 \leq D \leq d_2$ for some real numbers a_1, a_2, b_1, b_2, c_1, c_2, d_1, and d_2, then for every unitarily invariant norm $|||\cdot|||$,

$$|||SX - XS||| \leq \sqrt{(a_2 - a_1)^2 + (c_2 - c_1)^2}\, |||X|||$$

and

$$\|ST - TS\| \leq \frac{1}{2}\sqrt{(a_2 - a_1)^2 + (c_2 - c_1)^2}\sqrt{(b_2 - b_1)^2 + (d_2 - d_1)^2},$$

where $\|\cdot\|$ is the usual operator norm. Applications of these norm inequalities are given, and generalizations of these inequalities to a larger class of nonnormal operators are also obtained.

Mathematics Subject Classification (2000). 47A30, 47B15, 47B47.

Keywords. Commutator, normal operator, positive operator, unitarily invariant norm, norm inequality, Cartesian decomposition.

1. Introduction

The commutator of two bounded linear operators S and T acting on a Hilbert space $\mathcal{H}$ is the operator $ST - TS$. For the usual operator norm $\|\cdot\|$ and for any two operators S and T, we have

$$\|ST - TS\| \leq 2\,\|S\|\,\|T\|\,. \tag{1}$$

If S or T is positive, then

$$\|ST - TS\| \leq \|S\|\,\|T\|\,. \tag{2}$$

Moreover, if S and T are positive, then

$$\|ST - TS\| \leq \frac{1}{2} \|S\| \|T\| . \tag{3}$$

The inequality (1) is an immediate consequence of the triangle inequality and the submultiplicativity of the norm $\|\cdot\|$. More general forms of the improved inequalities (2) and (3), which are the stimulants of this work, have been recently given in [8] and [7], respectively. These inequalities can be also concluded from a general theorem on the norms of derivations in [10]. Related commutator estimates for the Hilbert-Schmidt norm have been given in [4], and related singular value inequalities and unitarily invariant norm inequalities for commutators of positive operators have been recently given in [7]. The connections between norms of commutators, pinchings, and spectral variation have been recently clarified in [3].

It should be mentioned here that the inequalities (1)–(3) are sharp.

For $S = \begin{bmatrix} 1 & 0 \\ 0 & -1 \end{bmatrix}$ and $T = \begin{bmatrix} 0 & 1 \\ 0 & 0 \end{bmatrix}$, the inequality (1) becomes an equality.

For $S = \begin{bmatrix} 1 & 0 \\ 0 & 0 \end{bmatrix}$ and $T = \begin{bmatrix} 0 & 1 \\ 0 & 0 \end{bmatrix}$, the inequality (2) becomes an equality.

For $S = \begin{bmatrix} 1 & 0 \\ 0 & 0 \end{bmatrix}$ and $T = \begin{bmatrix} 1 & 1 \\ 1 & 1 \end{bmatrix}$, the inequality (3) becomes an equality.

In this paper, we give considerable generalizations of the inequalities (2) and (3) to commutators of normal operators. Our analysis in this paper is completely different from and much simpler than those in [7] and [8], and our generalized inequalities are given in terms of the spectral bounds of the Cartesian parts of operators. The generalized versions of (2) are extended to the wider class of unitarily invariant norms.

Recall that if S is an operator, then the Cartesian decomposition of S is $S = A + iC$, where A and C are the self-adjoint operators $A = \text{Re } S$ and $C = \text{Im } S$. Note that $\|S\|^2 = \|S^*S\|$, and that if S is normal, then $S^*S = A^2 + C^2$ and $S - z$ is normal for all complex numbers z.

We will make a repeated use of the triangle inequality, the submultiplicativity of the norm $\|\cdot\|$, and the fact that for a self-adjoint operator A and for a positive real number a, $\|A\| \leq a$ if and only if $-a \leq A \leq a$. This is also equivalent to the condition that $\sigma(A) \subseteq [-a, a]$, where $\sigma(A)$ is the spectrum of A. Another fact that will be repeatedly used asserts that if S, T, and X are operators such that X belongs to a norm ideal associated with a unitarily invariant (or symmetric) norm $|||\cdot|||$, then $|||SXT||| \leq \|S\| \|T\| |||X|||$ (see, e.g., [1, p. 94] or [5, p. 79]). Recall that the usual operator norm and the Schatten p-norms (in particular, the Hilbert-Schmidt norm) are unitarily invariant.

In the sequel, the symbol $|||\cdot|||$ denotes any unitarily invariant norm, and for the sake of brevity, we will make no explicit mention of this norm. Thus, when we talk of $|||X|||$, we are assuming that the operator X belongs to the norm ideal associated with $|||\cdot|||$.

2. Main results

Our first main result can be stated as follows.

Theorem 1. *Let S be a normal operator with the Cartesian decomposition $S = A + iC$ such that $a_1 \le A \le a_2$ and $c_1 \le C \le c_2$ for some real numbers a_1, a_2, c_1, and c_2. Then, for every operator X,*

$$|||SX - XS||| \le \sqrt{(a_2 - a_1)^2 + (c_2 - c_1)^2}\, |||X||| . \tag{4}$$

Proof. Let $a = \frac{a_1+a_2}{2}$, $c = \frac{c_1+c_2}{2}$, and $z = a + ic$. Then

$$\begin{aligned} |||SX - XS||| &= |||(S - z)X - X(S - z)||| \\ &\le 2\, \|S - z\|\, |||X||| . \end{aligned} \tag{5}$$

But

$$\begin{aligned} \|S - z\|^2 &= \|A - a + i(C - c)\|^2 \\ &= \left\|(A - a)^2 + (C - c)^2\right\| \text{ (by the normality of } S - z) \\ &\le \|A - a\|^2 + \|C - c\|^2 . \end{aligned}$$

Since $-\left(\frac{a_2-a_1}{2}\right) \le A - a \le \frac{a_2-a_1}{2}$ and $-\left(\frac{c_2-c_1}{2}\right) \le C - c \le \frac{c_2-c_1}{2}$, it follows that $\|A - a\| \le \frac{a_2-a_1}{2}$ and $\|C - c\| \le \frac{c_2-c_1}{2}$, and so

$$\|S - z\|^2 \le \left(\frac{a_2 - a_1}{2}\right)^2 + \left(\frac{c_2 - c_1}{2}\right)^2 . \tag{6}$$

Now, it follows from the inequalities (5) and (6) that

$$|||SX - XS||| \le \sqrt{(a_2 - a_1)^2 + (c_2 - c_1)^2}\, |||X||| ,$$

as required. □

A generalized commutator version of Theorem 1 is given in the following result.

Theorem 2. *Let S and T be normal operators with the Cartesian decompositions $S = A + iC$ and $T = B + iD$ such that $a_1 \le A \le a_2$, $b_1 \le B \le b_2$, $c_1 \le C \le c_2$, and $d_1 \le D \le d_2$ for some real numbers a_1, a_2, b_1, b_2, c_1, c_2, d_1, and d_2. Then, for every operator X,*

$$\begin{gathered} |||SX - XT||| \le \\ \sqrt{(\max(a_2, b_2) - \min(a_1, b_1))^2 + (\max(c_2, d_2) - \min(c_1, d_1))^2}\, |||X||| . \end{gathered} \tag{7}$$

Proof. Let $R = \begin{bmatrix} S & 0 \\ 0 & T \end{bmatrix}$ and $Y = \begin{bmatrix} 0 & X \\ 0 & 0 \end{bmatrix}$. Then, as operators on $\mathcal{H} \oplus \mathcal{H}$, R is normal, $\min(a_1, b_1) \le \operatorname{Re} R \le \max(a_2, b_2)$, $\min(c_1, d_1) \le \operatorname{Im} R \le \max(c_2, d_2)$, $RY - YR = \begin{bmatrix} 0 & SX - XT \\ 0 & 0 \end{bmatrix}$, $|||RY - YR||| = |||(SX - XT) \oplus 0|||$, and $|||Y||| = |||X \oplus 0|||$. Applying Theorem 1 to the operators R and Y, and utilizing the Fan dominance principle (see, e.g., [1, p. 93] or [5, p. 82]), we obtain the desired inequality (7). □

In the following result, we obtain a norm inequality for commutators of normal operators.

Theorem 3. *Let S and T be normal operators with the Cartesian decompositions $S = A + iC$ and $T = B + iD$ such that $a_1 \le A \le a_2$, $b_1 \le B \le b_2$, $c_1 \le C \le c_2$, and $d_1 \le D \le d_2$ for some real numbers a_1, a_2, b_1, b_2, c_1, c_2, d_1, and d_2. Then*

$$\|ST - TS\| \le \frac{1}{2}\sqrt{(a_2 - a_1)^2 + (c_2 - c_1)^2}\sqrt{(b_2 - b_1)^2 + (d_2 - d_1)^2}. \tag{8}$$

Proof. Let $a = \frac{a_1+a_2}{2}$, $b = \frac{b_1+b_2}{2}$, $c = \frac{c_1+c_2}{2}$, $d = \frac{d_1+d_2}{2}$, $z = a + ic$, and $w = b + id$. Then

$$\begin{aligned} \|ST - TS\| &= \|(S - z)(T - w) - (T - w)(S - z)\| \\ &\le 2\,\|S - z\|\,\|T - w\|\,. \end{aligned} \tag{9}$$

But, as in the proof of Theorem 1, the normality of S and T implies that

$$\|S - z\| \le \sqrt{\left(\frac{a_2 - a_1}{2}\right)^2 + \left(\frac{c_2 - c_1}{2}\right)^2} \tag{10}$$

and

$$\|T - w\| \le \sqrt{\left(\frac{b_2 - b_1}{2}\right)^2 + \left(\frac{d_2 - d_1}{2}\right)^2}. \tag{11}$$

The desired inequality (8) now follows from the inequalities (9)–(11). □

For general (i.e., not necessarily normal) operators S and T, we have the following weaker estimate.

Theorem 4. *Let S and T be operators with the Cartesian decompositions $S = A + iC$ and $T = B + iD$ such that $a_1 \le A \le a_2$, $b_1 \le B \le b_2$, $c_1 \le C \le c_2$, and $d_1 \le D \le d_2$ for some real numbers a_1, a_2, b_1, b_2, c_1, c_2, d_1, and d_2. Then*

$$\|ST - TS\| \le \frac{1}{2}((a_2 + c_2) - (a_1 + c_1))((b_2 + d_2) - (b_1 + d_1)). \tag{12}$$

Proof. As in the proof of Theorem 3, we have

$$\begin{aligned} \|ST - TS\| &= \|(S - z)(T - w) - (T - w)(S - z)\| \\ &\le 2\,\|S - z\|\,\|T - w\| \\ &= 2\,\|A - a + i(C - c)\|\,\|B - b + i(D - d)\| \\ &\le 2\,(\|A - a\| + \|C - c\|)\,(\|B - b\| + \|D - d\|) \\ &\le 2\left(\frac{a_2 - a_1}{2} + \frac{c_2 - c_1}{2}\right)\left(\frac{b_2 - b_1}{2} + \frac{d_2 - d_1}{2}\right) \\ &= \frac{1}{2}((a_2 + c_2) - (a_1 + c_1))((b_2 + d_2) - (b_1 + d_1)), \end{aligned}$$

as required. □

3. Applications

As applications of Theorem 1, we have the following corollaries, which include generalizations of the inequality (2).

Corollary 1. *Let S be a normal operator with the Cartesian decomposition $S = A + iC$ such that A is positive. Then, for every operator X,*

$$|||SX - XS||| \leq \sqrt{\|A\|^2 + 4\|C\|^2}\, |||X|||. \tag{13}$$

Proof. The inequality (13) follows from the inequality (4) by letting $a_1 = 0$, $c_1 = -\|C\|$, $a_2 = \|A\|$, and $c_2 = \|C\|$. □

Remark 1. *In Corollary 1, if instead of the assumption that A is positive, we assume that C is positive, then we obtain the inequality*

$$|||SX - XS||| \leq \sqrt{4\|A\|^2 + \|C\|^2}\, |||X|||. \tag{14}$$

Corollary 2. *Let S be a normal operator with the Cartesian decomposition $S = A + iC$ such that A and C are positive. Then, for every operator X,*

$$|||SX - XS||| \leq \sqrt{\|A\|^2 + \|C\|^2}\, |||X|||. \tag{15}$$

Proof. The inequality (15) follows from the inequality (4) by letting $a_1 = c_1 = 0$, $a_2 = \|A\|$, and $c_2 = \|C\|$. □

Using arguments similar to those used in the proofs of Corollaries 1 and 2, we obtain the following norm inequalities for commutators of normal operators as consequences of Theorem 3. These inequalities include generalizations of the inequality (3).

Corollary 3. *Let S and T be normal operators with the Cartesian decompositions $S = A + iC$ and $T = B + iD$ such that A and B are positive. Then*

$$\|ST - TS\| \leq \frac{1}{2}\sqrt{\|A\|^2 + 4\|C\|^2}\sqrt{\|B\|^2 + 4\|D\|^2}. \tag{16}$$

Remark 2. *In Corollary 3, if instead of the assumption that A and B are positive, we assume that C and D are positive, then we obtain the inequality*

$$\|ST - TS\| \leq \frac{1}{2}\sqrt{4\|A\|^2 + \|C\|^2}\sqrt{4\|B\|^2 + \|D\|^2}. \tag{17}$$

Corollary 4. *Let S and T be normal operators with the Cartesian decompositions $S = A + iC$ and $T = B + iD$ such that A, B, C, and D are positive. Then*

$$\|ST - TS\| \leq \frac{1}{2}\sqrt{\|A\|^2 + \|C\|^2}\sqrt{\|B\|^2 + \|D\|^2}. \tag{18}$$

For a positive invertible operator A, we have $\|A^{-1}\|^{-1} \leq A \leq \|A\|$. Using this fact, the reasoning above can be invoked to establish improvements of the inequalities (2) and (3) for positive invertible operators, and improvements of the inequalities (13)–(18) for normal operators with positive invertible Cartesian parts.

Based on Theorem 2, we obtain natural generalizations and improvements of norm inequalities concerning differences of positive invertible operators in [6]. We leave the details to the interested reader.

4. Generalizations

Using an analysis employed in [10], it has been recently shown in [3] that if S and T are normal operators, then for every operator X,

$$|||SX - XS||| \leq c(S)\, |||X||| \tag{19}$$

and

$$\|ST - TS\| \leq \frac{1}{2} c(S)c(T), \tag{20}$$

where $c(S) = 2\inf_{z\in\mathbb{C}} \|S - z\|$ is the diameter of the smallest disk in the complex plane containing $\sigma(S)$.

Based on the spectral theorem for normal operators, one can easily show that if S is normal, then $\sigma(\operatorname{Re} S) = \operatorname{Re}\sigma(S)$ and $\sigma(\operatorname{Im} S) = \operatorname{Im}\sigma(S)$. Consequently, if a_1, a_2, c_1, and c_2 are as in Theorem 1, then

$$c(S) \leq \sqrt{(a_2 - a_1)^2 + (c_2 - c_1)^2}. \tag{21}$$

Thus, in view of the inequality (21), the inequalities (4) and (8) can be also concluded from the inequalities (19) and (20), respectively. However, in addition to the simple proofs of the inequalities (4) and (8) given in Section 2, the bounds obtained there are much more easily computable and they seem natural enough and applicable to be widely useful. Moreover, our methods used in Section 2 enable us to generalize Theorems 1 and 3 to a larger class of nonnormal operators.

It is evident that if S is a normal operator, then $\|S^2\| = \|S\|^2$. However, the converse is not true if $\dim\mathcal{H} > 2$. To see this, consider the three-dimensional example $S = \begin{bmatrix} 0 & 0 & 0 \\ 1 & 0 & 0 \\ 0 & 1 & 0 \end{bmatrix}$. Then $\|S^2\| = \|S\|^2 = 1$, but S is not normal. Tedious computations show that if $\dim\mathcal{H} = 2$, then an operator S is normal if and only if $\|S^2\| = \|S\|^2$.

If S is an operator with the Cartesian decomposition $S = A + iC$, then $S^*S + SS^* = 2(A^2 + C^2)$. So, by the fact that $\|S^*S\| = \|SS^*\| = \|S\|^2$ and the triangle inequality, we have

$$\|A^2 + C^2\| \leq \|S\|^2. \tag{22}$$

Also, by the fact that $\|S^2\| = \|\,|S|\,|S^*|\,\|$ and the arithmetic-geometric mean inequality for positive operators (see, e.g., [2] or [9]), we have

$$\begin{aligned} \|S^2\| &= \|\,|S|\,|S^*|\,\| \leq \frac{1}{2}\left\| |S|^2 + |S^*|^2 \right\| \\ &= \frac{1}{2}\|\, S^*S + SS^*\,\| = \|A^2 + C^2\|. \end{aligned} \tag{23}$$

Combining the inequalities (22) and (23), we conclude that if S is an operator with the Cartesian decomposition $S = A + iC$, then

$$\left\|S^2\right\| \le \left\|A^2 + C^2\right\| \le \|S\|^2 . \tag{24}$$

Thus, if $\left\|S^2\right\| = \|S\|^2$, then

$$\left\|S^2\right\| = \left\|A^2 + C^2\right\| . \tag{25}$$

This assertion allows us to replace the normality conditions on the operators S and T in Theorems 1 and 3 by the weaker conditions that $\left\|(S-z)^2\right\| = \|S-z\|^2$ and $\left\|(T-w)^2\right\| = \|T-w\|^2$, where z and w are as in Theorems 1 and 3. These generalizations of Theorems 1 and 3 yield analogous generalizations of Corollaries 1–4.

Finally, we remark that our commutator inequalities presented in this paper are sharp. Moreover, except for the inequality (12), these inequalities do not hold for general operators S and T. The example

$$S = \begin{bmatrix} a_1 + ic_1 & 0 \\ 0 & a_2 + ic_2 \end{bmatrix} \quad \text{and} \quad X = \begin{bmatrix} 0 & 1 \\ 0 & 0 \end{bmatrix}$$

shows that the inequality (4) is sharp, and the example

$$S = \begin{bmatrix} 0 & 2i \\ 0 & 0 \end{bmatrix} \quad \text{and} \quad X = \begin{bmatrix} 1 & 0 \\ 0 & -1 \end{bmatrix}$$

shows that it is not possible to drop the condition that $\left\|(S-z)^2\right\| = \|S-z\|^2$ in the generalized version of Theorem 1. Analogous two-dimensional examples can be constructed to demonstrate that the other commutator inequalities are sharp, and that the assumptions that $\left\|(S-z)^2\right\| = \|S-z\|^2$ and $\left\|(T-w)^2\right\| = \|T-w\|^2$ in the generalized versions of Theorems 1 and 3 are indispensable.

References

[1] R. Bhatia, *Matrix Analysis*, Springer-Verlag, New York, 1997.

[2] R. Bhatia and F. Kittaneh, *On the singular values of a product of operators*, SIAM J. Matrix Anal. Appl. **11** (1990), 272–277.

[3] R. Bhatia and F. Kittaneh, *Commutators, pinchings, and spectral variation*, Oper. Matrices **2** (2008), 143–151.

[4] A. Böttcher and D. Wenzel, *How big can the commutator of two matrices be and how big is it typically?* Linear Algebra Appl. **403** (2005), 216–228.

[5] I.C. Gohberg and M.G. Krein, *Introduction to the Theory of Linear Nonselfadjoint Operators*, Transl. Math. Monographs **18**, Amer. Math. Soc., Providence, RI, 1969.

[6] F. Kittaneh, *Norm inequalities for sums and differences of positive operators*, Linear Algebra Appl. **383** (2004), 85–91.

[7] F. Kittaneh, *Inequalities for commutators of positive operators*, J. Funct. Anal. **250** (2007), 132–143.

[8] F. Kittaneh, *Norm inequalities for commutators of positive operators and applications*, Math. Z. **258** (2008), 845–849.

[9] A. McIntosh, *Heinz inequalities and perturbation of spectral families*, Macquarie Mathematical Reports **79-006** (1979).

[10] J.G. Stampfli, *The norm of a derivation*, Pacific J. Math. **33** (1970), 737–747.

Fuad Kittaneh
Department of Mathematics
University of Jordan
Amman, Jordan
e-mail: fkitt@ju.edu.jo

International Series of Numerical Mathematics, Vol. 157, 155–166

Uniformly Continuous Superposition Operators in the Spaces of Differentiable Functions and Absolutely Continuous Functions

Janusz Matkowski

Abstract. Let $I, J \subset \mathbb{R}$ be intervals. We prove that if a superposition operator H generated by a two place $h : I \times J \to \mathbb{R}$,

$$H(\varphi)(x) := h(x, \varphi(x)),$$

maps the set $C^r(I, J)$ of all r-times continuously differentiable functions $\varphi : I \to J$ into the Banach space $C^r(I, \mathbb{R})$ and is uniformly continuous with respect to C^r-norm, then

$$h(x, y) = a(x)y + b(x), \qquad x \in I, \ y \in J,$$

for some $a, b \in C^r(I, \mathbb{R})$.

For the Banach space of absolutely continuous functions an analogous result is proved.

Mathematics Subject Classification (2000). Primary 47H30; Secondary 39B22.

Keywords. Superposition operator, Nemytskij operator, Lipschitzian operator, uniformly continuous operator, Banach space of r-times continuously differentiable functions, Banach space of absolutely continuous functions.

1. Introduction

Let $I, J \subset \mathbb{R}$ be intervals. By J^I denote the set of all functions $\varphi : I \to J$. For a given function $h : I \times J \to \mathbb{R}$, the mapping $H : J^I \to \mathbb{R}^I$ defined by

$$H(\varphi)(x) := h(x, \varphi(x)), \qquad \varphi \in J^I,$$

is called a superposition (composition or Nemytskij) operator of a generator h.

The superposition operators play important role in the theory of differential equations, integral equations and functional equations. It is known that every locally defined operator mapping the set $C(I, J)$ of continuous functions $\varphi : I \to J$ into $C(I, \mathbb{R})$ must be a superposition operator. Moreover H maps $C(I, J)$ into

$C(I,\mathbb{R})$ if, and only if, its generator h is continuous (Krasnoselskij). At this background it is surprising enough that there are discontinuous functions $h : I\times\mathbb{R} \to \mathbb{R}$ generating the superpositions operators H which map the space of continuously differentiable functions $C^1(I,\mathbb{R})$ into itself (cf. [2], p. 209).

Let $\mathcal{F}(I,\mathbb{R}) \subset \mathbb{R}^I$ be a function Banach space with a norm $\|\cdot\|_{\mathcal{F}}$. The fixed point methods applied in examination of the existence and uniqueness of a solution $\varphi \in \mathcal{F}(I,\mathbb{R})$ of the functional equation

$$\varphi(x) = h(x, \varphi(f(x))),$$

where f and h are he given functions (which strongly depends on the space $\mathcal{F}(I,\mathbb{R})$ cf. M. Kuczma [3]) in a natural way lead to the question: when is the superposition operator H Lipschitz continuous with respect to the norm $\|\cdot\|_{\mathcal{F}}$? The same question arises in connection with the problem of existence and uniqueness of the implicit function in the class $\mathcal{F}(I,\mathbb{R})$.

In [5] it has been proved that if a superposition operator maps the Banach space $\mathrm{Lip}(I,\mathbb{R})$ into itself and is globally Lipschitzian with respect to Lip-norm, that is, there is a $c \geq 0$ such that

$$\|H(\varphi) - H(\psi)\|_{\mathrm{Lip}} \leq c\,\|\varphi - \psi\|_{\mathrm{Lip}}\,, \qquad \varphi,\psi \in \mathrm{Lip}(I,\mathbb{R}),$$

then its generator h must be of the form

$$h(x,y) = a(x)y + b(x), \qquad x \in I,\ y \in \mathbb{R},$$

where $a, b \in \mathrm{Lip}(I,\mathbb{R})$. Then this result has been extended to some other function Banach spaces (cf. [5]–[7], cf. also J. Appell & P.P. Zabrejko [2]).

Given a positive integer number r, denote by $C^r(I,\mathbb{R})$ the Banach space of all r-times continuously differentiable functions $\varphi : I \to R$ with the norm $\|\cdot\|_r$. In Section 2 of the present paper, using the theory of Jensen functional equation, we prove the following (Theorem 1): *if the operator H mapping the set $C^r(I,J)$ into $C^r(I,R)$ satisfies the inequality*

$$\|H(\varphi) - H(\psi)\|_{C^r} \leq \gamma\left(\|\varphi - \psi\|_{C^r}\right), \qquad \varphi,\psi \in C^r(I,J),$$

for a function $\gamma : [0,\infty) \to [0,\infty)$ that is continuous at 0 and such that $\gamma(0) = 0$, then its generator h must be of the form

$$h(x,y) = a(x)y + b(x), \qquad x \in I,\ y \in J. \tag{$*$}$$

As a conclusion we obtain the main result of this section (Theorem 2) stating that the generator h of any superposition operator mapping $C^r(I,J)$ into $C^r(I,\mathbb{R})$ and uniformly continuous with respect to the norm $\|\cdot\|_r$ must of the form $(*)$.

Moreover, in the case $J = \mathbb{R}$ the assumptions in both results can be significantly weakened (cf. Proposition 1 and Corollary 1).

Denote by $AC(I,\mathbb{R})$ the Banach space of all absolutely continuous functions $\varphi : I \to \mathbb{R}$ and by $AC(I,J)$ the set of all functions $\varphi \in AC(I,\mathbb{R})$ such that $\varphi(I) \subset J$. Assume that the operator H maps $AC(I,J)$ into the space $AC(I,\mathbb{R})$.

In Section 3 we prove (Theorem 3): if the operator H satisfies the inequality

$$\|H(\varphi) - H(\psi)\|_{AC} \leq \gamma\left(\|\varphi - \psi\|_{AC}\right), \qquad \varphi, \psi \in AC(I, J),$$

where a function $\gamma : [0, \infty) \to [0, \infty)$ is continuous at 0 and such that $\gamma(0) = 0$, then its generator h must be of the form $(*)$ with a, b in $AC(I, \mathbb{R})$. Taking $\gamma(t) = ct$ for some $c \geq 0$ we obtain a result of [6].

Applying Theorem 3 we conclude that the generator h of any superposition operator mapping $AC(I, J)$ into $AC(I, \mathbb{R})$ and uniformly continuous with respect to the norm $\|\cdot\|_{AC}$ must of the form $(*)$ with a, b in $AC(I, \mathbb{R})$ (Theorem 4).

2. Uniformly continuous superposition operators in the space of r-times continuously differentiable functions

Let $x_0 \in I$ be fixed. By $C^r(I, \mathbb{R})$ we denote the Banach space of all r-times continuously differentiable functions $\varphi : I \to \mathbb{R}$ with the norm

$$\|\varphi\|_r := \sum_{i=0}^{r-1} \left|\varphi^{(i)}(x_0)\right| + \sup_{x \in I} \left|\varphi^{(r)}(x)\right|.$$

For an interval $J \subset \mathbb{R}$ we put

$$C^r(I, J) := \{\varphi \in C^r(I, \mathbb{R}) : \varphi(I) \subset J\}.$$

Theorem 1. *Let $I, J \subset \mathbb{R}$ be intervals and let $h : I \times J \to \mathbb{R}$. Suppose that $\gamma : [0, \infty) \to [0, \infty)$ is continuous at 0 and $\gamma(0) = 0$. If the superposition operator H of the generator h maps the set $C^r(I, J)$ into the Banach space $C^r(I, \mathbb{R})$ and satisfies the inequality*

$$\|H(\varphi) - H(\psi)\|_r \leq \gamma\left(\|\varphi - \psi\|_r\right), \qquad \varphi, \psi \in C^r(I, J), \tag{1}$$

then there are $a, b \in C^r(I, \mathbb{R})$ such that

$$h(x, y) = a(x)y + b(x), \qquad x \in I,\ y \in J.$$

Proof. Without any loss of generality we can assume that $I = [0, 1]$ and that $x_0 = 0$. Note that for arbitrary $y \in J$ the constant function $\varphi(t) = y$, $(t \in I)$, belongs to $C^r(I, J)$. Since H maps $C^r(I, J)$ into $C^r(I, \mathbb{R})$, the function $H(\varphi) = h(\cdot, y) \in C^r(I, \mathbb{R})$ and, consequently, h is continuous with respect to the first variable.

For arbitrarily fixed $y, \bar{y} \in J$ take $\varphi, \psi : I \to J$ defined by

$$\varphi(t) = y, \qquad \psi(t) = \bar{y}, \qquad t \in I.$$

Then, of course, $\varphi, \psi \in C^r(I, J)$ and, by the assumption, the functions $H(\varphi) = h(\cdot, y)$, $H(\psi) = h(\cdot, \bar{y})$ belong to $C^r(I, \mathbb{R})$ and

$$\|\varphi - \psi\|_r = |y - \bar{y}|.$$

Hence, applying the Lagrange mean-value theorem, the definition of the norm $\|\cdot\|_r$ and (1) we get, for all $x \in [0,1]$,

$$\begin{aligned}|h(x,y) - h(x,\bar{y})| &\le |h(0,y) - h(0,\bar{y})| + |h(x,y) - h(x,\bar{y}) - h(0,y) - h(0,\bar{y})| \\ &= |H(\varphi)(0) - H(\psi)(0)| \\ &\quad + |[H(\varphi)(x) - H(\psi)(x)] - [H(\varphi)(0) - H(\psi)(0)]| \\ &= |[H(\varphi) - H(\psi)](0)| + |[H(\varphi) - H(\psi)]'(c)| \\ &\le \|[H(\varphi) - H(\psi)]\|_r \le \gamma\left(\|\varphi - \psi\|_r\right) \le \gamma\left(|y - \bar{y}|\right).\end{aligned}$$

This inequality, the continuity of γ at 0 and the equality $\gamma(0) = 0$ imply that h is continuous with respect to the second variable.

Take arbitrary $x, \bar{x} \in I$, $x < \bar{x}$; and y_1, y_2, $\bar{y}_1$, $\bar{y}_2 \in J$. Let $\alpha : [0,1] \to \mathbb{R}$ be defined by

$$\alpha(t) := \begin{cases} 0 & \text{for } \ t < x \\ 2\frac{t-x}{\bar{x}-x} & \text{for } \ x \le t \le \frac{x+\bar{x}}{2} \\ 2\frac{\bar{x}-t}{\bar{x}-x} & \text{for } \ \frac{x+\bar{x}}{2} \le t \le \bar{x} \\ 0 & \text{for } \ t > \bar{x}. \end{cases}$$

Of course α is continuous and $\int_0^1 \alpha(s)ds = \int_x^{\bar{x}} \alpha(s)ds = 1$. Now define the functions $\varphi_1, \varphi_2 : [0,1] \to \mathbb{R}$ by

$$\varphi_i(t) := y_i + \frac{1}{(r-1)!}\frac{\bar{y}_i - y_i}{\bar{x} - x}\int_0^t (t-s)^{r-1}\alpha(s)ds, \qquad \text{for } \quad t \in [0,1], \ \ i = 1,2.$$

Note that $\varphi_i \in C^r(I,J)$ and

$$\varphi_i^{(k)}(t) = \frac{1}{(r-k-1)!}\frac{\bar{y}_i - y_i}{\bar{x} - x}\int_0^t (t-s)^{r-k-1}\alpha(s)ds, \quad k = 1, \dots, r-1,$$

$$\varphi_i^{(r)}(t) = \frac{\bar{y}_i - y_i}{\bar{x} - x}\alpha, \qquad\qquad i = 1,2, \ \ (t \in [0,1]).$$

It follows that

$$\varphi_i(0) = y_i, \qquad \varphi_i^{(k)}(0) = 0, \quad k = 1, \dots, r{-}1, \quad \varphi_i^{(r)} = \frac{\bar{y}_i - y_i}{\bar{x} - x}\alpha, \qquad\qquad i = 1,2,$$

whence, by the definitions of α and the norm $\|\cdot\|_r$,

$$\|\varphi_1 - \varphi_2\|_r = |y_1 - y_2| + \frac{|y_1 - y_2 - \bar{y}_1 + \bar{y}_2|}{|x - \bar{x}|}.$$

Since

$$\varphi_i(x) = y_i, \qquad \varphi_i(\bar{x}) = \bar{y}_i, \qquad\qquad i = 1,2,$$

by the Lagrange mean-value theorem, for some $c \in (x, \bar{x})$, we have

$$\begin{aligned}&\frac{|h(x,y_1)-h(x,y_2)-h(\bar{x},\bar{y}_1)+h(\bar{x},\bar{y}_2)|}{|x-\bar{x}|}\\&=\frac{|[H(\varphi_1)-H(\varphi_2](x)-[H(\varphi_1)-H(\varphi_2](\bar{x})|}{|x-\bar{x}|}=|[H(\varphi_1)-H(\varphi_2]'(c)|\\&\le \|H(\varphi_1)-H(\varphi_2)\|_r,\end{aligned}$$

whence, applying inequality (1) with $\varphi = \varphi_1$, $\psi = \varphi_2$, we obtain

$$\begin{aligned}&\frac{|h(x,y_1)-h(x,y_2)-h(\bar{x},\bar{y}_1)+h(\bar{x},\bar{y}_2)|}{|x-\bar{x}|}\\&\qquad\le \gamma\left(|y_1-y_2|+\frac{|y_1-y_2-\bar{y}_1+\bar{y}_2|}{|x-\bar{x}|}\right)\end{aligned}$$

for all $x, \bar{x} \in I$, $x < \bar{x}$; $y_1, y_2, \bar{y}_1, \bar{y}_2 \in J$.

Taking arbitrary $u, v \in J$ and setting here

$$y_1 := \frac{u+v}{2}, \qquad y_2 := u, \qquad \bar{y}_1 := v, \qquad \bar{y}_2 := \frac{u+v}{2}$$

we obtain

$$\frac{\left|h(x,\frac{u+v}{2})-h(x,u)-h(\bar{x},v)+h(\bar{x},\frac{u+v}{2})\right|}{|x-\bar{x}|^{\alpha}} \le \gamma\left(\frac{|u-v|}{2}\right)$$

whence

$$\left|h\left(x,\frac{u+v}{2}\right)-h(x,u)-h(\bar{x},v)+h\left(\bar{x},\frac{u+v}{2}\right)\right| \le |x-\bar{x}|^{\alpha}\gamma\left(\frac{|u-v|}{2}\right)$$

for all $x, \bar{x} \in I$, $x < \bar{x}$; $u, v \in J$.

Letting here $\bar{x}$ tend to x and making use of the continuity of h with respect to the first variable, we hence get

$$2h\left(x,\frac{u+v}{2}\right) = h(x,v)+h(x,u), \qquad x \in I;\ u \in J,$$

which proves that, for every fixed $x \in I$, the function $h(x, \cdot)$ satisfies the Jensen functional equation in the interval J. The continuity of h with respect to the second variable implies that (cf. J. Aczél [1], p. 43, Theorem 1, or M. Kuczma [3], p. 315, Theorem 1) for every $x \in I$ there exist $a(x), b(x) \in \mathbb{R}$ such that

$$h(x,y) = a(x)y + b(x), \qquad x \in I,\ y \in \mathbb{R}.$$

If J is a nontrivial interval, there are $y_1, y_2 \in J$, $y_1 \neq y_2$. By assumption the functions $h(\cdot, y_1) = ay_1 + b$ and $h(\cdot, y_2) = ay_2 + b$ belong to $\mathrm{Lip}^{\alpha}(I, \mathbb{R})$. It follows that $a, b \in \mathrm{Lip}^{\alpha}(I, \mathbb{R})$. If J is trivial, the result is obvious. □

Taking $J = \mathbb{R}$ and $\gamma(t) = ct$, $t \in \mathbb{R}$, we obtain the main result of [6].

In the case $J = \mathbb{R}$ we have the following stronger result.

Proposition 1. *Let $I \subset \mathbb{R}$ be an interval, $h : I \times \mathbb{R} \to \mathbb{R}$, and $\gamma : [0, \infty) \to [0, \infty)$ be continuous at 0 and such that $\gamma(0) = 0$. Denote by $\mathcal{A}$ be the set of all functions $\varphi : I \to \mathbb{R}$ of the form*

$$\varphi(t) = \alpha t + \beta, \qquad t \in I.$$

Suppose the superposition operator H of the generator h maps the set $\mathcal{A}$ into the Banach space $C^r(I, \mathbb{R})$. If H satisfies the inequality

$$\|H(\varphi) - H(\psi)\|_r \le \gamma \left(\|\varphi - \psi\|_r\right), \qquad \varphi, \psi \in \mathcal{A},$$

then there exist $a, b \in C^r(I, \mathbb{R})$ such that

$$h(x, y) = a(x)y + b(x), \qquad x \in I,\ y \in \mathbb{R}.$$

Proof. Without any loss of generality we may assume that $I = [0, 1]$.

Since all the constant functions in I belong to $\mathcal{A}$, in a similar way as in the previous theorem we can show that h is continuous with respect to both variables. Take arbitrary $x, \bar{x} \in I$, $x < \bar{x}$, p, q, $k, l \in \mathbb{R}$ and

$$\varphi(t) = pt + k, \qquad \psi(t) = qt + l, \qquad t \in I.$$

Of course

$$\|\varphi - \psi\|_r = |k - l| + |p - q| .$$

By the Lagrange mean-value theorem there is $c \in (x, \bar{x})$ such that

$$\begin{aligned} &\frac{|h(x, px + k) - h(x, qx + l) - h(\bar{x}, p\bar{x} + k) + h(\bar{x}, q\bar{x} + l)|}{|x - \bar{x}|} \\ &\quad = \frac{|[H(\varphi) - H(\psi)](x) - [H(\varphi) - H(\psi)](\bar{x})|}{|x - \bar{x}|} = |[H(\varphi) - H(\psi)]'(c)| \\ &\quad \le \|H(\varphi) - H(\psi)\|_r \end{aligned}$$

whence, by the assumption,

$$\begin{aligned} &\frac{|h(x, px + k) - h(x, qx + l) - h(\bar{x}, p\bar{x} + k) + h(\bar{x}, q\bar{x} + l)|}{|x - \bar{x}|} \\ &\quad \le \gamma \left(|k - l| + |p - q|\right), \end{aligned}$$

which can be written in the form

$$\begin{aligned} &|h(x, px + k) - h(x, qx + l) - h(\bar{x}, p\bar{x} + k) + h(\bar{x}, q\bar{x} + l)| \\ &\quad \le |x - \bar{x}| \, \gamma \left(|k - l| + |p - q|\right). \end{aligned}$$

Take arbitrary $u, v \in \mathbb{R}$. Putting here

$$p = q = \frac{u - v}{2(x - \bar{x})}, \qquad k = \frac{(2x - \bar{x})v - \bar{x}u}{2(x - \bar{x})}, \qquad l = \frac{(x - 2\bar{x})u + xv}{2(x - \bar{x})}$$

we obtain

$$\left| h\left(x, \frac{u + v}{2}\right) - h\left(x, u\right) - h\left(\bar{x}, v\right) + h\left(\bar{x}, \frac{u + v}{2}\right) \right| \le |x - \bar{x}| \, \gamma \left(\left|\frac{u - v}{2}\right|\right)$$

for all $x, \bar{x} \in I$, $x \neq \bar{x}$, and $u, v \in \mathbb{R}$. Letting $\bar{x}$ tend to x, by the continuity of h, we hence get

$$2h\left(x, \frac{u+v}{2}\right) = h(x,u) + h(x,v), \qquad x \in I,\ u,v \in \mathbb{R}.$$

Now we can argue similarly as in the proof of Theorem 1. □

The main result of this section reads as follows.

Theorem 2. *Let $I, J \subset \mathbb{R}$ be intervals and let $h : I \times J \to \mathbb{R}$. If the superposition operator H of the generator h maps the class $C^r(I,J)$ into the class $C^r(I,\mathbb{R})$ and is uniformly continuous with respect to the norm $\|\cdot\|_r$, then there are $a, b \in C^r(I,\mathbb{R})$ such that*

$$h(x,y) = a(x)y + b(x), \qquad x \in I,\ y \in J.$$

Proof. The uniform continuity of H implies that the function $\gamma : [0,\infty) \to [0,\infty)$,

$$\gamma(t) := \sup\left\{\|H(\varphi) - H(\psi)\|_r : \|\varphi - \psi\|_r \le t\right\}, \qquad t \ge 0,$$

is correctly defined, continuous at 0, $\gamma(0) = 0$ and, clearly, we have

$$\|H(\varphi) - H(\psi)\|_r \le \gamma\left(\|\varphi - \psi\|_r\right), \qquad \varphi, \psi \in \mathrm{Lip}^\alpha(I,J).$$

Now the result is a consequence of Theorem 1. □

Similarly, applying Proposition 1, we obtain

Corollary 1. *Let $I \subset \mathbb{R}$ be an interval, $h : I \times \mathbb{R} \to \mathbb{R}$, and $\gamma : [0,\infty) \to [0,\infty)$ be continuous at 0 and such that $\gamma(0) = 0$. Denote by $\mathcal{A}$ the set of all functions $\varphi : I \to \mathbb{R}$ of the form*

$$\varphi(t) = \alpha t + \beta, \qquad t \in I.$$

Suppose the superposition operator H of the generator h maps the set $\mathcal{A}$ into the Banach space $C^r(I,\mathbb{R})$. If H is uniformly continuous with respect to the norm $\|\cdot\|_r$ in $\mathcal{A}$ then there exist $a, b \in C^r(I,\mathbb{R})$ such that

$$h(x,y) = a(x)y + b(x), \qquad x \in I,\ y \in \mathbb{R}.$$

3. Uniformly continuous superposition operators in the space absolutely continuous functions

Let $I \subset \mathbb{R}$ be an interval and let $x_0 \in I$. Then the set $AC(I,\mathbb{R})$ of all absolutely continuous functions $\varphi : I \to \mathbb{R}$ with the norm

$$\|\varphi\|_{AC} := |\varphi(x_0)| + \int_I |\varphi'(t)|\, dt$$

is a Banach space. For an interval $J \subset \mathbb{R}$ denote by $AC(I,J)$ the set of all $\varphi \in AC(I,\mathbb{R})$ such that $\varphi(I) \subset J$.

Theorem 3. *Let $I, J \subset \mathbb{R}$ be intervals and $h : I \times J \to \mathbb{R}$. Suppose that $\gamma : [0,\infty) \to [0,\infty)$ is continuous at 0 and $\gamma(0) = 0$. If the superposition operator H of the generator h maps the set $AC(I,J)$ into the Banach space $AC(I,\mathbb{R})$ and satisfies the inequality*

$$\|H(\varphi) - H(\psi)\|_{AC} \leq \gamma\left(\|\varphi - \psi\|_{AC}\right), \qquad \varphi, \psi \in AC(I,J), \tag{2}$$

then there exist $a, b \in AC(I,\mathbb{R})$ such that

$$h(x,y) = a(x)y + b(x), \qquad x \in I,\ y \in \mathbb{R}.$$

Proof. Without any loss of generality we can assume that $I = [0,1]$ and that

$$\|\varphi\|_{AC} := |\varphi(0)| + \int_0^1 |\varphi'(t)|\, dt.$$

For arbitrary $y, \bar{y} \in J$ take $\varphi, \psi : I \to J$ defined by

$$\varphi(t) = y, \qquad \psi(t) = \bar{y}, \qquad t \in I.$$

Then, of course, $\varphi, \psi \in AC(I,J)$ and, by the assumption, $H(\varphi) = h(\cdot, y)$ and $H(\psi) = h(\cdot, \bar{y})$ belong to $AC(I,\mathbb{R})$ and

$$\|\varphi - \psi\|_{AC} = |y - \bar{y}|\,.$$

Hence, making use of (2), we have, for all $x \in I$,

$$\begin{aligned}
|h(x,y) - h(x,\bar{y})| &\leq |h(0,y) - h(0,\bar{y})| + |h(x,y) - h(x,\bar{y}) - h(0,y) - h(0,\bar{y})| \\
&= |h(0,y) - h(0,\bar{y})| + \left| \int_0^x \frac{d}{dt}[h(t,y) - h(t,\bar{y})] dt \right| \\
&\leq |h(0,y) - h(0,\bar{y})| + \int_0^x \left| \frac{d}{dt}[h(t,y) - h(t,\bar{y})] \right| dt \\
&\leq |h(0,y) - h(0,\bar{y})| + \int_0^1 \left| \frac{d}{dt}[h(t,y) - h(t,\bar{y})] \right| dt \\
&= |[H(\varphi) - H(\psi)](0)| + \int_0^1 \left| \frac{d}{dt}[H(\varphi) - H(\psi)](t) \right| dt \\
&= \|H(\varphi) - H(\psi)\|_{AC} \leq \gamma\left(|y - \bar{y}|\right),
\end{aligned}$$

whence, for all $x, \bar{x} \in I$, $y, \bar{y} \in J$,

$$\begin{aligned}
|h(x,y) - h(\bar{x},\bar{y})| &\leq |h(x,y) - h(\bar{x},y)| + |h(\bar{x},y) - h(\bar{x},\bar{y})| \\
&\leq |H(\varphi)(x) - H(\varphi)(\bar{x})| + \gamma\left(|y - \bar{y}|\right),
\end{aligned}$$

which implies the continuity of h in $I \times J$.

Take arbitrary $n \in \mathbb{N}$, $x \in I$, y_1, y_2, $\bar{y}_1$, $\bar{y}_2 \in J$, and a finite sequence $x_1, x_2, \ldots, x_{2n}$ such that

$$0 < x_1 < x_2 < \cdots < x_{2n} < 1.$$

Let $\varphi : I \to J$ be the polygonal function the graph of which is uniquely determined by the vertices

$$(0, y_1),\ (x_1, y_1),\ \ (x_2, y_2), \ldots, (x_{2k-1}, y_1),\ (x_{2k}, y_2), \ldots, (x_{2n}, y_2),\ (1, y_2)$$

and, similarly, let $\psi : I \to J$ be the polygonal function the graph of which is uniquely determined by the vertices

$$(0, \bar{y}_1),\ (x_1, \bar{y}_1),\ \ (x_2, \bar{y}_2), \ldots, (x_{2k-1}, \bar{y}_1),\ (x_{2k}, \bar{y}_2), \ldots, (x_{2n}, \bar{y}_2),\ (1, \bar{y}_2).$$

Clearly, $\varphi, \psi \in AC(I, J)$. Since φ and ψ are constant in the intervals $[0, x_1]$ and $[x_{2n}, 1]$, and affine in each of the intervals $[x_k, x_{k+1}]$, $k = 1, \ldots, 2n-1$, by the definition of the norm$\|\cdot\|_{AC}$, we have

$$\begin{aligned}
\|\varphi - \psi\|_{AC} &= |\varphi(0) - \psi(0)| + \int_0^1 |\varphi'(t) - \psi'(t)|\, dt \\
&= |y_1 - \bar{y}_1| + \sum_{k=1}^{2n-1} \int_{x_k}^{x_{k+1}} |\varphi'(t) - \psi'(t)|\, dt \\
&= |y_1 - \bar{y}_1| + \sum_{k=1}^{2n-1} |y_1 - \bar{y}_1 - y_2 + \bar{y}_2| \\
&= |y_1 - \bar{y}_1| + (2n-1)\, |y_1 - \bar{y}_1 - y_2 + \bar{y}_2|\, .
\end{aligned}$$

Moreover

$$\begin{aligned}
&\|H(\varphi) - H(\psi)\|_{AC} \\
&\quad = |h(0, \varphi(0)) - h(0, \psi(0))| + \int_0^1 \left| \frac{d}{dt} \left[h(t, \varphi(t)) - h(t, \psi(t)) \right] \right| dt \\
&\quad \geq \int_0^1 \left| \frac{d}{dt} \left[h(t, \varphi(t)) - h(t, \psi(t)) \right] \right| dt \\
&\quad = \sum_{k=1}^{2n-1} \int_{x_k}^{x_{k+1}} \left| \frac{d}{dt} \left[h(t, \varphi(t)) - h(t, \psi(t)) \right] \right| dt \\
&\quad \geq \sum_{k=1}^{2n-1} \left| \int_{x_k}^{x_{k+1}} \frac{d}{dt} \left[h(t, \varphi(t)) - h(t, \psi(t)) \right] dt \right| \\
&\quad = \sum_{k=1}^{2n-1} \left| h(x_{k+1}, \varphi(x_{k+1})) - h(x_{k+1}, \psi(x_{k+1})) - h(x_k, \varphi(x_k)) + h(x_k, \psi(x_k)) \right| .
\end{aligned}$$

Note that for each $k \in \{1, 2, \ldots, 2n\}$ either $\varphi(x_k) = y_1$ or $\varphi(x_k) = y_2$ and

$$\varphi(x_k) = y_1 \Longleftrightarrow \varphi(x_{k+1}) = y_2, \qquad k \in \{1, 2, \ldots, 2n-1\}.$$

It is also true if we replace φ by ψ, y_1 by $\bar{y}_1$ and y_2 by $\bar{y}_2$. Therefore, letting x_k tend to x for all $k \in \{1, 2, \ldots, 2n\}$ and making use of the continuity of h in $I \times J$, we hence get

$$\begin{aligned}\|H(\varphi) - H(\psi)\|_{AC} &\geq \sum_{k=1}^{2n-1} |h(x, y_1) - h(x, \bar{y}_1) - h(x, y_2) + h(x, \bar{y}_2)| \\ &= (2n-1)\,|h(x, y_1) - h(x, \bar{y}_1) - h(x, y_2) + h(x, \bar{y}_2)|\,.\end{aligned}$$

Hence, applying inequality (2), we obtain

$$\begin{aligned}&(2n-1)\,|h(x, y_1) - h(x, \bar{y}_1) - h(x, y_2) + h(x, \bar{y}_2)| \\ &\qquad \leq \gamma\,(|y_1 - \bar{y}_1| + (2n-1)\,|y_1 - \bar{y}_1 - y_2 + \bar{y}_2|)\end{aligned}$$

for all $n \in \mathbb{N}$, $x \in I$, $y_1, y_2, \bar{y}_1, \bar{y}_2 \in J$.

Taking $r \in (0, 1)$, $u, v \in J$ and substituting here

$$y_1 := (1-r)u + rv, \qquad y_2 := u, \qquad \bar{y}_1 := v, \qquad \bar{y}_2 := ru + (1-r)v$$

we obtain

$$\begin{aligned}&(2n-1)\,|h(x, (1-r)u + rv) - h(x, u) - h(\bar{x}, v) + h(\bar{x}, ru + (1-r)v)| \\ &\qquad \leq \gamma\,(r\,|u - v|)\end{aligned}$$

for all $x, \bar{x} \in I$, $x < \bar{x}$; $u, v \in J$.

Letting here $\bar{x}$ tend to x and making use of the continuity of h with respect to the first variable, we hence get

$$\begin{aligned}&(2n-1)\,|h(x, (1-r)u + rv) - h(x, u) - h(x, v) + h(x, ru + (1-r)v)| \\ &\qquad \leq \gamma\,(r\,|u - v|)\end{aligned}$$

for all $n \in \mathbb{N}$, $x \in I$; $r \in (0, 1)$, $u, v \in J$. Since $n \in \mathbb{N}$ is arbitrary, it follows that

$$h(x, (1-r)u + rv) - h(x, u) - h(x, v) + h(x, ru + (1-r)v) = 0$$

for all $x \in I$; $r \in (0, 1)$, $u, v \in J$ (which means that, for each $x \in I$, the function $h(x, \cdot)$ is Wright-affine). For $r = \frac{1}{2}$ we hence get

$$2h\left(x, \frac{u+v}{2}\right) = h(x, v) + h(x, u), \qquad x \in I;\ u \in J,$$

that is, for every $x \in I$, the function $h(x, \cdot)$ satisfies the Jensen functional equation. The continuity of h implies that for every $x \in I$ there are $a(x), b(x) \in \mathbb{R}$ such that

$$h(x, y) = a(x)y + b(x), \qquad x \in I,\ y \in J,$$

(cf. J. Aczél [1], p. 43, Theorem 1, or M. Kuczma [3], p. 315, Theorem 1). Since $h(\cdot, c) \in AC(I, \mathbb{R})$ for every $c \in J$, the functions a and b are absolutely continuous. □

Taking $\gamma(t) = kt$ ($t \geq 0$) for some $k \geq 0$ we get a result of [6].

Remark 1. *Theorem 3 remains valid if we replace the norm* $\|\cdot\|_{AC}$ *by the following one*

$$\|\varphi\| := \sup_{x \in I} |\varphi(x)| + \int_I |\varphi'(t)|\, dt.$$

Since

$$\|\varphi\|_{AC} \leq \|\varphi\| \leq 2\,\|\varphi\|_{AC}\,,$$

both these norms are equivalent. It easy to check the following: if $I \subset \mathbb{R}$ *is a compact interval and* $a, b \in AC(I, \mathbb{R})$ *then the superposition operator* H *of the generator* $h(x, y) = a(x)y + b(x)$, $(x \in I,\ y \in \mathbb{R})$ *maps the Banach space* $AC(I, \mathbb{R})$ *into itself, and*

$$\|H(\varphi) - H(\psi)\|_{AC} \leq \|a\|\,\|\varphi - \psi\|_{AC}\,, \qquad \varphi, \psi \in AC(I, \mathbb{R}),$$

that is H *is Lipschitzian.*

Now we can prove the main result of this section which reads as follows.

Theorem 4. *Let* $I, J \subset \mathbb{R}$ *be intervals and* $h : I \times J \to \mathbb{R}$*. Suppose that the superposition operator* H *of the generator* h *maps the set* $AC(I, J)$ *into the Banach space* $AC(I, \mathbb{R})$*. Then* H *is uniformly continuous if, and only if, there exist* $a, b \in AC(I, \mathbb{R})$ *such that*

$$h(x, y) = a(x)y + b(x), \qquad x \in I,\ y \in \mathbb{R}.$$

Proof. Suppose that H is uniformly continuous. Then for every $\varepsilon > 0$ there is a $\delta > 0$ such that for all $\varphi, \psi \in AC(I, J)$,

$$\|\varphi - \psi\|_{AC} \leq \delta \Longrightarrow \|H(\varphi) - H(\psi)\|_{AC} \leq \varepsilon.$$

It follows that the function $\gamma : [0, \infty) \to [0, \infty)$,

$$\gamma(t) := \sup\left\{\|H(\varphi) - H(\psi)\|_{AC} : \|\varphi - \psi\|_{AC} = t\right\}, \qquad t \geq 0,$$

is correctly defined, γ is continuous at 0 and $\gamma(0) = 0$. Since

$$\|H(\varphi) - H(\psi)\|_{AC} \leq \gamma\left(\|\varphi - \psi\|_{AC}\right), \qquad \varphi, \psi \in AC(I, J),$$

the "only if" part of the theorem follows from the previous result. As the "if" part is obvious, the proof is completed. □

Remark 2. *Clearly, without any loss of generality, one can assume that the function* γ *is increasing. It follows that Theorem* 3 *and Theorem* 4 *remain valid on replacing the* AC*-norm by equivalent ones.*

Final Remark

The suitable results for the spaces of Hölder functions and bounded variation functions will be considered in our next paper.

References

[1] J. Aczél, *Lectures on functional equations and their applications*, Academic Press, New York and London, 1966.

[2] J. Appell, P.P. Zabrejko, *Nonlinear Superposition Operators*, Cambridge University Press, Cambridge, 1990.

[3] M. Kuczma, *Functional equations in a single variable*, Monografie Matematyczne **46**, Polish Scientific Publishers, Warszawa, 1968.

[4] M. Kuczma, *An Introduction to the Theory of Functional Equations and Inequalities*, Polish Scientific Publishers and Silesian University, Warszawa – Kraków – Katowice, 1985.

[5] J. Matkowski, *Functional equations and Nemytskij operators*, Funkc. Ekvacioj Ser. Int. **25** (1982), 127–132.

[6] J. Matkowski, *Form of Lipschitz operators of substitution in Banach spaces of differentiable functions*, Zeszyty Nauk. Politch. Lódz. Mat. **17** (1984), 5–10.

[7] J. Matkowski, *Lipschitzian composition operators in some function spaces*, Nonlinear Analysis, Theory, Methods & Applications **30** (1997), 719–726.

Janusz Matkowski
Faculty of Mathematics
Computer Science and Econometry
University of Zielona Góra
Podgórna 50
PL-65246 Zielona Góra, Poland

and

Institute of Mathematics
Silesian University
Bankowa 14
PL-40007 Katowice, Poland
e-mail: J.Matkowski@wmie.uz.zgora.pl

International Series of Numerical Mathematics, Vol. 157, 167–178

Tight Enclosures of Solutions of Linear Systems

Takeshi Ogita and Shin'ichi Oishi

Abstract. This paper is concerned with the problem of verifying the accuracy of an approximate solution of a linear system. A fast method of calculating both lower and upper error bounds of the approximate solution is proposed. By the proposed method, it is possible to obtain the error bounds which are as tight as needed. As a result, it can be verified that the obtained error bounds are of high quality. Numerical results are presented elucidating properties and efficiencies of the proposed verification method.

Mathematics Subject Classification (2000). 65G20, 65G50.

Keywords. Verified numerical computation, linear system, tight enclosure.

1. Introduction

We are concerned with the problem of verifying the accuracy of an approximate solution $\widetilde{x}$ of a linear system

$$Ax = b, \tag{1}$$

where A is a real $n \times n$ matrix and b is a real n-vector. If A is nonsingular, there exists a unique solution $x^* := A^{-1}b$. We aim on verifying the nonsingularity of A and calculating some $\underline{\epsilon}, \overline{\epsilon} \in \mathbb{R}^n$ such that

$$\mathbf{o} \le \underline{\epsilon} \le |x^* - \widetilde{x}| \le \overline{\epsilon} \quad \text{with} \quad \mathbf{o} := (0, \ldots, 0)^T \in \mathbb{R}^n. \tag{2}$$

Here, for a real vector $v = (v_1, \ldots, v_n)^T \in \mathbb{R}^n$, we denote by $|v| = (|v_1|, \ldots, |v_n|)^T \in \mathbb{R}^n$ the nonnegative vector consisting of entrywise absolute values.

A number of fast self-validating algorithms (cf., for example, [6, 8, 14]) have been proposed to verify the nonsingularity of A and to compute $\overline{\epsilon}$ in (2). This paper also considers to compute $\underline{\epsilon}$. If $\underline{\epsilon}_i \approx \overline{\epsilon}_i$, then we can *verify* that the error bounds (and the verification) are of high quality. A geometric image of the inclusions for the exact solution x^* such as (2) can be depicted as in Figure 1.

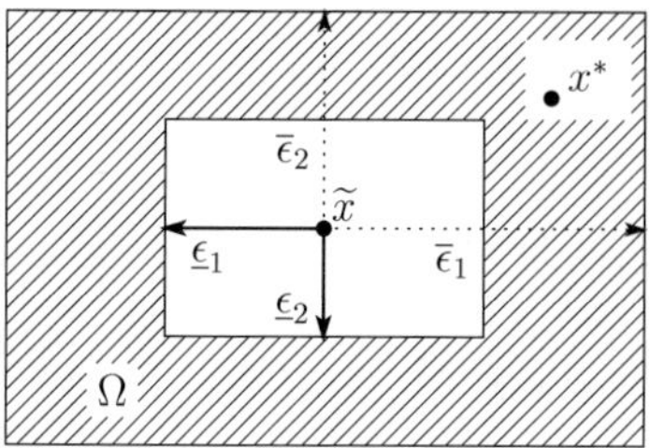

FIGURE 1. Inner and outer enclosures of the exact solution (two-dimensional case). The exact solution x^* exists in Ω.

A main point of this paper is to develop a method of calculating both $\underline{\epsilon}$ and $\overline{\epsilon}$ satisfying (2), which are as tight as we need. For the purpose, the iterative refinement is used. If we obtain tight error bounds, we can set an appropriate criterion for improving an approximate solution $\widetilde{x}$ by the iterative refinement method.

We assume that the floating-point system used in this paper follows IEEE standard 754 for floating-point arithmetic [1]. Moreover, we suppose that all floating-point operations are executed according to the prescribed rounding mode defined in IEEE standard 754. Under such conditions, we will propose a fast algorithm of calculating a verified solution of (1) in terms of (2).

The rest of the paper is organized as follows: In Section 2, we state notations and definitions used in this paper. In Section 3, we analyze the behavior of the iterative refinement. In Section 4, we briefly review previous verification methods of calculating a componentwise error bound of an approximate solution $\widetilde{x}$ of (1). After that, we propose the verification theory for calculating tight error bounds of $\widetilde{x}$. In Section 5, we present a concrete algorithm of calculating tight error bounds of $\widetilde{x}$. In Section 6, some numerical results are presented elucidating properties and efficiencies of the proposed verification method. Finally in Section 7, we conclude the paper.

2. Notation and definitions

Let $\mathbb{R}$ denote the set of real numbers. Let $\mathbb{F}$ be a set of floating-point numbers following IEEE standard 754. Let $\mathbf{u}$ be the unit-roundoff. In IEEE 754 double precision arithmetic, $\mathbf{u} = 2^{-53}$. Throughout this paper, we assume that the operations in $\mathrm{fl}(\cdots)$ is all executed by floating-point arithmetic in given rounding mode (default is round-to-nearest). Let $\mathbb{IR}$ denote the set of interval real numbers and $\mathbb{IF}$ denote a set of interval floating-point numbers. Note that $\mathbb{IF} \subset \mathbb{IR}$. For a real matrix $A = (a_{ij}) \in \mathbb{R}^{n\times n}$, we denote by $|A| = (|a_{ij}|) \in \mathbb{R}^{n\times n}$ the nonnegative matrix consisting of entrywise absolute values. For real $n \times n$ matrices $A = (a_{ij}), B = (b_{ij})$, an inequality $A \le B$ is understood entrywise, *i.e.*, $a_{ij} \le b_{ij}$

for all (i,j). We express an interval matrix including A by $[A] := [\underline{A}, \overline{A}] \in \mathbb{IR}^{n\times n}$ where $\underline{A}$ and $\overline{A}$ is a lower and an upper bound of A, respectively. For real vectors, we apply these definitions similarly.

The *magnitude* and the *mignitude*[1] of an interval quantity $[a] \in \mathbb{IR}$, which are the largest and the smallest absolute values in $[a]$, are defined by

$$\mathrm{mag}([a]) := \max_{a\in[\underline{a},\overline{a}]} |a| \quad \text{and} \quad \mathrm{mig}([a]) := \min_{a\in[\underline{a},\overline{a}]} |a|,$$

respectively. For an interval vector and an interval matrix, they are applied entrywise.

Throughout this paper, n-vectors e and $\mathbf{o}$ are defined by $e := (1,\dots,1)^T$ and $\mathbf{o} := (0,\dots,0)^T$, respectively. For $p \in \{1,2,\infty\}$ we denote p-norm of a real $m\times n$ matrix $A = (a_{ij})$ by

$$\|A\|_1 := \max_{1\le j\le n} \sum_{i=1}^{m} |a_{ij}|, \qquad \|A\|_2 := \sigma_{\max}(A), \qquad \|A\|_\infty := \max_{1\le i\le m} \sum_{j=1}^{n} |a_{ij}|,$$

where $\sigma_{\max}(A)$ denotes the largest singular value of A. Moreover, the condition number of A is defined by

$$\mathrm{cond}_p(A) := \|A\|_p \|A^{-1}\|_p \quad \text{for } p \in \{1,2,\infty\}.$$

3. Iterative refinement

To obtain a tight enclosure of an approximate solution $\widetilde{x}$ of a linear system $Ax = b$, we shall show some properties of the iterative refinement (cf., e.g., [2]).

Using the iterative refinement for the approximate solution $\widetilde{x} \in \mathbb{F}^n$, we may improve $\widetilde{x}$ by $\widetilde{x}+y$ where $y := \sum_{k=1}^{K} z^{(k)}$ with $z^{(k)} \in \mathbb{F}^n$ for $1 \le k \le K$. Then $z^{(k)}$ is called the staggered correction for $\widetilde{x}$. This approach seems to be already used in [12]. If a good approximate inverse R of A has been calculated, we can obtain $\widetilde{x}+y$ with arbitrarily high precision using the iterative refinements:

$y^{(0)} = \mathbf{o}$
for $k = 1,2,\dots,K$
  $r^{(k)} = \mathtt{AccDot}(b - A(\widetilde{x} + y^{(k-1)}))$ % computing an accurate residual
  $z^{(k)} = R * r^{(k)}$ % approximation of $A^{-1}r^{(k)}$
  $y^{(k)} = y^{(k-1)} + z^{(k)}$ % updating y

This makes only sense for calculating the residual $b - A(\widetilde{x}+y^{(k-1)})$ when an accurate dot product "AccDot" is available. Fortunately, a fast and portable method of implementing the accurate dot product has been developed in [7, 9] and it is available in INTLAB [11]. We can use them for this purpose. For detail, see [7, 9]. Of course, LU factors of A can be used for calculating $z^{(k)}$ instead of R by forward and backward substitutions.

[1]This seems to be a technical term used in interval analysis.

Assume that an approximate inverse $R \in \mathbb{F}^{n \times n}$ of A is computed by a backward stable algorithm, e.g., LU factorization with partial pivoting. Then, the following is known as a rule of thumb: Let $\mathbf{u}$ denote the unit-roundoff. For $\mu := \mathrm{cond}_\infty(A) < \mathbf{u}^{-1}$ and $G := I - RA$,

$$\alpha := \|G\|_\infty = \mathcal{O}(n\mathbf{u})\mu. \tag{3}$$

Let $\widetilde{x} = Rb$ and $e := (1, \ldots, 1)^T$. Since

$$|A^{-1}b - \widetilde{x}| = |A^{-1}b - Rb| = |(I - RA)A^{-1}b| \le |G||A^{-1}b|,$$

it holds that

$$|A^{-1}b - \widetilde{x}| \le \|A^{-1}b\|_\infty |G|e. \tag{4}$$

After an iterative refinement by using $y^{(1)} = R(b - A\widetilde{x})$, it follows that

$$\begin{aligned} |A^{-1}b - (\widetilde{x} + y^{(1)})| &= |A^{-1}b - \widetilde{x} - R(b - A\widetilde{x})| = |(I - RA)(A^{-1}b - \widetilde{x})| \\ &\le |G||A^{-1}b - \widetilde{x}|. \end{aligned} \tag{5}$$

Inserting (3) and (4) into (5) yields

$$|A^{-1}b - (\widetilde{x} + y^{(1)})| \le \|A^{-1}b\|_\infty |G|^2 e.$$

For $k \ge 2$, it can inductively be proved for $y^{(k)} = y^{(k-1)} + R(b - A(\widetilde{x} + y^{(k-1)}))$ that

$$|A^{-1}b - (\widetilde{x} + y^{(k)})| \le \|A^{-1}b\|_\infty |G|^{k+1} e$$

and

$$|A^{-1}b - (\widetilde{x} + y^{(k)})| \le \alpha^{k+1} \|A^{-1}b\|_\infty e. \tag{6}$$

Therefore, if $\alpha < 1$, then the iterative refinement converges with the factor $\alpha = \mathcal{O}(n\mathbf{u})\mu$ for each iteration. In practice, due to the rounding error, we have $\widetilde{x}^{(k)} = \mathrm{fl}(\widetilde{x} + y^{(k)})$ with $\widetilde{x}^{(0)} = \widetilde{x}$ and

$$|A^{-1}b - \widetilde{x}^{(k)}| \le \mathbf{u}|A^{-1}b| + \mathcal{O}(\alpha^{k+1})\|A^{-1}b\|_\infty e. \tag{7}$$

This is a *componentwise* error bound and explains the behavior of the iterative refinement. Namely, we can grasp the following tendency of the iterative refinement: Let $x^* := A^{-1}b$, $x_{\max} := \max_{1 \le i \le n} |x_i^*| = \|A^{-1}b\|_\infty$ and $x_{\min} := \min_{1 \le i \le n} |x_i^*|$. Suppose $x_{\min} \ne 0$. Let also $\widetilde{x}$ be an approximate solution of $Ax = b$. If $x_{\max}/x_{\min}$ is very large, then a component of $\widetilde{x}$ corresponding to $x_{\min}$ is fairly less accurate than that corresponding to $x_{\max}$ in the sense of relative errors, which is due to the second term $\mathcal{O}(\alpha^{k+1})\|A^{-1}b\|_\infty e$ of the right-hand side in (7). An extreme case is that $x_{\min} = 0$ and then the iterative refinements for $\widetilde{x}_{\min}$ can not converge until entering the underflow range. To avoid it, an additional stopping criterion ((15) in Section 4) for the iterative refinement is needed.

For example, consider the case where $A \in \mathbb{F}^{5 \times 5}$ is generated by an algorithm proposed in [4] with $\mathrm{cond}_\infty(A) \approx 10^{10}$ and the *exact* solution $A^{-1}b = (1, 10^3, 10^6, 10^9, 134217728)^T$. Here, the last component 134217728 is not important but necessary only for generating the part of the exact solution $(1, 10^3, 10^6, 10^9)^T$, so that we omit to consider it. All computations are done in double precision arithmetic on Matlab, so that $\mathbf{u} = 2^{-53} \approx 10^{-16}$. An approxi-

TABLE 1. History of iterative refinement for $k = 0, 1, 2$; Approximate solutions $\widetilde{x}^{(k)}$ and their true absolute errors $|x^* - \widetilde{x}^{(k)}|$.

i	$\widetilde{x}^{(0)}$	$\widetilde{x}^{(1)}$	$\widetilde{x}^{(2)}$
1	$-1.711885408678072 \cdot 10^2$	$\underline{0.9999}35694692795$	$\underline{1.00000000000}2729$
2	$\underline{1.0}21301738732815 \cdot 10^3$	$\underline{1.00000001}1437893 \cdot 10^3$	$\underline{1.000000000000000} \cdot 10^3$
3	$\underline{1.0000}55792398647 \cdot 10^6$	$\underline{0.9999999999}79213 \cdot 10^6$	$\underline{1.000000000000000} \cdot 10^6$
4	$\underline{1.0000000}83648967 \cdot 10^9$	$\underline{0.9999999999999}80 \cdot 10^9$	$\underline{1.000000000000000} \cdot 10^9$

i	$\lvert x^* - \widetilde{x}^{(0)}\rvert$	$\lvert x^* - \widetilde{x}^{(1)}\rvert$	$\lvert x^* - \widetilde{x}^{(2)}\rvert$
1	$1.7218\cdots \times 10^2$	$6.4305\cdots \times 10^{-5}$	$2.7284\cdots \times 10^{-12}$
2	$2.1301\cdots \times 10^1$	$1.1437\cdots \times 10^{-5}$	$3.4106\cdots \times 10^{-13}$
3	$5.5792\cdots \times 10^1$	$2.0787\cdots \times 10^{-5}$	0
4	$8.3648\cdots \times 10^1$	$2.0265\cdots \times 10^{-5}$	0

mate inverse R of A is computed by the Matlab's function `inv`, which uses BLAS and LAPACK routines. Then, $\alpha = \|I - RA\|_\infty \approx 1.8 \cdot 10^{-7}$, which is almost consistent with the fact that $\mathbf{u} \cdot \mathrm{cond}_\infty(A) \approx 10^{-6}$. An initial approximate solution $\widetilde{x}^{(0)}$ is computed by $\widetilde{x}^{(0)} = \mathrm{fl}(Rb)$. The results of the iterative refinements are displayed in Table 1. As expected, each component is gradually improved with the factor α, in this case about 6 or 7 decimal digits, for each iteration until achieving the maximum accuracy.

4. Verification theory

In this section, we will briefly review some previous methods of calculating a componentwise error bound of an approximate solution $\widetilde{x}$ of a linear system $Ax = b$.

First, we present in the following a linearized version of Yamamoto's theorem [14].

Theorem 4.1 (Yamamoto [14]). *Let A be a real $n \times n$ matrix and b be a real n-vector. Let $\widetilde{x}$ be an approximate solution of $Ax = b$ and $r := b - A\widetilde{x}$. Suppose R is an approximate inverse of A and $G := I - RA$ with I denoting the $n \times n$ identity matrix. If $\|G\|_\infty < 1$, then A is nonsingular and*

$$|A^{-1}b - \widetilde{x}| \le |Rr| + \frac{\|Rr\|_\infty}{1 - \|G\|_\infty}|G|e. \tag{8}$$

On the other hand, the following alternative approach for calculating a componentwise error bound is known in [6].

Theorem 4.2 (Ogita et al. [6]). *Let $A, b, \widetilde{x}$ and r be as in Theorem 4.1. Let $\widetilde{y}$ be an approximate solution of $Ay = r$. If A is nonsingular, then it holds that*

$$|A^{-1}b - \widetilde{x}| \le |\widetilde{y}| + \|A^{-1}\|_p \|r - A\widetilde{y}\|_p e \tag{9}$$

for $p \in \{1, 2, \infty\}$.

The advantages of this approach are as follows:

- Although it needs an upper bound ρ of $\|A^{-1}\|_p$, it does not necessarily need to compute an approximate inverse R of A. For example, $\|A^{-1}\|_2 = 1/\sigma_{\min}(A)$, where $\sigma_{\min}(A)$ denotes the smallest singular value of A. Therefore, ρ can be estimated from $\sigma_{\min}(A)$ (cf., e.g., [13]). Moreover, Theorem 4.2 can be applied to verification for solutions of sparse linear systems if it is not so difficult to compute ρ (cf., e.g., [5]).
- If $\widetilde{y}$ is accurate enough, then the reminder term $\|A^{-1}\|_p\|r - A\widetilde{y}\|_p e$ becomes almost negligible.
- It is compatible with iterative refinement (see Section 3).

The main point of Theorem 4.2 is that $\widetilde{y}$ can arbitrarily be improved for a fixed approximate solution $\widetilde{x}$. To obtain tight error bounds of $\widetilde{x}$ utilizing Theorem 4.1, we will modify it as in Proposition 4.3.

We shall extend Theorem 4.1. Suppose that an approximate inverse $R \in \mathbb{F}^{n\times n}$ of A is obtained. Then a main part of computational effort to obtain the error bounds of $\widetilde{x}$ is to calculate an upper bound of $\|I - RA\|_\infty$. To do this, a possibility is to calculate $[G] \in \mathbb{IF}^{n\times n}$ such that $I - RA \subseteq [G]$. It is known (e.g., [8]) that if $\|\mathrm{mag}([G])\|_\infty < 1$, then an upper bound ρ of $\|A^{-1}\|_\infty$ can be obtained by

$$\|A^{-1}\|_\infty \le \frac{\|R\|_\infty}{1 - \|I - RA\|_\infty} \le \frac{\|R\|_\infty}{1 - \|\mathrm{mag}([G])\|_\infty} =: \rho. \tag{10}$$

Using a usual matrix multiplication for including $I - RA$ with directed rounding requires $4n^3$ flops [8, 10]. Faster methods of calculating an upper bound of $\|I - RA\|_\infty$ have also been presented in [8].

We now assume that $\|G\|_\infty < 1$ for $G := I - RA$. Then A is nonsingular. For an arbitrary $\widetilde{y} \in \mathbb{R}^n$, it holds that

$$A^{-1}b - \widetilde{x} = A^{-1}b - (\widetilde{x} + \widetilde{y}) + \widetilde{y}$$

and

$$|\widetilde{y}| - \epsilon \le |A^{-1}b - \widetilde{x}| \le |\widetilde{y}| + \epsilon \quad \text{with} \quad \epsilon := |A^{-1}b - (\widetilde{x} + \widetilde{y})|.$$

By regarding $\widetilde{x} + \widetilde{y}$ as an approximate solution of $Ax = b$ (or $\widetilde{y}$ as that of $Ay = r$ where $r := b - A\widetilde{x}$), Theorem 4.1 implies

$$\begin{aligned}\epsilon &\le |R(b - A(\widetilde{x} + \widetilde{y}))| + \frac{\|R(b - A(\widetilde{x} + \widetilde{y}))\|_\infty}{1 - \|G\|_\infty}|G|e \\ &\le |R(r - A\widetilde{y})| + \frac{\|R(r - A\widetilde{y})\|_\infty}{1 - \|G\|_\infty}|G|e =: \epsilon_1.\end{aligned}$$

From this, we have the following proposition.

Proposition 4.3. *Let $A, R, G, b, e, \widetilde{x}$ and r be as in Theorem 4.1. Let $\widetilde{y}$ be an approximate solution of $Ay = r$. If $\|G\|_\infty < 1$, then A is nonsingular and*

$$\max(|\widetilde{y}| - \epsilon_1, \mathbf{o}) \le |A^{-1}b - \widetilde{x}| \le |\widetilde{y}| + \epsilon_1, \tag{11}$$

where

$$\epsilon_1 := |R(r - A\widetilde{y})| + \frac{\|R(r - A\widetilde{y})\|_\infty}{1 - \|G\|_\infty}|G|e.$$

Note that the validity of the proposition is independent of the quality of $\widetilde{y}$.

From Proposition 4.3, we can obtain tight componentwise lower and upper error bounds of $\widetilde{x}$ by updating $\widetilde{y}$ using the iterative refinement until satisfying

$$|\widetilde{y}_i| \geq (\epsilon_1)_i \quad \text{for all } i, \widetilde{y}_i \neq 0, \tag{12}$$

which becomes an appropriate stopping criterion for the iterations.

Next, we shall extend Theorem 4.2. Suppose $\|A^{-1}\|_p \leq \rho$. Then it follows for $p \in \{1, 2, \infty\}$ that

$$\begin{aligned} \epsilon &= |A^{-1}(b - A(\widetilde{x} + \widetilde{y}))| = |A^{-1}(r - A\widetilde{y})| \\ &\leq \|A^{-1}(r - A\widetilde{y})\|_\infty e \leq \|A^{-1}(r - A\widetilde{y})\|_p e \\ &\leq \rho\|r - A\widetilde{y}\|_p e =: \epsilon_2. \end{aligned}$$

From this, we also have the following proposition.

Proposition 4.4. *Let $A, b, \widetilde{x}$ and r be as in Theorem 4.1. Let $\widetilde{y}$ be an approximate solution of $Ay = r$. Assume that A is nonsingular and ρ satisfies $\|A^{-1}\|_p \leq \rho$ for any $p \in \{1, 2, \infty\}$. Then*

$$\max(|\widetilde{y}| - \epsilon_2, \mathbf{o}) \leq |A^{-1}b - \widetilde{x}| \leq |\widetilde{y}| + \epsilon_2, \tag{13}$$

where $\epsilon_2 := \rho\|r - A\widetilde{y}\|_p e$.

From Proposition 4.4 after obtaining an upper bound ρ of $\|A^{-1}\|_p$, we can also set an appropriate stopping criterion

$$\min_{1 \leq i \leq n,\ \widetilde{y}_i \neq 0} |\widetilde{y}_i| \geq \rho\|r - A\widetilde{y}\|_p \tag{14}$$

for the iterative refinement. To treat the case $\widetilde{y}_j = 0$ for some j, we add the following criterion for improving $\widetilde{y}$:

$$|\widetilde{x}_j| \geq \mathbf{u} \cdot \rho\|r - A\widetilde{y}\|_p \quad \text{for all } j \text{ such that } \widetilde{y}_j = 0, \tag{15}$$

which ensures the maximum accuracy of $\widetilde{x}_j$ in the working precision.

Remark 4.5. *In general, it is difficult for this kind of verification method to prove $x_j = 0$ for some j, which means $\widetilde{x}_j + \widetilde{y}_j = 0$ and $(\epsilon_1)_j = 0$ (or $(\epsilon_2)_j = 0$). If $x_j = 0$, then the best possible inclusion by floating-point numbers (except zero) of x_j is $[-\underline{\mathbf{u}}, \underline{\mathbf{u}}]$, where $\underline{\mathbf{u}}$ denotes the underflow unit. In IEEE 754 double precision arithmetic, $\underline{\mathbf{u}} = 2^{-1074}$. Therefore, if $x_j = \widetilde{x}_j = 0$, then the iterative refinement tries to improve $\widetilde{y}$ until $\widetilde{y}_j$ enters the underflow range, and finally it falls into infinite loops. To avoid it, the maximum number of loops should be set for the iterative refinement.*

5. Algorithm for tight enclosures

Based on Proposition 4.4 and the discussions in the previous sections, we now present a fast algorithm of calculating tight and componentwise error bounds of an approximate solution of a linear system. Here, we express the algorithm in Matlab-style, which is an almost executable INTLAB code.

Algorithm 5.1. *Let A be a real nonsingular $n \times n$ matrix. Suppose a preconditioner R for A, an upper bound α of $\|I - RA\|_\infty$, an upper bound ρ of $\|A^{-1}\|_\infty$, and an approximate solution $\widetilde{x}$ of a linear system $Ax = b$ are given. Then the following algorithm calculates componentwise lower and upper error bounds $\underline{\epsilon}$ and $\overline{\epsilon}$ of $\widetilde{x}$ such that $\mathbf{o} \le \underline{\epsilon} \le |A^{-1}b - \widetilde{x}| \le \overline{\epsilon}$.*

```
function [x̃, ϵ_, ϵ̄] = vclss(A, b, R, x̃, α, ρ)
  r = AccDot(b − Ax̃);                      % accurate residual b − Ax̃
  y = R ∗ r;                               % y = y^(1): initial estimated error
  for loop = 1 : mloop                     % mloop: maximum number of loops
    [r_s] = AccDot(b − A(x̃ + y), [ ]);     % accurate inclusion of b − A(x̃ + y)
    setround(+1)                           % rounding upwards
    β = ρ ∗ norm(mag([r_s]), p);           % ρ‖b − A(x̃ + y)‖_p ≤ β
    setround(0)                            % rounding to nearest
    q = find(abs(y) < 10^2 ∗ β);           % find all indeces where |y_i| < 10^2 β
    if isempty(q)                          % check whether |y_i| ≥ 10^2 β for all i
      break
    elseif all(u ∗ abs(x(q)) ≥ 10^2 ∗ β)   % check whether u|x_j| ≥ 10^2 β
      break                                % for all j ∈ q
    end
    z = R ∗ mid([r_s]);                    % correction term for y
    y = stag(y, z, α);                     % staggered correction of y
  end
  [t] = midrad(0, β ∗ e);                  % [t] = [−β, β] · e
  m = size(y, 2);                          % y = Σ_{j=1}^m y^(j)
  for j = m : −1 : 1                       % inclusion of y + [−β, β] · e
    [t] = [t] + y^(j);
  end
  ϵ̄ = mag([t]);                            % upper bound of |A^{-1}b − x̃|
  ϵ_ = max(mig([t]), o);                   % lower bound of |A^{-1}b − x̃|
```

In Algorithm 5.1, the instruction `stag` is to be assumed as the staggered correction. Let us briefly explain how it works, although the details are not shown in this paper; Suppose a preconditioner R for A satisfies $\|I - RA\|_\infty \le \alpha$. Let $y = \sum_{j=1}^{m} y^{(j)}$ with $y^{(j)} \in \mathbb{F}^n$ and $\{y_i^{(1)}, y_i^{(2)}, \ldots, y_i^{(m)}\}$ being a non-overlapping sequence[2] for all i. Let $z \in \mathbb{F}^n$ be a correction term obtained by an iterative refinement for improving y. Then, $w = \mathtt{stag}(y, z)$ updates y to $w = \sum_{j=1}^{M} w^{(j)} \approx y + z$

[2] Basically this means $y^{(j+1)} \le 2\mathbf{u}y^{(j)}$. For detail, see [7, 9].

with $\{w_i^{(1)}, w_i^{(2)}, \ldots, w_i^{(M)}\}$, $M \le m+1$ being also a non-overlapping sequence for all i. Since z is obtained by the iterative refinement, the normwise relative error of z is basically less than α. Assume that $\mathbf{u}|y_i^{(m)}| > \alpha|z_i|$ with $z_i \neq 0$ for some i is satisfied. If we only update $y^{(m)}$ to $w^{(m)} = \mathrm{fl}(y^{(m)} + z)$, then the rounding error $\psi := (y_i^{(m)} + z_i) - \mathrm{fl}(y_i^{(m)} + z_i)$ is discarded, whereas ψ still contains a useful information as the correction term. In that case, we should use the information as $w^{(m+1)} := (y^{(m)} + z) - \mathrm{fl}(y^{(m)} + z)$.

We stress that all the computations in Algorithm 5.1 can be done in $\mathcal{O}(n^2)$ flops, so that the algorithm works fast compared with the verification of the non-singularity of A.

6. Numerical examples

In this section, we present some results of numerical experiments showing the performance of our proposed method. We use a PC with an Intel Core Duo 1.06GHz CPU and Matlab 7.4.0 (R2007a) with INTLAB 5.3. All computations are done in IEEE 754 double precision, so that $\mathbf{u} = 2^{-53} \approx 10^{-16}$.

We compare the quality of componentwise error bounds of approximate solutions of linear systems by the following methods:

(Method-P) The proposed method based on Proposition 4.4 (Algorithm 5.1)

(Method-Y) The method based on Theorem 4.1

(Method-O) The method based on Theorem 4.2

We denote componentwise lower and upper error bounds obtained by Method-P as $\underline{e}$ and $\overline{e}$, respectively. We also denote componentwise (upper) error bounds obtained by Method-Y and Method-O as e_{Y} and e_{O}, respectively. The exact error is denoted as e^*. For $\widetilde{e} \in \{\overline{e}, e_{\mathrm{Y}}, e_{\mathrm{O}}\}$, we define the maximum ratio between e^* and $\widetilde{e}$ by

$$\mathit{Ratio}(e^*, \widetilde{e}) := \max_{1 \le i \le n, e_i^* \neq 0} \frac{\widetilde{e}_i}{|e_i^*|} \ge 1.$$

If $\mathit{Ratio}(e^*, \widetilde{e}) \approx 1$, then it can be said that $\widetilde{e}$ is a tight error bound. Moreover, we define the maximum relative distance between $\underline{e}$ and $\overline{e}$ by

$$\mathit{RelDist}(\overline{e}, \underline{e}) := \max_{1 \le i \le n, \overline{e}_i \neq 0} \frac{\overline{e}_i - \underline{e}_i}{\overline{e}_i}.$$

For example, if $\mathit{RelDist}(\overline{e}, \underline{e}) < 0.1$, then it can be said that $\underline{e}$ and $\overline{e}$ are tight error bounds. In practice, we do not normally know the exact error e^*, so that $\mathit{RelDist}(\overline{e}, \underline{e})$ is useful to know how tight the error bounds are.

For all examples, we compute approximate inverses by the Matlab's function `inv`. In all the methods, the accurate dot product [7, 9] is used for calculating the residuals $b - A\widetilde{x}$ and $b - A(\widetilde{x} + y)$.

TABLE 2. Error bounds of approximate solutions

Case	n	cond(A)	$Ratio(e^*, e_Y)$	$Ratio(e^*, e_O)$	$Ratio(e^*, \overline{e})$	$RelDist(\overline{e}, \underline{e})$	# iter
1-1	100	10^6	≈ 1	≈ 1	≈ 1	1.03×10^{-5}	1
1-2	100	10^{12}	≈ 1	≈ 1	≈ 1	2.43×10^{-7}	3
2-1	10	10^6	≈ 1	≈ 1	≈ 1	2.18×10^{-16}	5
2-2	10	10^6	≈ 1	8.99×10^2	≈ 1	4.04×10^{-16}	4
3-1	10	10^{12}	≈ 1	2.84×10^7	≈ 1	1.97×10^{-16}	11
3-2	10	10^{12}	8.28×10^4	5.39×10^{14}	≈ 1	4.04×10^{-16}	9
4	100	10^{12}	7.26×10^{95}	7.35×10^{104}	≈ 1	4.40×10^{-16}	28

We treat the following cases:

Case 1-1. A is generated as a random matrix with $n = 100$ and $\mathrm{cond}_2(A) \approx 10^6$ by a Higham's test matrix `randsvd` and $b := (1, 1, \ldots, 1)^T$. Set $\widetilde{x} := \mathrm{fl}(R \cdot b)$.

Case 1-2. Similar to Case 1-1 except $\mathrm{cond}_2(A) \approx 10^{12}$.

Case 2-1. A and b are generated by the algorithm in [4] with $n = 10$, $\mathrm{cond}_2(A) \approx 10^6$ and a part of the exact solution $x^*(1 : n) = t + \delta t$ where

$$t = (1, 10, 10^2, \ldots, 10^{n-1})^T \quad \text{and} \quad |\delta t_i| \le \mathbf{u}|t_i|.$$

Set $\widetilde{x} := \mathrm{fl}(R \cdot b)$. Note that $\dim(A)$ becomes a little greater than n because of setting the desired exact solution.

Case 2-2. Similar to Case 2-1 except $\widetilde{x} := \mathrm{fl}(x^*)$, which is the best possible approximate solution in double precision.

Case 3-1. Similar to Case 2-1 except $\mathrm{cond}_2(A) \approx 10^{12}$. In this case, $\widetilde{x} := \mathrm{fl}(R \cdot b)$, so that $\widetilde{x}$ becomes a poor approximate solution in the sense of relative accuracy.

Case 3-2. Similar to Case 3-1 except $\widetilde{x} := \mathrm{fl}(x^*)$.

Case 4. Similar to Case 3-2 except $n = 100$.

The results are displayed in Table 2. The notation "# iter" means the number of iterations for the iterative refinement in Method-P.

From the results, we can observe the following facts:

- In usual cases such as Cases 1-1 and 1-2, all the methods can normally give tight error bounds.
- Method-Y seems to be more robust than Method-O.
- In the case where there is a big difference in the order of magnitude in the exact solution, the relative error bounds obtained by Method-Y and Method-O become poor if an approximate solution is accurate.
- Method-P can always give tight error bounds. The number of iterations for the iterative refinement in Method-P increases depending on the condition number of A and the difference in the order of magnitude in the approximate solution.

7. Conclusions

We proposed a fast method of calculating both lower and upper error bounds of an approximate solution of a linear system. The proposed method is based on the iterative refinement and the staggered correction. Using the proposed method, we can obtain the error bounds which are as tight as needed. As a result, we can verify that the obtained error bounds are of high quality. By the numerical results, we confirmed that the proposed verification method worked as expected.

References

[1] ANSI/IEEE, IEEE *Standard for Binary Floating Point Arithmetic*, Std 754–1985 edition, IEEE, New York, 1985.

[2] G.H. Golub, C.F. Van Loan, *Matrix Computations*, 3rd ed., The Johns Hopkins University Press, Baltimore and London, 1996.

[3] N.J. Higham, *Accuracy and Stability of Numerical Algorithms*, 2nd ed., SIAM Publications, Philadelphia, PA, 2002.

[4] S. Miyajima, T. Ogita, S. Oishi, *A method of generating linear systems with an arbitrarily ill-conditioned matrix and an arbitrary solution*, Proceedings of 2005 International Symposium on Nonlinear Theory and its Applications, Bruges (2005), 741–744.

[5] T. Ogita, S. Oishi, Y. Ushiro, *Fast verification of solutions for sparse monotone matrix equations*, Topics in Numerical Analysis: With Special Emphasis on Nonlinear Problems (Computing, Supplement 15, G. Alefeld and X. Chen eds.), Springer Wien, New York, Austria (2001), 175–187.

[6] T. Ogita, S. Oishi, Y. Ushiro, *Computation of sharp rigorous componentwise error bounds for the approximate solutions of systems of linear equations*, Reliable Computing, **9**(3) (2003), 229–239.

[7] T. Ogita, S.M. Rump, S. Oishi, *Accurate Sum and Dot Product*, SIAM J. Sci. Comput. **26**(6) (2005), 1955–1988.

[8] S. Oishi, S.M. Rump, *Fast verification of solutions of matrix equations*, Numer. Math. **90**(4) (2002), 755–773.

[9] S.M. Rump, T. Ogita, S. Oishi *Accurate floating-point summation, Part I and II*, SIAM J. Sci. Comput., to appear.

[10] S.M. Rump, *Fast and parallel interval arithmetic*, BIT **39**(3) (1999), 534–554.

[11] S.M. Rump, *INTLAB – INTerval LABoratory*, Developments in Reliable Computing (T. Csendes ed.), Kluwer Academic Publishers, Dordrecht, 1999, 77–104. `http://www.ti3.tu-harburg.de/rump/intlab/`

[12] S.M. Rump, *Kleine Fehlerschranken bei Matrixproblemen*, Universität Karlsruhe, PhD thesis, 1980.

[13] S.M. Rump, *Verification methods for dense and sparse systems of equations*, Topics in Validated Computations – Studies in Computational Mathematics (J. Herzberger ed.), Elsevier, Amsterdam, 1994, 63–136.

[14] T. Yamamoto, *Error bounds for approximate solutions of systems of equations*, Japan J. Appl. Math. **1**(1) (1984), 157–171.

Takeshi Ogita
Department of Mathematics
Tokyo Woman's Christian University
2-6-1 Zempukuji, Suginami-ku
Tokyo 167-8585, Japan

and

Faculty of Science and Engineering
Waseda University
3-4-1 Okubo Shinjuku-ku
Tokyo 169-8555, Japan
e-mail: `ogita@lab.twcu.ac.jp`

Shin'ichi Oishi
Faculty of Science and Engineering
Waseda University
3-4-1 Okubo Shinjuku-ku
Tokyo 169-8555, Japan
e-mail: `oishi@waseda.jp`

Part IV

Inequalities in Approximation Theory

PHB '07

International Series of Numerical Mathematics, Vol. 157, 181–192

Operators of Bernstein-Stancu Type and the Monotonicity of Some Sequences Involving Convex Functions

Ioan Gavrea

Abstract. By using the operators introduced by D.D. Stancu in 1969, we show that the results obtained in the papers [1], [2] and [5] follow from the properties of these operators. We also present some improvements and generalizations of the results obtained in the above mentioned papers.

Mathematics Subject Classification (2000). 26D15, 26A51.

Keywords. Inequality, convex functions, Bernstein-Stancu operators, monotonicity.

1. Introduction

Throughout this paper we denote by n a natural number greater than 1. The divided difference $[x_1, x_2, x_3; f]$ of a function $f \in \mathbb{R}^{[0,1]}$ on the distinct nodes, $x_1, x_2, x_3 \in [0, 1]$ is defined by

$$[x_1, x_2, x_3; f] = \frac{f(x_1)}{(x_1 - x_2)(x_1 - x_3)} + \frac{f(x_2)}{(x_2 - x_3)(x_2 - x_1)} + \frac{f(x_3)}{(x_3 - x_1)(x_3 - x_2)}.$$

A function $f \in \mathbb{R}^{[0,1]}$ is said be convex (concave) if

$$[x_1, x_2, x_3; f] \geq 0 \quad ([x_1, x_2, x_3; f] \leq 0)$$

for every distinct nodes $x_1, x_2, x_3 \in [0, 1]$.

If $f \in \mathbb{R}^{[0,1]}$ is a convex (concave) function then

$$[x_1, x_2; f] \leq [x_2, x_3; f] \quad ([x_1, x_2; f] \geq [x_2, x_3; f])$$

for every distinct nodes $x_1, x_2, x_3 \in [0, 1]$ such that

$$x_1 < x_2 < x_3,$$

where

$$[x_1, x_2; f] = \frac{f(x_2) - f(x_1)}{x_2 - x_1}.$$

J.-Ch. Kuang ([2]) proved the following result:

Theorem 1.1. ([2]) *Let f be a strictly increasing convex (concave) function on $[0,1]$. Then:*

$$\frac{1}{n}\sum_{k=1}^{n} f\left(\frac{k}{n}\right) > \frac{1}{n+1}\sum_{k=1}^{n+1} f\left(\frac{k}{n+1}\right) > \int_0^1 f(x)dx. \tag{1.1}$$

G. Bennett and G. Jameson [1] investigate the monotonicity of various averages of the values of a convex (concave) function at n equally spaced points. More precisely, they defined

$$A_n(f) = \frac{1}{n-1}\sum_{r=1}^{n-1} f\left(\frac{r}{n}\right) \quad (n \geq 2)$$

$$B_n(f) = \frac{1}{n+1}\sum_{r=0}^{n} f\left(\frac{r}{n}\right) \quad (n \geq 1)$$

$$s_n(f) = \frac{1}{n}\sum_{r=0}^{n-1} f\left(\frac{r}{n}\right) \quad (n \geq 1)$$

$$S_n(f) = \frac{1}{n}\sum_{r=1}^{n} f\left(\frac{r}{n}\right) \quad (n \geq 1)$$

and proved the following theorems:

Theorem 1.2. ([1]) *If f is a convex function on the open interval $(0,1)$, then $A_n(f)$ increases with n. If f is concave, $A_n(f)$ decreases with n.*

Theorem 1.3. ([1]) *If f is convex on $[0,1]$, then $B_n(f)$ decreases with n. If f is concave, $B_n(f)$ increases with n.*

Theorem 1.4. ([1]) *Suppose that f is monotonic and either convex or concave on $[0,1]$. Then, with the above notation, $s_n(f)$ increases with n, and $S_n(f)$ decreases.*

We remark that Theorem 1.1 ([2]) is the same with Theorem 1.4 ([1]).

F. Qi [4] proves the following result.

Theorem 1.5. ([4]) *Let f be a strictly increasing convex (concave) function on $[0,1]$. Then*

$$\frac{1}{n}\sum_{i=k+1}^{n+k} f\left(\frac{i}{n+k}\right) > \frac{1}{n+1}\sum_{i=k+1}^{n+k+1} f\left(\frac{i}{n+k+1}\right) > \int_0^1 f(x)dx, \tag{1.2}$$

where k is a fixed natural number.

Recently, F. Qi and B.-N. Guo ([5]) obtain the following result:

Theorem 1.6. ([5]) *Let f be an increasing (concave) function defined on $[0,1]$ and $\{a_i\}_{i\in\mathbb{N}}$ be an increasing positive sequence such that the sequence*

$$\left\{i\left(\frac{a_i}{a_{i+1}}-1\right)\right\}_{i\in\mathbb{N}}$$

decreases ($\{i\left((a_{i+1}/a_i)-1\right)\}_{i\in\mathbb{N}}$ *increases*), *then*

$$\frac{1}{n}\sum_{i=1}^{n} f\left(\frac{a_i}{a_n}\right) \geq \frac{1}{n+1}\sum_{i=1}^{n+1} f\left(\frac{a_i}{a_{n+1}}\right) \geq \int_0^1 f(t)dt. \tag{1.3}$$

Let $B_n(f;x)$ be the well-known Bernstein polynomial of degree n defined by

$$B_n(f;x) = \sum_{k=0}^{n} f(k/n)p_{n,k}(x), \tag{1.4}$$

where

$$p_{n,k}(x) = \binom{n}{k}x^k(1-x)^{n-k}, \quad k=0,\ldots,n.$$

The following identity can be found in [3] (p. 309, Th. 4.1).

For $n=1,2,\ldots$ we have

$$B_n(f;x) - B_{n+1}(f;x) = \frac{x(1-x)}{n(n+1)}\sum_{k=0}^{n-1}\left[\frac{k}{n},\frac{k+1}{n+1},\frac{k+1}{n};f\right]p_{n-1,k}(x). \tag{1.5}$$

From (1.5) we obtain

$$\begin{aligned}\frac{x(1-x)}{n(n+1)}\min_{k=\overline{0,n-1}}\left[\frac{k}{n},\frac{k+1}{n+1},\frac{k+1}{n};f\right] &\leq B_n(f;x) - B_{n+1}(f;x) \\ &\leq \frac{x(1-x)}{n(n+1)}\max_{k=\overline{0,n-1}}\left[\frac{k}{n},\frac{k+1}{n+1},\frac{k+1}{n};f\right].\end{aligned} \tag{1.6}$$

Integrating between 0 and 1 both sides of (1.6) we get

$$\begin{aligned}&\frac{1}{6n(n+1)}\min_{k=\overline{0,n-1}}\left[\frac{k}{n},\frac{k+1}{n+1},\frac{k+1}{n};f\right] \\ &\leq \frac{1}{n+1}\sum_{k=0}^{n} f\left(\frac{k}{n}\right) - \frac{1}{n+2}\sum_{k=0}^{n+1} f\left(\frac{k}{n+1}\right) \\ &\leq \frac{1}{6n(n+1)}\max_{k=\overline{0,n-1}}\left[\frac{k}{n},\frac{k+1}{n+1},\frac{k+1}{n};f\right].\end{aligned} \tag{1.7}$$

Certainly, inequalities (1.6) and (1.7) improve the result from the Theorem 1.4 relative to $S_n(f)$.

The aim of this paper is to improve the results from the Theorems 1.1–1.6, using operators of Bernstein-Stancu type.

2. Main results

In 1969 D.D. Stancu [6] considered the following linear positive operators, $B_n^{(\alpha,\beta)}$ defined by:

$$B_n^{(\alpha,\beta)}(f;x) = \sum_{k=0}^{n} f\left(\frac{k+\alpha}{n+\beta}\right) p_{n,k}(x), \tag{2.1}$$

where $0 \le \alpha \le \beta$ and $p_{n,k}(x) = \binom{n}{k} x^k (1-x)^{n-k}$, $k = 0, 1, \ldots, n$.

In the following we will establish some properties of the operators defined by (2.1) for some particular values of the parameters α, β and $n \ge 2$.

Theorem 2.1. *Let $f : (0,1) \to \mathbb{R}$ be a convex function. Then*

$$B_{n-1}^{(1,2)}(f;x) - B_{n-2}^{(1,2)}(f;x) \ge \sum_{k=0}^{n-2} p_{n-2,k}(x) \frac{nx-k-1}{n(n+1)} \left[\frac{k+1}{n}, \frac{k+1}{n+1}; f\right] \tag{2.2}$$

$$\begin{aligned} B_{n-1}^{(1,2)}(f;x) - B_{n-2}^{(1,2)}(f;x) \le{}& \frac{x(n-1)-(n-2)x^2}{n(n+1)} \left[\frac{n-1}{n}, \frac{n}{n+1}; f\right] \\ &- \frac{(1-x)((n-2)x+1)}{n(n+1)} \left[\frac{1}{n}, \frac{1}{n+1}; f\right]. \end{aligned} \tag{2.3}$$

Proof. Using the definition (2.1) of the operator $B_n^{(1,2)}$ we get

$$\begin{aligned} &B_{n-1}^{(1,2)}(f;x) - B_{n-2}^{(1,2)}(f;x) \\ &= \sum_{k=0}^{n-1} \binom{n-1}{k} x^k (1-x)^{n-k-1} f\left(\frac{k+1}{n+1}\right) \\ &\quad - \sum_{k=0}^{n-2} \binom{n-2}{k} x^k (1-x)^{n-k-2} ((1-x)+x) f\left(\frac{k+1}{n}\right) \\ &= \sum_{k=0}^{n-1} \binom{n-1}{k} x^k (1-x)^{n-k-1} f\left(\frac{k+1}{n+1}\right) \\ &\quad - \sum_{k=0}^{n-2} \binom{n-2}{k} x^k (1-x)^{n-k-1} f\left(\frac{k+1}{n}\right) \\ &\quad - \sum_{k=1}^{n-1} \binom{n-2}{k-1} x^k (1-x)^{n-k-1} f\left(\frac{k}{n}\right). \end{aligned} \tag{2.4}$$

Using the equality

$$\binom{n}{k} = \binom{n-1}{k-1} + \binom{n-1}{k}$$

(2.4) becomes

$$B_{n-1}^{(1,2)}(f;x) - B_{n-2}^{(1,2)}(f;x)$$
$$= \sum_{k=1}^{n-1} \binom{n-2}{k-1} x^k (1-x)^{n-k-1} \left(f\left(\frac{k+1}{n+1}\right) - f\left(\frac{k}{n}\right)\right)$$
$$- \sum_{k=1}^{n} \binom{n-2}{k-1} x^{k-1} (1-x)^{n-k} \left(f\left(\frac{k}{n}\right) - f\left(\frac{k}{n+1}\right)\right)$$

or, in terms of divided differences

$$B_{n-1}^{(1,2)}(f;x) - B_{n-2}^{(1,2)}(f;x) \tag{2.5}$$
$$= \frac{1}{n(n+1)} \left(\sum_{k=0}^{n-2} \binom{n-2}{k} x^{k+1} (1-x)^{n-k-2} (n-k-1) \left[\frac{k+1}{n}, \frac{k+2}{n+1}; f\right]\right)$$
$$- \sum_{k=0}^{n-2} \binom{n-2}{k} x^k (1-x)^{n-k-1} (k+1) \left[\frac{k+1}{n}, \frac{k+1}{n+1}; f\right].$$

But

$$\left[\frac{k}{n}, \frac{k+1}{n+1}; f\right] \geq \left[\frac{k}{n}, \frac{k}{n+1}; f\right]$$

and so, from (2.5) we have

$$B_{n-1}^{(1,2)}(f;x) - B_{n-2}^{(1,2)}(f;x) \geq \frac{1}{n(n+1)} \sum_{k=0}^{n-2} p_{n-2,k}(x)(n-x-k-1) \left[\frac{k+1}{n}, \frac{k+1}{n+1}; f\right].$$

Inequality (2.2) is proved.

Since f is a convex function we have

$$\begin{aligned} \left[\frac{k}{n}, \frac{k+1}{n+1}; f\right] &\leq \left[\frac{n-1}{n}, \frac{n}{n+1}; f\right] \\ \left[\frac{k}{n}, \frac{k}{n+1}; f\right] &\geq \left[\frac{1}{n}, \frac{1}{n+1}; f\right] \end{aligned} \qquad k = \overline{1, n-1} \tag{2.6}$$

From (2.5) and (2.6) we get (2.3).

Corollary 2.2. *Let $f : (0,1) \to \mathbb{R}$ be a convex function on $(0,1)$ and $\alpha \geq 0$ a fixed number. Then the following inequalities hold:*

$$\frac{\Gamma(n)}{\Gamma(n+\alpha+1)} \sum_{k=0}^{n-1} \frac{\Gamma(k+\alpha+1)}{k!} f\left(\frac{k+1}{n+1}\right) - \frac{\Gamma(n-1)}{\Gamma(n+\alpha)} \sum_{k=0}^{n-2} \frac{\Gamma(k+\alpha+1)}{k!} f\left(\frac{k+1}{n}\right)$$
$$\geq \frac{\alpha}{n(n+1)} \cdot \frac{\Gamma(n-1)}{\Gamma(n+\alpha+1)} \sum_{k=0}^{n-2} (n-k-1) \frac{\Gamma(k+\alpha+1)}{k!} \left[\frac{k+1}{n}, \frac{k+1}{n+1}; f\right] \tag{2.7}$$

$$\frac{1}{n}\sum_{k=0}^{n-1}\frac{\Gamma(k+\alpha+1)}{k!}f\left(\frac{k+1}{n+1}\right)-\frac{n+\alpha}{n(n-1)}\sum_{k=0}^{n-2}\frac{\Gamma(k+\alpha+1)}{k!}f\left(\frac{k+1}{n}\right)$$
$$\geq\frac{\alpha\left[\frac{1}{n},\frac{1}{n+1};f\right]}{(\alpha+1)(\alpha+2)}\cdot\frac{\Gamma(\alpha+n+1)}{(n+1)!} \tag{2.8}$$

$$\frac{1}{n}\sum_{k=0}^{n-1}\frac{\Gamma(k+\alpha+1)}{k!}f\left(\frac{k+1}{n+1}\right)-\frac{n+\alpha}{n(n-1)}\sum_{k=0}^{n-2}\frac{\Gamma(k+\alpha+1)}{k!}f\left(\frac{k+1}{n}\right)$$
$$\leq\left(\frac{\alpha+n+1}{(\alpha+2)(\alpha+3)}\left[\frac{n-1}{n},\frac{n}{n+1};f\right]\right.$$
$$\left.-\left(\frac{(n-1)\alpha+n+1}{(\alpha+1)(\alpha+2)(\alpha+3)}\left[\frac{1}{n},\frac{1}{n+1};f\right]\right)\frac{\Gamma(\alpha+n+1)}{(n+1)!}\right). \tag{2.9}$$

Proof. If we multiply (2.2) and (2.3) by x^α and then integrate between 0 and 1 we get (2.7) and (2.9). The relation (2.8) follows from (2.7) and from the inequality

$$\left[\frac{k+1}{n},\frac{k+1}{n+1};f\right]\geq\left[\frac{1}{n},\frac{1}{n+1};f\right],\quad k=\overline{0,n-2}.$$

Remark 2.3. If $\alpha=0$ then, from inequalities (2.8) and (2.9) we obtain

$$0\leq\frac{1}{n}\sum_{k=0}^{n-1}f\left(\frac{k+1}{n+1}\right)-\frac{1}{n-1}\sum_{k=0}^{n-2}f\left(\frac{k+1}{n}\right) \tag{2.10}$$
$$\leq\frac{1}{6}\left(\left[\frac{n-1}{n},\frac{n}{n+1};f\right]-\left[\frac{1}{n},\frac{1}{n+1};f\right]\right)$$

for every convex function f on $(0,1)$.

Theorem 2.4. *For any function $f:[0,1]\to\mathbb{R}$, the following equality holds:*

$$B_{n-1}^{(\beta+1,\beta+1)}(f;x)-B_n^{(\beta+1,\beta+1)}(f;x)=(1-x)\sum_{i=0}^{n-1}p_{n-1,i}(x)E_{i,n,\beta}(f) \tag{2.11}$$

where

$$E_{i,n,\beta}(f)=\frac{1}{(n+\beta)(n+\beta+1)}\left((i+\beta+1)\left[\frac{i+\beta+1}{n+\beta+1},\frac{i+\beta+1}{n+\beta};f\right]\right. \tag{2.12}$$
$$\left.-i\left[\frac{i+\beta}{n+\beta},\frac{i+\beta+1}{n+\beta+1};f\right]\right).$$

Proof. Using the definition of $B_n^{(\alpha,\beta)}(f;x)$ we obtain successively:

$$B_{n-1}^{(\beta+1,\beta+1)}(f;x)-B_n^{(\beta+1,\beta+1)}(f;x)$$
$$=\sum_{i=0}^{n-1}p_{n-1,i}(x)f\left(\frac{i+\beta+1}{n+\beta}\right)-\sum_{i=0}^{n}p_{n,i}(x)f\left(\frac{i+\beta+1}{n+\beta+1}\right)$$

$$
\begin{aligned}
&= \sum_{i=0}^{n-1} \binom{n-1}{i} x^i (1-x)^{n-i-1} [(1-x)+x] f\left(\frac{i+\beta+1}{n+\beta}\right) \\
&\quad - \sum_{i=0}^{n} \binom{n}{i} x^i (1-x)^{n-i} f\left(\frac{i+\beta+1}{n+\beta+1}\right) \\
&= \sum_{i=0}^{n-1} \binom{n-1}{i} x^i (1-x)^{n-i} f\left(\frac{i+\beta+1}{n+\beta}\right) \\
&\quad + \sum_{i=0}^{n-1} \binom{n-1}{i} x^{i+1} (1-x)^{n-i-1} f\left(\frac{i+\beta+1}{n+\beta}\right) \\
&\quad - \sum_{i=0}^{n} \left[\binom{n-1}{i} + \binom{n-1}{i-1}\right] x^i (1-x)^{n-i} f\left(\frac{i+\beta+1}{n+\beta+1}\right) \\
&= \sum_{i=0}^{n-1} \binom{n-1}{i} x^i (1-x)^{n-i} f\left(\frac{i+\beta+1}{n+\beta}\right) + \sum_{i=1}^{n} \binom{n-1}{i-1} x^i (1-x)^{n-i} f\left(\frac{i+\beta}{n+\beta}\right) \\
&\quad - \sum_{i=0}^{n} \binom{n-1}{i} x^i (1-x)^{n-i} f\left(\frac{i+\beta+1}{n+\beta+1}\right) \\
&\quad - \sum_{i=0}^{n} \binom{n-1}{i-1} x^i (1-x)^{n-i} f\left(\frac{i+\beta+1}{n+\beta+1}\right) \\
&= \sum_{i=0}^{n-1} \binom{n-1}{i} x^i (1-x)^{n-i} \left[f\left(\frac{i+\beta+1}{n+\beta}\right) - f\left(\frac{i+\beta+1}{n+\beta+1}\right)\right] \\
&\quad - \sum_{i=0}^{n-1} \binom{n-1}{i-1} x^i (1-x)^{n-i} \left[f\left(\frac{i+\beta+1}{n+\beta+1}\right) - f\left(\frac{i+\beta}{n+\beta}\right)\right] \\
&= (1-x) \sum_{i=0}^{n-1} p_{n-1,i}(x) \left\{ f\left(\frac{1+\beta+1}{n+\beta}\right) - f\left(\frac{i+\beta+1}{n+\beta+1}\right) \right. \\
&\quad \left. - \frac{i}{n-i} \left[f\left(\frac{i+\beta+1}{n+\beta+1}\right) - f\left(\frac{i+\beta}{n+\beta}\right)\right] \right\} \\
&= (1-x) \sum_{i=0}^{n-1} p_{n-1,i}(x) \left\{ \frac{i+\beta+1}{(n+\beta)(n+\beta+1)} \left[\frac{i+\beta+1}{n+\beta+1}, \frac{i+\beta+1}{n+\beta}; f\right] \right. \\
&\quad \left. - \frac{i}{(n+\beta)(n+\beta+1)} \left[\frac{i+\beta}{n+\beta}, \frac{i+\beta+1}{n+\beta+1}; f\right] \right\} \\
&= (1-x) \sum_{i=0}^{n-1} p_{n-1,i}(x) E_{i,n,\beta}(f).
\end{aligned}
$$

Corollary 2.5. *Let f be a convex function on $[0,1]$. Then*

$$B_{n-1}^{(\beta+1,\beta+1)}(f;x) - B_n^{(\beta+1,\beta+1)}(f;x) \tag{2.13}$$
$$\geq \frac{(1-x)(\beta+1)}{(n+\beta)(n+\beta+1)} \sum_{i=0}^{n-1} \left[\frac{i+\beta}{n+\beta}, \frac{i+\beta+1}{n+\beta+1}; f\right] p_{n-1,i}(x)$$
$$\geq \frac{(1-x)(\beta+1)}{(n+\beta)(n+\beta+1)} \left[\frac{\beta}{n+\beta}, \frac{\beta+1}{n+\beta+1}; f\right]$$

and

$$B_{n-1}^{(\beta+1,\beta+1)}(f;x) - B_n^{(\beta+1,\beta+1)}(f;x) \tag{2.14}$$
$$\leq \frac{1}{(n+\beta)(n+\beta+1)} \left\{(n+\beta)x^{n-1}(1-x)\left[\frac{n+\beta}{n+\beta+1}, 1; f\right]\right\}$$
$$+ \sum_{i=0}^{n-2} p_{n-1,i}(x)((i+\beta+1-(n+\beta)x)\left[\frac{i+\beta+1}{n+\beta}, \frac{i+\beta+2}{n+\beta+1}; f\right].$$

Proof. The convexity of the function f implies that

$$\left[\frac{i+\beta+1}{n+\beta+1}, \frac{i+\beta+1}{n+\beta}; f\right] \geq \left[\frac{i+\beta}{n+\beta}, \frac{i+\beta+1}{n+\beta+1}; f\right]. \tag{2.15}$$

From (2.11) and (2.15) we obtain (2.13). Relation (2.11) can be written in the following form:

$$B_{n-1}^{(\beta+1,\beta+1)}(f;x) - B_n^{(\beta+1,\beta+1)}(f;x)$$
$$= \frac{1-x}{(n+\beta)(n+\beta+1)} \left(\sum_{i=1}^{n}(i+\beta)\left[\frac{i+\beta}{n+\beta+1}, \frac{i+\beta}{n+\beta}; f\right] p_{n-1,i-1}(x)\right.$$
$$\left. - \sum_{i=0}^{n-1} i \left[\frac{i+\beta}{n+\beta}, \frac{i+\beta+1}{n+\beta+1}; f\right] p_{n-1,i}(x)\right)$$

or

$$B_{n-1}^{(\beta+1,\beta+1)}(f;x) - B_n^{(\beta+1,\beta+1)}(f;x) \tag{2.16}$$
$$= \frac{1-x}{(n+\beta)(n+\beta+1)} \left\{(n+\beta)\left[\frac{n+\beta}{n+\beta+1}, 1; f\right] x^{n-1}\right.$$
$$+ \sum_{i=1}^{n-1}(i+\beta)\left[\frac{i+\beta}{n+\beta+1}, \frac{i+\beta}{n+\beta}; f\right] p_{n-1,i-1}(x)$$
$$\left. - \sum_{i=1}^{n-1} i \left[\frac{i+\beta}{n+\beta}, \frac{i+\beta+1}{n+\beta+1}; f\right] p_{n-1,i}(x)\right\}.$$

Inequality (2.14) follows from (2.16) and from the inequality

$$\left[\frac{i+\beta}{n+\beta+1}, \frac{i+\beta}{n+\beta}; f\right] \le \left[\frac{i+\beta}{n+\beta}, \frac{i+\beta+1}{n+\beta+1}; f\right], \quad i=\overline{1, n-1}.$$

Corollary 2.6. *Let f be a convex function on $[0,1]$. For every $\alpha, \beta \ge 0$ we have*

$$A_n \le \frac{(n-1)!}{\Gamma(n+\alpha+1)} \sum_{i=0}^{n-1} \frac{\Gamma(i+\alpha+1)}{i!} f\left(\frac{i+\beta+1}{n+\beta}\right) - \frac{n!}{\Gamma(n+\alpha+2)} \sum_{i=0}^{n} \frac{\Gamma(i+\alpha+1)}{i!} f\left(\frac{i+\beta+1}{n+\beta+1}\right) \le B_n \tag{2.17}$$

where

$$A_n = \frac{\beta+1}{(\alpha+1)(\alpha+2)} \cdot \frac{1}{(n+\beta)(n+\beta+1)} \left[\frac{\beta}{n+\beta}, \frac{\beta+1}{n+\beta+1}; f\right]$$

$$B_n = \frac{\beta+1}{(\alpha+1)(\alpha+2)} \cdot \frac{1}{(n+\beta)(n+\beta+1)} \left[\frac{n+\beta}{n+\beta+1}, 1; f\right].$$

Proof. We note that

$$\int_0^1 x^\alpha (B_{n-1}^{(\beta+1,\beta+1)}(f;x) - B_n^{(\beta+1,\beta+1)}(f;x))dx = \frac{(n-1)!}{\Gamma(n+\alpha+1)} \sum_{i=0}^{n-1} \frac{\Gamma(i+\alpha+1)}{i!} f\left(\frac{i+\beta+1}{n+\beta}\right) - \frac{n!}{\Gamma(n+\alpha+2)} \sum_{i=0}^{n} \frac{\Gamma(i+\alpha+1)}{i!} f\left(\frac{i+\beta+1}{n+\beta+1}\right). \tag{2.18}$$

Now, if we multiply (2.13) by x^α and then integrate between 0 and 1 we get

$$\int_0^1 x^\alpha (B_{n-1}^{(\beta+1,\beta+1)}(f;x) - B_n^{(\beta+1,\beta+1)}(f;x))dx \ge A_n. \tag{2.19}$$

From (2.18) and (2.19) we obtain the first inequality from (2.17).

For the proof of the second inequality from (2.17) we note that

$$\int_0^1 x^\alpha p_{n-1,i}(x)(i+\beta+1-(n+\beta)x)dx > 0, \quad i=0,1,\ldots,n-1. \tag{2.20}$$

If we multiply (2.13) by x^α and then integrate between 0 and 1 we get:

$$\int_0^1 x^\alpha (B_{n-1}^{(\beta+1,\beta+1)}(f;x) - B_n^{(\beta+1,\beta+1)}(f;x))dx \le \frac{1}{(n+\beta)(n+\beta+1)} \Bigg\{(n+\beta)\left[\frac{n+\beta}{n+\beta+1}, 1; f\right] \int_0^1 x^{n+\alpha-1}(1-x)dx + \sum_{i=0}^{n-2} \int_0^1 x^\alpha p_{n-1,i}(x)(i+\beta+1-(n+\beta)x)dx \left[\frac{i+\beta+1}{n+\beta}, \frac{i+\beta+2}{n+\beta+1}; f\right]\Bigg\}. \tag{2.21}$$

By the inequality

$$\left[\frac{i+\beta+1}{n+\beta}, \frac{i+\beta+2}{n+\beta+1}; f\right] \le \left[\frac{n+\beta}{n+\beta+1}, 1; f\right], \quad i = 0, 1, \ldots, n-2$$

and from inequalities (2.21) and (2.20), we obtain

$$\int_0^1 x^\alpha (B_{n-1}^{(\beta+1,\beta+1)}(f;x) - B_n^{(\beta+1,\beta+1)}(f;x))dx \le B_n$$

the proof is completed.

Remark 2.7. For $\alpha = 0$, (2.17) becomes

$$\begin{aligned} &\frac{\beta+1}{2(n+\beta)(n+\beta+1)} \left[\frac{\beta}{n+\beta}, \frac{\beta+1}{n+\beta+1}; f\right] \\ &\le \frac{1}{n}\sum_{i=1}^{n} f\left(\frac{i+\beta}{n+\beta}\right) - \frac{1}{n+1}\sum_{i=1}^{n+1} f\left(\frac{i_\beta}{n+\beta+1}\right) \\ &\le \frac{\beta+1}{2(n+\beta)(n+\beta+1)} \left[\frac{n+\beta}{n+\beta+1}, 1; f\right] \end{aligned} \tag{2.22}$$

which is an improvement of Theorem 1.5 ([4]), in the case when f is a convex function.

Let us suppose that f is a concave function. Then $g = -f$ is a convex function and, so the inequalities obtained change sign. For example, if f is a concave function then (2.22) becomes

$$\begin{aligned} &\frac{\beta+1}{2(n+\beta)(n+\beta+1)} \left[\frac{\beta}{n+\beta}, \frac{\beta+1}{n+\beta+1}; f\right] \\ &\ge \frac{1}{n}\sum_{i=1}^{n} f\left(\frac{i+\beta}{n+\beta}\right) - \frac{1}{n+1}\sum_{i=1}^{n} f\left(\frac{i+\beta}{n+\beta+1}\right) \\ &\ge \frac{\beta+1}{2(n+\beta)(n+\beta+1)} \left[\frac{n+\beta}{n+\beta+1}, 1; f\right]. \end{aligned} \tag{2.23}$$

Inequality (2.23) is an improvement of Theorem 1.5 ([4]) in the case when f is a concave function.

In the following we will consider more general Bernstein-Stancu type operators. More precisely, let us consider a fixed natural number n, $n \ge 1$ and let $x_{i,n}$, $i = 0, 1, \ldots, n, n+1$ be nodes on $[0, 1]$ such that

$$0 \le x_{0,n} \le x_{1,n} < \cdots < x_{n,n} \le 1. \tag{2.24}$$

We also assume that, for every $k = 1, 2, \ldots, n$ we have

$$x_{k,n+1} < x_{k,n} < x_{k+1,n+1}, \quad n \in \mathbb{N}^*. \tag{2.25}$$

Now, we consider the operator $S_n : \mathbb{R}^{[0,1]} \to \Pi_n$ defined by

$$S_n(f;x) = \sum_{k=0}^{n} p_{n,k}(x) f(x_{k,n}). \tag{2.26}$$

The operator S_n defined by (2.26) is called an operator of Bernstein-Stancu type relative to the nodes $x_{k,n}$.

Theorem 2.8. *The following cquality is true:*

$$\begin{aligned} &S_{n-1}(f;x) - S_n(f;x) \qquad\qquad (2.27)\\ &= (1-x)\sum_{k=1}^{n-1} p_{n-1,k}(x)\Big\{(x_{k,n-1}-x_{k,n})[x_{k,n},x_{k,n-1};f]\\ &\quad - \frac{k(x_{k,n}-x_{k-1,n-1})}{n-k}[x_{k-1,n-1},x_{k,n};f]\Big\}\\ &\quad + (1-x)^n(f(x_{0,n-1})-f(x_{0,n})) + x^n(f(x_{n-1,n-1})-f(x_{n,n})). \end{aligned}$$

Proof. The proof of this theorem is similar to Theorem 2.4.

Corollary 2.9. *If the nodes $x_{k,n}$ satisfy the additional requirement*

$$\begin{aligned} x_{0,n-1} &\geq x_{0,n}\\ x_{n-1,n-1} &\geq x_{n,n} \qquad\qquad (2.28)\\ x_{k,n-1} - x_{k,n} &\geq \frac{k}{n-k}(x_{k,n}-x_{k-1,n-1}), \end{aligned}$$

$k = 1, 2, \ldots, n-1$, $n \in \mathbb{N}^*$, then for any increasing convex function f we have

$$\frac{1}{n}\sum_{k=1}^{n} f(x_{k-1,n-1}) \geq \frac{1}{n+1}\sum_{k=1}^{n+1} f(x_{k-1,n}). \tag{2.29}$$

Proof. By (2.27) and (2.28) we obtain

$$S_{n-1}(f;x) \geq S_n(f;x) \tag{2.30}$$

for any increasing convex function. If we integrate both sides of (2.30) between 0 and 1 we obtain (2.29).

The following result follows from (2.29).

Corollary 2.10. *Let $(a_n)_{n\in\mathbb{N}^*}$ be an increasing sequence, $a_n \in [0,1]$, $n \in \mathbb{N}^*$, such that*

$$\left\{n\left(1-\frac{a_n}{a_{n+1}}\right)\right\}_{n\in\mathbb{N}^*}$$

is an increasing sequence. Then for any increasing convex function on $[0,1]$ we have:

$$\frac{1}{n}\sum_{k=1}^{n} f\left(\frac{a_k}{a_n}\right) \geq \frac{1}{n+1}\sum_{k=1}^{n+1} f\left(\frac{a_k}{a_{n+1}}\right).$$

Acknowledgements

We would like to thank the anonymous reviewers for their important observations that led to the improvement of this work.

References

[1] G. Bennett, G. Jameson, *Monotonic averages of convex functions*, J. Math. Anal. Appl. **252**(2000), 410–430.

[2] J.-Ch. Kuang, *Some extensions and refinements of Minc-Sathre inequality*, Math. Gaz. **83**(1999), 123–127.

[3] R.A. DeVore, G.G. Lorentz, *Constructive Approximation*, Springer, 1993.

[4] F. Qi, *Generalizations of Alzer's and Kuang's inequality*, Tamkang J. Math. **31**(3) (2000), 223–227.

[5] F. Qi, B.-N. Guo, *Monotonicity of sequences involving convex function and sequence*, Math. Inequal. Appl. **2**(2006), 247–254.

[6] D.D. Stancu, *On a generalization of Bernstein polynomials* (Romanian), Studia Univ. Babeş-Bolyai **14**(2) (1969), 31–45.

Ioan Gavrea
Technical University of Cluj-Napoca
Department of Mathematics
Str. C. Daicoviciu 15
RO-3400 Cluj-Napoca, Romania
e-mail: Ioan.Gavrea@math.utcluj.ro

International Series of Numerical Mathematics, Vol. 157, 193–200

Inequalities Involving the Superdense Unbounded Divergence of Some Approximation Processes

Alexandru Ioan Mitrea and Paulina Mitrea

Abstract. Estimations concerning the norm of the approximating functionals associated to some approximation procedures are given, in order to deduce their superdense unbounded divergence.

Mathematics Subject Classification (2000). 41A05.

Keywords. Superdense set, unbounded divergence, Bernoulli's inequality, Jackson's inequality, approximation error.

1. Introduction

Given an integer $s \geq 0$, denote by C^s the Banach space of all functions $f : [-1, 1] \to \mathbb{R}$ which are continuous together with their derivatives up to the order s, endowed with the norm:

$$\|f\|_s = \|f^{(s)}\| + \sum_{j=0}^{s-1} |f^{(j)}(0)|, \text{ if } s \geq 1$$
$$\|f\|_0 = \|f^{(0)}\| = \|f\|,$$

where $\|\cdot\|$ means the uniform norm.

Let $\mathcal{M} = \{x_n^k : \ n \geq 1, \ 1 \leq k \leq i_n\}$ be a triangular node matrix, where $(i_n)_{n\geq 1}$ is a strictly increasing sequence of natural numbers and $-1 \leq x_n^1 < x_n^2 < \cdots < x_n^{i_n} \leq 1$.

Let us consider, too, a given linear continuous functional $A : C^s \to \mathbb{R}$ and the approximating functionals $D_n : C^s \to \mathbb{R}$, $n \geq 1$

$$D_n f = \sum_{j=0}^{m} \sum_{i=1}^{j_n} a_n^{kj} f^{(j)}(x_n^k), \quad f \in C^s, \tag{1.1}$$

where m is a given integer, with $0 \leq m \leq s$ and a_n^{kj} are real coefficients.

In this paper, we shall deal with *approximation procedures* described by the formulas

$$Af = D_n f + R_n f; \quad n \geq 1, \ f \in C^s, \tag{1.2}$$

where $R_n f$, $n \geq 1$, are the *approximation errors*.

We shall assume that these procedures are of *interpolatory type*, i.e.,

$$AP = D_n P \tag{1.3}$$

for each polynomial P whose degree does not exceed $q_n = (m+1)i_n - 1$.

Many approximation procedures are comprised within the framework of the scheme described by the relations (1.2) and (1.1): *pointwisely Lagrange interpolation* ($Af = f(x_0)$, with a given $x_0 \in [-1,1]$), *numerical differentiation* (with $s \geq 1$, $m \leq s-1$ and $Af = f^{(m+1)}(x_0)$, usually $x_0 = 0$), *quadrature procedures* ($Af = \displaystyle\int_{-1}^{1} w(x)f(x)dx$, with a given weight-function $w(x)$).

The aim of this paper is to establish various types of inequalities regarding the norm of the approximating functionals D_n corresponding to the numerical differentiation, in order to characterize the topological structure of the set of unbounded divergence of the family $\{D_n : n \geq 1\}$, using the following *principle of the condensation of singularities* of Functional Analysis:

1.1. Theorem. [2], [10]. *If X is a Banach space, Y is a normed space and $(A_n)_{n\geq 1}$, $A_n : X \to Y$ is a sequence of continuous linear operators with $\limsup\limits_{n\to\infty} \|A_n\| = \infty$, then the set of singularities of the family $\{A_n : n \geq 1\}$,*

$$\mathcal{S}(A_n) = \{x \in X : \limsup_{n\to\infty} \|A_n x\| = \infty\}$$

is superdense in X.

We recall that a subset S of a topological space T is said to be *superdense* in T if it is residual, uncountable and dense in T.

In what follows, we denote by M_i, $i \geq 1$, some positive constants, which do not depend on n.

2. Numerical differentiation on Chebyshev node matrix

Firstly, take $m = 0$ and $Af = f'(0)$, so (1.1) and (1.2) become the numerical differentiation formulas (shortly n.d.f.):

$$f'(0) = \sum_{k=1}^{i_n} a_n^k f(x_n^k) + R_n f, \quad f \in C^s, \ s \geq 1, \ n \geq 1. \tag{2.1}$$

Let us consider, for each $\alpha > -1$, the Jacobi node matrix $\mathcal{M}^\alpha$, whose nth row contains the roots of the ultraspherical Jacobi polynomial $P_n^{(\alpha)}$, $n \geq 1$.

R.A. Lorentz [3] has established the convergence of the n.d.f. (2.1) in the case of the matrix of extreme Chebyshev nodes (i.e., $\alpha = 1/2$) and their divergence for equidistant nodes in $[-1,1]$. In fact, the set of all functions in C^1 for which the n.d.f. (2.1) unboundedly diverge is superdense in the Banach space $(C^1, \|\cdot\|_1)$ in the case of equidistant nodes, while the convergence in the case of Jacobi nodes hold for each $\alpha \geq -\frac{1}{2}$, [4]. On the same topic, we remark the results of K. Balázs, [1], J. Szabados and P. Vértesi [7], [9].

In this section, we shall take $m = 1$, $s = 2$ and $Af = f''(0)$, $f \in C^s$. The approximation procedures described by (1.2) and (1.1) lead to the n.d.f.:

$$\begin{cases} f''(0) = D_n f + R_n f;\ f \in C^2,\ n \geq 1 \\ D_n f = \sum_{k=1}^{i_n} a_n^k f(x_n^k) + \sum_{k=1}^{i_n} b_n^k f'(x_n^k) \end{cases} \tag{2.2}$$

Further, suppose that the node matrix $\mathcal{M}^\alpha$ is the even Chebyshev node matrix $\mathcal{M}_T$, i.e., $\alpha = -\frac{1}{2}$ and $i_n = 2n$, so x_n^k, $1 \leq k \leq 2n$, are the roots of the Chebyshev polynomial $T_{2n}(x) = \cos(2n \arccos x)$, $-1 \leq x \leq 1$, $n \geq 1$. Putting $t_{2n}^k = \sin \frac{2k-1}{4n}\pi$, $1 \leq k \leq n$, it is easy to see that $x_n^{n+k} = t_{2n}^k$, $x_n^{n-k+1} = -t_{2n}^k = t_{2n}^{-k}$, $1 \leq k \leq n$, so $\mathcal{M}_T = \{\pm t_{2n}^k :\ n \geq 1,\ 1 \leq k \leq n\}$. Writing a_{2n}^2, a_{2n}^{-k} instead of a_n^{n+k}, a_n^{n-k+1}, $1 \leq k \leq n$, respectively, and b_{2n}^k, b_{2n}^{-k} instead of b_{2n}^k, b_{2n}^{-k}, $1 \leq k \leq n$, respectively, the n.d.f. (2.2) become:

$$f''(0) = D_{2n} f + R_{2n} f, \quad f \in C^2,\ n \geq 1 \tag{2.3}$$

$$D_{2n} f = \sum_{|k| \leq n,\ k \neq 0} a_{2n}^k f(t_{2n}^k) + \sum_{|k| \leq n,\ k \neq 0} b_{2n}^k f'(f_{2n}^k). \tag{2.4}$$

Now, let $h_{2n,0}^j(t)$ and $h_{2n,1}^j(t)$ be the fundamental polynomials of Hermite interpolation, [6]. Taking into account the interpolatory condition (1.3) with respect to n.d.f. (2.3) and (2.4), we get

$$a_{2n}^k = A(h_{2n,0}^k) = (h_{2n,0}^k)''(0),$$

$$b_{2n}^k = A(h_{2n,1}^k) = (h_{2n,1}^k)''(0), \quad |k| \leq n, \quad k \neq 0,$$

i.e., (see also [5]):

$$\begin{cases} a_{2n}^k = 2 \cdot \dfrac{T_{2n}''(t_{2n}^k)}{(T_{2n}'(t_{2n}^k))^3} \left[\dfrac{T_{2n}(0) T_{2n}''(0)}{t_{2n}^k} + \dfrac{T_{2n}^2(0)}{(t_{2n}^k)^3} \right] \\ \qquad + \dfrac{2}{(T_{2n}'(t_{2n}^k))^2} \left[\dfrac{T_{2n}(0) T_{2n}''(0)}{(t_{2n}^k)^2} + 3 \cdot \dfrac{T_{2n}^2(0)}{(t_{2n}^k)^4} \right], \ 1 \leq k \leq n \\ a_{2n}^{-k} = a_{2n}^k,\ 1 \leq k \leq n \end{cases} \tag{2.5}$$

$$\begin{cases} b_{2n}^k = -\dfrac{2}{(T_{2n}'(t_{2n}^k))^2} \left[\dfrac{T_{2n}(0) T_{2n}''(0)}{t_{2n}^k} + \dfrac{T_{2n}^2(0)}{(t_{2n}^k)^3} \right], \ 1 \leq k \leq n \\ b_{2n}^{-k} = -b_{2n}^k,\ 1 \leq k \leq n. \end{cases} \tag{2.6}$$

Let $\alpha_{2n} = \sum_{k=1}^{n} |a_{2n}^k|$ and $\beta_{2n} = \sum_{k=1}^{n} |b_{2n}^k|$, $n \geq 1$.

In order to establish estimations concerning the approximation error $R_n f$ and to characterize the topological structure of the unbounded set of n.d.f. (2.3), we need two-sided estimates for α_{2n} and β_{2n}.

2.1. Lemma. *The following inequalities*

$$M_1 n^2 \leq \alpha_{2n} \leq M_2 n^2 \tag{2.7}$$

$$M_3 n \ln n \leq \beta_{2n} \leq M_4 n \ln n \tag{2.8}$$

hold for sufficiently large n's, i.e., $\alpha_{2n} \sim n^2$ *and* $\beta_{2n} \sim n \ln n$.

Proof. It is easy to verify the relations

$$T'_{2n}(t^k_{2n}) = (-1)^{n+k} \cdot 2n \cdot \left(\cos \frac{2k-1}{4n}\pi\right)^{-1},$$

$$T''_{2n}(t^k_{2n}) = (-1)^{n+k} \cdot 2nt^k_{2n} \left(\cos \frac{2k-1}{4n}\pi\right)^{-3},$$

so, according to (2.5) and (2.6) we obtain:

$$\begin{cases} a^k_{2n} &= \dfrac{-1}{2n^2(t^k_{2n})^n}(4n^2(t^k_{2n})^2 - 3 + 2(t^k_{2n})^2) \\ b^k_{2n} &= \dfrac{1-(t^k_{2n})^2}{2n^2(t^k_{2n})^3}(4n^2(t^k_{2n})^2 - 1). \end{cases} \tag{2.9}$$

The classic inequalities

$$\frac{2}{\pi}x \leq \sin x \leq x, \quad 0 \leq x \leq \frac{\pi}{2} \tag{2.10}$$

show that

$$\begin{cases} a^1_{2n} > 0;\ a^k_{2n} < 0,\ \forall\ k \geq 2, \text{ if } n \geq 2 \\ b^k_{2n} > 0,\ \forall\ n \geq 1,\ \forall\ k \in \{1,2,3,\dots,n\}. \end{cases} \tag{2.11}$$

It follows from (2.9) and (2.11):

$$|a^k_{2n}| = \frac{2}{(t^k_{2n})^2} + \frac{1}{n^2(t^k_{2n})^2} - \frac{3}{2n^2(t^k_{2n})^4}. \tag{2.12}$$

Relation (2.10), written as $\dfrac{1}{x} \leq \dfrac{1}{\sin x} \leq \dfrac{\pi}{2x}$, $0 < x < \dfrac{\pi}{2}$, combined with (2.12), gives:

$$\begin{aligned} &\frac{n^2}{\pi^2}\left(\frac{32}{(2k-1)^2} + \frac{16}{(2k-1)^2 n^2} - \frac{24\pi^2}{(2k-1)^4}\right) \\ &\qquad \leq |a^k_{2n}| \leq n^2\left(\frac{8}{(2k-1)^2} + \frac{4}{n^2(2k-1)^2}\right); \quad k \geq 2. \end{aligned}$$

Adding, we obtain, for sufficiently large n's:

$$\sum_{k=2}^{n} |a_{2n}^k| \leq 8n^2 \sum_{k=2}^{n} \frac{1}{(2k-1)^2} + 4 \sum_{k=2}^{n} \frac{1}{(2k-1)^2} \leq M_5 n^2; \tag{2.13}$$

$$\sum_{k=2}^{n} |a_{2n}^k| \geq 8n^2 \sum_{k=2}^{2n} \frac{1}{(2k-1)^2} \left(\frac{4}{\pi^2} - 3\frac{1}{(2k-1)^2} \right) \tag{2.14}$$

$$\geq 8n^2 \left(\frac{4}{\pi^2} - \frac{1}{3} \right) \sum_{k=2}^{n} \frac{1}{(2k-1)^2} \geq M_6 n^2.$$

On the other hand, we have:

$$a_{2n}^1 \geq \frac{1}{2n^2} \cdot \frac{256n^4}{\pi^4} \left[3 - (2n^2+1)\frac{\pi^2}{16n^2} \right] \geq \frac{8n^2}{\pi^4}(48 - 3\pi^2) \tag{2.15}$$

$$a_{2n}^1 \leq \frac{1}{2n^2} \left[3 - (2n^2+1)^2 \cdot \frac{1}{4n^2} \right] 16n^4 = 2(10n^2 - 1) \leq 20n^2. \tag{2.16}$$

Inequalities (2.11), (2.13), (2.14), (2.15) and (2.16) prove (2.7). Similarly, taking into account (2.11) and writing b_{2n}^k of (2.9) as

$$b_{2n}^k = \frac{2}{t_{2n}^k} - 2t_{2n}^k + \frac{1}{2n^2 t_{2n}^k} - \frac{1}{2n^2 (t_{2n}^k)^3}, \quad 1 \leq k \leq n, \tag{2.17}$$

we get, in accordance with (2.10):

$$\sum_{k=1}^{n} |b_{2n}^k| \leq 4n \sum_{k=1}^{n} \frac{1}{2k-1} + \frac{1}{2n} \sum_{k=1}^{n} \frac{1}{2k-1}$$

$$\leq M_7 n \ln n + M_8 \frac{\ln n}{n};$$

$$\sum_{k=1}^{n} |b_{2n}^k| \geq \frac{8n}{\pi} \sum_{k=1}^{n} \frac{1}{2k-1} - \frac{\pi}{n} \sum_{k=1}^{n} (2k-1) - n\pi \sum_{k=1}^{n} \frac{1}{(2k-1)^3}$$

$$\geq M_9 n \ln n - M_{10} n - M_{11} n.$$

So, (2.8) is valid, which completes the proof.

Now, we are in a position to prove the main result of this paper.

2.2. Theorem. *The set of all functions in C^2 for which the n.d.f. described by (2.3) and (2.4) unboundedly diverge, i.e.,*

$$\left\{ f \in C^2 : \limsup_{n\to\infty} |D_{2n} f| = \infty \right\},$$

is superdense in the Banach space $(C^2, \|\cdot\|_2)$.

Proof. Let $\delta_{2n}^k = \dfrac{1}{3} \min\{t_{2n}^k - t_{2n}^{k-1}; t_{2n}^{k+1} - t_{2n}^k\}$, $1 \leq k \leq n-1$, with $t_{2n}^0 = 0$, $\delta_{2n}^n = \dfrac{1}{3} \min\{t_{2n}^n - t_{2n}^{n-1}, 1 - t_{2n}^n\}$ and $\delta_{2n} = \max\{\delta_{2n}^k : 1 \leq k \leq n\}$.

Define the functions $g_{2n}^k : [-1,1] \to \mathbb{R}$ and $g_{2n} : [-1,1] \to \mathbb{R}$, $n \geq 1, 1 \leq k \leq n$ as follows:

$$g_{2n}^k(t) = (t - t_{2n}^k)[(t - t_{2n}^k)^3 - (\delta_{2n}^k)^3]$$
$$g_{2n}(t) = \begin{cases} g_{2n}^k(t)\operatorname{sign} b_{2n}^k, & \text{if } |t - t_{2n}^k| \leq \delta_{2n}^k,\ 1 \leq k \leq n \\ 0; & \text{otherwise.} \end{cases}$$

It is a simple exercise to show that $g_{2n} \in C^2$ and

$$|g_{2n}''(t)| \leq |(g_{2n}^k)''(t)| \leq 24(\delta_{2n}^k)^5, \quad |t| \leq 1,\ n \geq 1,$$

so:

$$\|g_{2n}\|_2 = |g_{2n}(0)| + |g_{2n}'(0)| + \|g_{2n}''\| \leq 24(\delta_{2n}^k)^5 \leq 24\delta_{2n}^5, \tag{2.18}$$

because $g_{2n}(0) = g_{2n}'(0) = 0$.

Now, introduce the functions $f_{2n} \in C^2$ by $f_{2n} = \dfrac{1}{24\delta_{2n}^5} g_{2n}$, $n \geq 1$.

The relations $g_{2n}(t_{2n}^k) = 0$, $g_{2n}'(t_{2n}^k) = -(\delta_{2n}^k)^6$, together with (2.18) give:

$$D_{2n}f_{2n} = -\frac{\delta_{2n}}{24}\sum_{k=1}^{n}\left(\frac{\delta_{2n}^k}{\delta_{2n}}\right)^6 |b_{2n}^k|. \tag{2.19}$$

The relations $t_{2n}^k - t_{2n}^{k-1} = 2\sin\dfrac{\pi}{4n}\sin\dfrac{n-k+1}{2n}\pi$, $1 \leq k \leq n$ and (2.10), combined with the definitions of δ_{2n}^k and δ_{2n} lead to:

$$M_{11}\frac{n-k+1}{n^2} \leq \delta_{2n}^k \leq M_{12}\frac{n-k+1}{n^2}, \quad 1 \leq k \leq n \tag{2.20}$$
$$M_{11} \leq n\delta_{2n} \leq M_{12}, \tag{2.21}$$

because $\delta_{2n}^k \geq \delta_{2n}^{k-1}$, $1 \leq k \leq n-1$, so $\delta_{2n} = \delta_{2n}^1$.

We deduce from (2.19), (2.20) and (2.21):

$$|D_{2n}f_{2n}| \geq M_{13}\frac{1}{n}\sum_{k=1}^{n}\left(\frac{n-k+1}{n}\right)^6 |b_{2n}^k| \tag{2.22}$$
$$= \frac{M_{13}}{n}\sum_{k=1}^{n}\left(1 - \frac{k-1}{n}\right)^6 |b_{2n}^k| \geq \frac{M_{13}}{n}\sum_{k=1}^{n}\left(1 - 6\frac{k-1}{n}\right)|b_{2n}^k|$$

taking into account Bernoulli's inequality $(1-x)^m \geq 1 - mx$, $x \in [0,1]$, $m \geq 1$.

The inequalities (2.22) and (2.8) give:

$$|D_{2n}f_{2n}| \geq \frac{M_{13}}{n}\beta_{2n} - \frac{M_{14}}{n^2}\sum_{k=1}^{n}(k-1)|b_{2n}^k|. \tag{2.23}$$

It follows from (2.17), (2.11) and (2.10):

$$|b_{2n}^k| = b_{2n}^k \leq \frac{2}{t_{2n}^2} + \frac{1}{2n^2 t_{2n}^k} \leq \frac{4n}{2k-1} + \frac{2}{n(2k-1)},$$

so:

$$\sum_{k=1}^{n}(k-1)|b_{2n}^{k}| \le \left(4n+\frac{2}{n}\right)\sum_{k=1}^{n}\frac{k-1}{2k-1} \le \left(4n+\frac{2}{n}\right)\cdot\frac{n}{2} \le M_{15}n^2,$$

which, combined with (2.23) and (2.8), leads to the inequality:

$$|D_{2n}f_{2n}| \ge \frac{M_{13}}{n}M_3 n\ln n - \frac{M_{14}}{n^2}M_{15}n^2 \ge M_{16}\ln n. \tag{2.24}$$

Using (2.24) we obtain

$$\|D_{2n}\| = \sup\{|D_{2n}f| : \ f\in C^2, \ \|f\|\le 1\} \ge |D_{2n}f_{2n}| \ge M_{16}\ln n, \tag{2.25}$$

for sufficiently large n's.

Finally, apply Theorem 1.1, with $X = C^2$, $Y = \mathbb{R}$, $A_n = D_{2n}$. In accordance with (2.25) we have:

$$\limsup_{n\to\infty}\|A_n\| = \limsup_{n\to\infty}\|D_{2n}\| = \infty,$$

which completes the proof.

2.3. Remark. It follows from Jackson's type inequalities, [8], that there exists a polynomial P of degree equal to q_n and a positive M_{17} so that the relations

$$\|f^{(j)} - p^{(j)}\| \le M_{17}q_n^{-r}\omega\left(f^{(r)};\frac{1}{q_n}\right); \quad f\in C^r;\ r\ge s;\ 0\le j\le r$$

hold for sufficiently large n's, where $\omega(g;\cdot)$ is the modulus of continuity of a function $g\in C$.

From here, we deduce, similarly to [5],

$$|R_nf| \le M_{18}n^{-r}(M_{19}n + \alpha_{2n} + \beta_{2n}\cdot n)\omega\left(f^{(r)};\frac{1}{n}\right),$$

which gives, in accordance with Lemma 2.1:

$$|R_nf| \le M_{20}n^{2-r}(\ln n)\omega\left(f^{(r)};\frac{1}{n}\right), \quad r\ge 2.$$

So, we can state:

2.3.1. The n.d.f. described by (2.3) and (2.4) are convergent on the class C^r, $r\ge 3$, i.e., $\lim\limits_{n\to\infty} D_{2n}f = Af = f''(0)$, $\forall\ f\in C^r$.

2.3.2. The n.d.f. described by (2.3) and (2.4) are convergent on the subset of all functions $f\in C^2$ whose second derivatives satisfy a Dini-Lipschitz condition

$$\lim_{\delta\searrow 0}\omega(f'';\delta)\ln\delta = 0.$$

References

[1] K. Balázs, *On the convergence of the derivatives of Lagrange interpolation polynomials*, Acta Math. Acad. Sci. Hungar. **55** (1990), 323–325.

[2] S. Cobzaş and I. Muntean, *Condensation of singularities for some interpolation processes*, Proc. of ICAOR (Approximation and Optimization), vol. 1, 227–232, Transilvania Press Publishers, Cluj-Napoca, 1997.

[3] R.A. Lorentz, *Convergence of Numerical Differentiation*, J. Approx. Theory **30** (1980), 59–70.

[4] A.I. Mitrea, *Convergence and Superdense Unbounded Divergence in Approximation Theory*, Transilvania Press Publishers, Cluj (Romania) (1998), [Zbl.0978.41001]

[5] A.I. Mitrea, *On the convergence of a class of approximation procedures*, PU.M.A. (Budapest-Siena), vol. 15 no. 2-3 (2004), 225–234.

[6] A. Schönhage, *Approximationstheorie*, Berlin, Walter de Gruyter, 1991.

[7] J. Szabados and P. Vertési, *Interpolation of Functions*, World Sci. Publ. Co., Singapore, 1990.

[8] G. Szegő, *Orthogonal Polynomials*, Amer. Math. Soc. Providence, R.I., 1975.

[9] P. Vertési, *A Panorama of Hungarian Mathematics in the Twentieth Century*, **14** Springer-János Bolyai Mathematical Society (ed. J. Horváth) (2005).

[10] L. Waelbroeck, *Topological Vector Spaces and Algebras*, Lecture Notes in Mathematics **230**, Springer-Verlag, Berlin/New York, 1971.

Alexandru Ioan Mitrea and Paulina Mitrea
Technical University of Cluj-Napoca
Department of Mathematics
str. C. Daicoviciu nr.15
Romania
e-mail: alexandru.ioan.mitrea@math.utcluj.ro
e-mail: paulina.mitrea@cs.utcluj.ro

International Series of Numerical Mathematics, Vol. 157, 201–214

An Overview of Absolute Continuity and Its Applications

Constantin P. Niculescu

Abstract. The aim of this paper is to illustrate the usefulness of the notion of absolute continuity in a series of fields such as Functional Analysis, Approximation Theory and PDE.

Mathematics Subject Classification (2000). Primary 41A36, 46B50, 47B07, 46E35; Secondary 47B38, 54E40.

Keywords. Absolute continuity, Korovkin's theorem, uniform continuity, compact inclusions.

1. Introduction

The basic idea of *absolute continuity* is to control the behavior of a function $f : X \to \mathbb{R}$ via an estimate of the form

$$|f| \leq \varepsilon q + \delta(\varepsilon)p, \quad \text{for every } \varepsilon > 0, \tag{1}$$

where $p, q : X \to \mathbb{R}$ are suitably chosen nonnegative functions. Technically, this means that for every $\varepsilon > 0$, one can find $\delta(\varepsilon) > 0$ such that $|f(x)| \leq \varepsilon q(x) + \delta(\varepsilon)p(x)$, for all $x \in X$. Thus the property of absolute continuity can be seen as a relaxation of the condition of domination

$$|f| \leq p.$$

In this respect (1) allows us to interpolate between two extreme cases: $|f| \leq q$ and $|f| \leq p$, one appearing as "too weak" and the other "too special".

Measure Theory offers us the important case of σ-additive measures defined on a σ-algebra $\mathcal{T}$ (of subsets of a set T). In this context, a measure $m : \mathcal{T} \to \mathbb{C}$ is said to be *absolutely continuous* with respect to a positive measure $\mu : \mathcal{T} \to \mathbb{R}$ (abbreviated, $m \ll \mu$) if for every $\varepsilon > 0$ there is a $\eta = \eta(\varepsilon) > 0$ such that for all $A \in \mathcal{T}$ with $\mu(A) \leq \eta$ we have

$$|m(A)| \leq \varepsilon.$$

Since m has finite variation $|m|$ (see [10], Theorem 19.13 (v)), the condition $m \ll \mu$ yields

$$|m(A)| \leq \varepsilon + \frac{|m|(T)}{\eta}\mu(A) \quad \text{for all } A \in \mathcal{T} \text{ and } \varepsilon > 0, \tag{2}$$

that represents the case of (1) when $X = \mathcal{T}$, $f = m$, $q = 1$, $p = \mu$ and $\delta = |m|(T)/\eta$. In turn, (2) yields the absolute continuity of m with respect to μ since for every $A \in \mathcal{T}$ with

$$\mu(A) \leq \frac{\varepsilon\eta(\varepsilon/2)}{2\,|m|(T)}$$

we have $|m(A)| \leq \varepsilon/2 + \varepsilon/2 = \varepsilon$.

The main criterion of absolute continuity in the above context is provided by the membership of negligible sets:

$$m \ll \mu \text{ if and only if } \mu(A) = 0 \text{ implies } m(A) = 0. \tag{3}$$

See [10], Exercise 19.67, p. 339.

The subject of absolutely continuous functions in Real Analysis can be covered by the above discussion since a function $f : [a, b] \to \mathbb{R}$ is absolutely continuous if and only if it is of the form

$$f(x) = f(a) + m([a, x]),$$

for a suitable Borel measure m which is absolutely continuous with respect to the Lebesgue measure.

The theory of inequalities offers many interesting applications where the concept of absolute continuity is instrumental. In particular this is the case of the famous Hardy-Landau-Littlewood inequalities. See [18].

The aim of this paper is to illustrate the usefulness of the notion of absolute continuity in other areas of mathematics such as Functional Analysis, Approximation Theory and PDE. In particular we show how this notion allows us to derive some quantitative facts from different qualitative properties.

Most of the results we discuss below are not in full generality, but it was our option to emphasize ideas rather than technical results.

2. Absolute continuity in Functional Analysis

Inspired by the case of Measure Theory, the author initiated in the early 70s an operator theoretical generalization of the concept of absolute continuity, which proved to be useful in understanding the properties of weakly compact operators defined on some special Banach spaces such as $C(K)$ and its relatives; as usually, $C(K)$ represents the Banach space (endowed with the sup norm) of all continuous real-valued functions defined on a compact Hausdorff space K.

The basic fact, which led to the concept of absolutely continuous operator, is as follows:

Theorem 1. (*C.P. Niculescu* [15], [16]). *Suppose that E is a Banach space. A bounded linear operator $T \in L(C(K), E)$ is weakly compact if and only if there exists a positive Borel measure μ on K such that for every $\varepsilon > 0$ one can find a $\delta(\varepsilon) > 0$ such that*

$$\|T(f)\| \leq \varepsilon \|f\| + \delta(\varepsilon) \int_K |f| \, d\mu, \tag{4}$$

whenever $f \in C(K)$.

Proof. If T is weakly compact, then the set

$$\mathcal{K} = \{|x' \circ T| : x' \in E', \ \|x'\| \leq 1\}$$

is relatively weakly compact in $C(K)'$ (see [19], p. 119); according to the Riesz representation theorem (see [10], p. 177), the functionals on a space $C(K)$ can be viewed as Borel regular measures, so here modulus means variation. By a classical result due to A. Grothendieck [9], the relative weak compactness of $\mathcal{K}$ means that for every bounded sequence of Borel measurable functions $f_n : K \to \mathbb{R}$ which is pointwise convergent to 0 we have

$$\lim_{n\to\infty} \int_K f_n d\nu = 0, \quad \text{uniformly for } \nu \in \mathcal{K}. \tag{5}$$

Claim: For every $\varepsilon > 0$ there exist a number $\eta(\varepsilon) > 0$ and a finite subset $\mathcal{K}_\varepsilon \subset \mathcal{K}$ such that every Borel measurable function $f : K \to \mathbb{R}$ with $0 \leq f \leq 1$ and $\sup_{\nu\in\mathcal{K}_\varepsilon} \int_K f d\nu \leq \eta(\varepsilon)$ verifies the inequality

$$\sup_{\nu\in\mathcal{K}} \int_K f d\nu \leq \varepsilon.$$

Once the claim is proved, we can easily infer that the measure

$$\mu = \sum_{n=1}^{\infty} \left(\frac{1}{2^n} \sup_{\nu\in\mathcal{K}_{1/n}} \nu \right)$$

verifies a condition of the following form,

$$f \in C(K), \ \|f\| \leq 1, \ \int_K |f| \, d\mu \leq \tilde{\eta}(\varepsilon) \Rightarrow \|T(f)\| \leq \varepsilon, \tag{6}$$

where $\tilde{\eta}(\varepsilon) > 0$ can be obtained from $\eta(\varepsilon)$ by rescaling. Now it is clear that T verifies the inequality (4) for $\delta(\varepsilon) = \|T\| / \tilde{\eta}(\varepsilon)$.

The *Claim* can be proved by reductio ad absurdum. In fact, if the contrary is true, then there are a number $\varepsilon_0 > 0$ and two sequences $(f_n)_n$ (of Borel measurable functions on K) and $(\nu_n)_n$ (of elements of $\mathcal{K}$) such that

i) $0 \leq f_n \leq 1$
ii) $\sup_{1\leq k\leq n} \int_K f_n d\nu_k \leq 2^{-n-1}$
iii) $\int_K f_n d\nu_{n+1} \geq \varepsilon_0$

for all n. Put $g_n = \sup\{f_k : k \geq n\}$ and $g = \inf\{g_n : n \geq 1\}$. Then

$$\sup_{1\leq k\leq n} \int_K g_n d\nu_k \leq 2^{-n},$$

so by (5) we infer that

$$\int_K g d\nu_k = \lim_{n\to\infty} \int_K g_n d\nu_k = 0$$

uniformly for $k \in \mathbb{N}$, a fact that contradicts the inequalities iii) above. Thus the proof of *Claim* is done.

Suppose now that T verifies the estimate (4). We shall show that T maps the weak Cauchy sequences of elements of $C(K)$ into norm convergent sequences in E (whence T is weakly compact by a result due to Grothendieck [9]). In fact, if $(f_n)_n$ is a weak Cauchy sequence in $C(K)$, then by Lebesgue's dominated convergence theorem we get

$$\lim_{m,n\to\infty} \int_K |f_m - f_n| \, d\mu = 0$$

and thus from (4) we conclude that $(Tf_n)_n$ is a norm Cauchy sequence. □

Since the inclusion $L^2(\mu) \subset L^1(\mu)$ is continuous, the inequality (4) yields the following one,

$$\|T(f)\| \leq \varepsilon \|f\| + \delta(\varepsilon) \left(\int_K |f|^2 \, d\mu \right)^{1/2}. \tag{7}$$

According to the Banach-Saks theorem, every bounded sequence in a Hilbert space has a Cesàro converging subsequence. Thus from Theorem 1 we infer the following interesting property of weakly compact operators defined on a $C(K)$ space:

Corollary 1. *If $T \in L(C(K), E)$ is weakly compact, then T maps every bounded sequence into a sequence with Cesàro converging subsequences.*

Another direct consequence of Theorem 1 is as follows:

Corollary 2. *Suppose that $T \in L(C(K), E)$ is an weakly compact operator and $(f_n)_n$ is a bounded sequence of functions in $C(K)$ which converges pointwise to a function $f \in C(K)$. Then $\|T(f_n) - T(f)\| \to 0$.*

For further developments related to Theorem 1 see our papers [14], [15], [16], [17], and the monograph of J. Diestel, H. Jarchow and A. Tonge [7], Ch. 15.

The property of absolute continuity is also instrumental in establishing the Radon-Riesz property for L^p-spaces with $1 \leq p < \infty$. See Corollary 3 below, which is a consequence of following result due to H. Brezis and E.H. Lieb [4], about the "missing term" in Fatou's Lemma:

Theorem 2. *Let $(f_n)_n$ be a sequence of functions in a space $L^p(\mu)$ with $p \in [1, \infty)$, which verifies the following conditions:*

i) $\sup \|f_n\| < \infty$; ii) $f_n \to f$ *almost everywhere.*

Then $f \in L^p(\mu)$ and $\lim_{n\to\infty} (\|f_n\|^p - \|f_n - f\|^p) = \|f\|^p$.

Corollary 3. *Assume that* $(f_n)_n$ *is a sequence of functions in a space* $L^p(\mu)$ ($p \in [1,\infty)$) *such that:*

i) $\|f_n\| \to \|f\|$;

ii) $f_n \to f$ *almost everywhere.*

Then $\|f_n - f\| \to 0$.

Proof of Theorem 2. We start by noticing the following inequality (that illustrates a property of absolute continuity): For each $\varepsilon > 0$ there is a $\delta = \delta(\varepsilon) > 0$ such that

$$||a+b|^p - |a|^p| \le \varepsilon\, |a|^p + \delta |b|^p \tag{8}$$

for all $a, b \in \mathbb{R}$.

This is clear for $p = 1$. For $p > 1$ we shall use the convexity of the function $|x|^p$. Indeed,

$$\begin{aligned} |a+b|^p \le (|a| + |b|)^p &= \left((1-\lambda)\frac{|a|}{(1-\lambda)} + \lambda \frac{|b|}{\lambda}\right)^p \\ &\le |a|^p + \left((1-\lambda)^{1-p} - 1\right)|a|^p + \lambda^{1-p}|b|^p \end{aligned}$$

for all $a, b \in \mathbb{R}$ and $\lambda \in (0,1)$. For $\lambda = 1-(1+\varepsilon)^{-1/(p-1)}$, this inequality yields (8).

The membership of f to the space $L^p(\mu)$ is motivated by Fatou's lemma. According to (8),

$$\begin{aligned} g_{n,\varepsilon} &= (||f_n|^p - |f_n - f|^p - |f|^p| - \varepsilon\, |f_n - f|^p)^+ \\ &\le (1+\delta)\, |f|^p \end{aligned}$$

so that by the dominated convergence theorem we get

$$\lim_{n\to\infty} \int g_{n,\varepsilon} d\mu = 0.$$

Taking into account the inequality

$$||f_n|^p - |f_n - f|^p - |f|^p| \le g_{n,\varepsilon} + \varepsilon\, |f_n - f|^p,$$

we infer that

$$\limsup_{n\to\infty} \int ||f_n|^p - |f_n - f|^p - |f|^p|\, d\mu \le \varepsilon \sup_{n\in\mathbb{N}} \|f_n - f\|^p,$$

whence $\lim_{n\to\infty} (\|f_n\|^p - \|f_n - f\|^p) = \|f\|^p$. □

In what follows we shall concentrate on the connection between absolute continuity and the Arzelà-Ascoli criterion of compactness. Roughly speaking, this criterion asserts that in a function space, the property of being relatively compact means the boundedness plus a certain kind of equi-membership.

If M is a metric space, then an estimate of the form

$$|f(s) - f(t)| \le C d(s,t) \quad \text{for all } s, t \in M$$

is characteristic for the Lipschitz functions $f : M \to \mathbb{R}$. The following relaxation in terms of absolute continuity

$$|f(s) - f(t)| \leq \varepsilon + \delta(\varepsilon)d(s,t) \quad \text{for all } s,t \in M \tag{9}$$

represents precisely the condition of uniform continuity. Indeed, a function $f : M \to \mathbb{R}$ is uniformly continuous if and only if there is a nonnegative function $\omega : [0,\infty) \to \mathbb{R}$ such that $\omega(0) = 0$, ω is continuous at $x = 0$ and

$$|f(s) - f(t)| \leq \omega\left(d(s,t)\right) \quad \text{for all } s,t \in M.$$

As a consequence of (9) we easily infer the well-known fact that every uniformly continuous function $f : \mathbb{R} \to \mathbb{R}$ verifies an estimate of the form

$$|f(x)| \leq a\,|x| + b.$$

A characterization of the metric spaces on which every continuous function is also uniformly continuous appeared in [11].

In the special case when M is also compact, the role of the distance function in (9) can be taken by any separating function for M. Recall that a *separating function* is a nonnegative continuous function $\gamma : M \times M \to \mathbb{R}$ such that

$$\gamma(s,t) = 0 \text{ implies } s = t.$$

If M is a compact subset of $\mathbb{R}^N$, and $f_1, \ldots, f_m \in C(M)$ is a family of functions which separates the points of M (in particular this is the case of the coordinate functions $\mathrm{pr}_1, \ldots, \mathrm{pr}_N$), then

$$\gamma(s,t) = \sum_{k=1}^{m} \left(f_k(s) - f_k(t)\right)^2 \tag{10}$$

is a separating function.

More generally, all separating parametric in General Topology (see [2]) are also separating functions.

The separating functions play an important role in Approximation Theory. This will be detailed in the next section.

Lemma 1. *If K is a compact metric space, and $\gamma : K \times K \to \mathbb{R}$ is a separating function, then any real-valued continuous function f defined on K verifies an estimate of the following form*

$$|f(s) - f(t)| \leq \varepsilon + \delta(\varepsilon)\gamma(s,t) \quad \textit{for all } s,t \in K.$$

Proof. In fact, if the estimate above doesn't work, then for a suitable $\varepsilon_0 > 0$ one can find two sequences $(s_n)_n$ and $(t_n)_n$ of elements of K such that

$$|f(s_n) - f(t_n)| \geq \varepsilon_0 + 2^n \gamma(s_n, t_n) \tag{11}$$

for all n. Without loss of generality we may assume (by passing to subsequences) that both sequences $(s_n)_n$ and $(t_n)_n$ are convergent, respectively to s and t. Since f is bounded, the inequality (11) forces $s = t$. Indeed,

$$\frac{|f(s_n) - f(t_n)|}{2^n} \geq \gamma(s_n, t_n) \to \gamma(s,t) \geq 0.$$

On the other hand, from (11) we can infer that $|f(s) - f(t)| \geq \varepsilon_0$ (and thus $s \neq t$). This contradiction shows that the conclusion of Lemma 1 is true. □

Lemma 1 is the topological counterpart of a well-known result in Measure Theory (precisely, of the equivalence (3)).

Now, a careful inspection of the proof of the Arzelà-Ascoli criterion of compactness in a space $C(K)$ shows that this criterion can be reformulated in terms of absolute continuity as follows:

Theorem 3. *If K is a compact metric space, then a bounded subset $\mathcal{A}$ of the Banach space $C(K)$ is relatively compact if and only if for every $\varepsilon > 0$ there is a number $\delta = \delta(\varepsilon) > 0$ such that*

$$|f(s) - f(t)| \leq \varepsilon + \delta(\varepsilon) d(s,t) \quad \text{for all } s, t \in K \text{ and } f \in \mathcal{A}.$$

Here the role of the distance function can be taken by any separating function for K.

We leave the details to the reader, as an exercise.

The above discussion can be easily extended to the case of functions with values in a complete metric space. Besides, the result of Theorem 3 remains valid for many other spaces, for example, for the space $C^r([a,b])$, of all functions $f : [a,b] \to \mathbb{R}$ which are r-times continuously differentiable, endowed with the norm

$$\|f\|_r = \sum_{k=0}^{r} \sup_{a \leq t \leq b} \left| f^{(k)}(t) \right|.$$

In fact, $C^r([a,b])$ is isomorphic to a subspace of $C([a,b] \times \{0, \ldots, r\})$. This remark can be used to prove that the canonical inclusion

$$j : C^{r+1}([a,b]) \to C^r([a,b]) \tag{12}$$

is compact.

A variant of Theorem 3 in the case of functions defined on a noncompact domain is as follows:

Theorem 4. *Given a bounded open subset Ω of $\mathbb{R}^N$, we denote by $BC(\Omega)$ the Banach space of all continuous bounded functions $f : \Omega \to \mathbb{R}$, endowed with the sup norm. A bounded subset $\mathcal{A}$ of $BC(\Omega)$ is relatively compact if and only if for every $\varepsilon > 0$ there is a number $\delta = \delta(\varepsilon) > 0$ such that*

$$|f(s) - f(t)| \leq \varepsilon + \delta(\varepsilon) d(s,t) \quad \text{for all } s, t \in \Omega \text{ and } f \in \mathcal{A}.$$

3. Absolute continuity and approximation theory

We start with the beautiful result of P.P. Korovkin [12], which put in a new perspective the whole subject of approximation in the case of continuous functions. In order to state this result we need a preparation.

Suppose that E is a Banach lattice. A linear operator $T : E \to E$ is called *positive* if

$$x \geq 0 \text{ implies } T(x) \geq 0.$$

Such an operator is always bounded. See [19], p. 84. For $E = C(K)$ this fact can be checked easily since

$$-\|f\| \cdot 1 \leq f \leq \|f\| \cdot 1 \text{ implies } -\|f\| \cdot T(1) \leq T(f) \leq \|f\| \cdot T(1)$$

and thus $\|T(f)\| \leq \|T(1)\| \cdot \|f\|$.

Theorem 5. (*P.P. Korovkin* [12]). *Consider the functions* $e_0(x) = 1$, $e_1(x) = x$, $e_2(x) = x^2$ *in* $C([0,1])$, *and suppose there is given a sequence*

$$T_n : C([0,1]) \to C([0,1]) \quad (n \in \mathbb{N})$$

of positive linear operators such that $T_n(f) \to f$ *uniformly on* $[0,1]$ *for* $f \in \{e_0, e_1, e_2\}$. *Then*

$$T_n(f) \to f \quad \text{uniformly on } [0,1]$$

for every $f \in C([0,1])$.

The proof is both simple and instructive, so we shall include here the details. The main ingredient is the fact that every function $f \in C([0,1])$ verifies an estimate of the form

$$|f(s) - f(t)| \leq \varepsilon + \delta(\varepsilon)|s - t|^2.$$

See Lemma 1. Then

$$|f - f(t)e_0| \leq \varepsilon e_0 + \delta(\varepsilon)\left(e_2 - 2te_1 + t^2 e_0\right)$$

which implies that $|T_n(f)(s) - f(t)T_n(e_0)(s)|$ is bounded above by

$$\varepsilon T_n(e_0)(s) + \delta(\varepsilon)[T_n(e_2)(s) - 2tT_n(e_1)(s) + t^2 T_n(e_0)(s)]$$

for every $s \in [0,1]$. Therefore

$$\begin{aligned}|T_n(f)(t) - f(t)| &\leq |T_n(f)(t) - f(t)T_n(e_0)(t)| + |f(t)| \cdot |T_n(e_0)(t) - 1| \\ &\leq \varepsilon T_n(e_0)(t) + \delta(\varepsilon)[T_n(e_2)(t) - 2tT_n(e_1)(t) + t^2 T_n(e_0)(t)] \\ &\qquad + \|f\| \cdot |T_n(e_0)(t) - 1|\end{aligned}$$

whence we conclude that $T_n(f) \to f$ uniformly on $[0,1]$.

The above argument (based on Lemma 1) is actually strong enough to cover a much more general result:

Theorem 6. *Suppose that* K *is a compact metric space and* γ *is a separating function for* M. *If* $T_n : C(K) \to C(K)$ $(n \in \mathbb{N})$ *is a sequence of positive linear operators such that* $T_n(1) \to 1$ *uniformly and*

$$T_n(\gamma(\cdot, t))(t) \to 0 \quad \text{uniformly in } t, \tag{13}$$

then $T_n(f) \to f$, *uniformly for each* $f \in C(K)$.

Proof. In fact, taking into account Lemma 1, we have

$$\begin{aligned}|T_n(f)(t) - f(t)| &\leq |T_n(f)(t) - f(t)T_n(1)(t)| + |f(t)| \cdot |T_n(1)(t) - 1| \\ &\leq T_n(|f - f(t)|)(t) + \|f\| \cdot |T_n(1)(t) - 1| \\ &\leq T_n(\varepsilon + \delta(\varepsilon)\gamma(\cdot,t))(t) + \|f\| \cdot |T_n(1)(t) - 1| \\ &\leq \varepsilon T_n(1)(t) + \delta(\varepsilon)T_n(\gamma(\cdot,t))(t) + \|f\| \cdot |T_n(1)(t) - 1|\end{aligned}$$

and the conclusion follows from our hypothesis. □

Theorem 6 is a variant of a recent result by H.E. Lomeli and C.L. Garcia [13] (based on a slightly different concept of separating function).

In order to understand how Theorem 6 extends the Theorem of Korovkin, let us consider the case were M is a compact subset of $\mathbb{R}^N$ and

$$\gamma(s,t) = \sum_{k=1}^{m} (f_k(s) - f_k(t))^2$$

is the separating function (associated to a family of functions $f_1, \ldots, f_m \in C(M)$ which separates the points of M). In this case the condition (13) of uniform convergence can be obtained by imposing that

$$T_n(f) \to f \quad \text{uniformly for } t \in M,$$

for each of the functions $f \in \{1, f_1, \ldots, f_m, f_1^2, \ldots, f_m^2\}$. For $M = [0,1]$ the identity separates the points of M, a fact that leads to the Theorem of Korovkin.

Corollary 4. (*Weierstrass Approximation Theorem*). *If f belongs to $C([a,b])$, then there exists a sequence of polynomials that converges to f uniformly on $[a,b]$.*

Proof. We can restrict to the case where $[a,b] = [0,1]$ (by performing the linear change of variable $t = (x-a)/(b-a)$). Then we apply Theorem 6 for $M = [0,1]$, $\gamma(s,t) = (s-t)^2$ and T_n the nth Bernstein operator,

$$T_n(f)(t) = \sum_{k=0}^{n} \binom{n}{k} t^k (1-t)^{n-k} f(k/n).$$

In fact,

$$T_n(\gamma(\cdot,t))(t) = \frac{t(1-t)}{n}$$

for all $t \in [0,1]$. This computation is part of Bernstein's classical proof of the Weierstrass Approximation Theorem. See [6], pp. 290–292. □

Corollary 5. (*Féjer Approximation Theorem*). *The Cesàro averages of the Fourier partial sums of a continuous function f of period 2π converge uniformly to f.*

Proof. We have to consider the Féjer kernels

$$K_n(t) = \begin{cases} \frac{1}{2n}\left(\dfrac{\sin\frac{nt}{2}}{\sin\frac{t}{2}}\right)^2 & \text{if } t \neq 2k\pi,\ k \in \mathbb{Z} \\ \frac{n}{2} & \text{if } t = 2k\pi,\ k \in \mathbb{Z}. \end{cases}$$

A direct computation shows that

$$K_n(t) = \frac{1}{2} + \frac{1}{n}\sum_{m=1}^{n-1}\sum_{k=1}^{m} \cos kt.$$

The result of Corollary 5 follows from Theorem 6, applied to $M = \mathbb{R} \bmod 2\pi$, $\gamma(s,t) = 1 - \cos(s-t)$ and the sequence of operators

$$T_n(f)(t) = \frac{1}{\pi}\int_{-\pi}^{\pi} K_n(t-s)f(s)ds. \qquad \square$$

Since Lemma 1 does not work for all metric spaces, we cannot use arbitrary separating functions in the case of noncompact metric spaces. However we can still formulate a Korovkin type criterion of convergence for operators acting on the Banach lattice $BUC(M)$ (of all uniformly continuous bounded functions on the metric space M, endowed with the sup norm).

Theorem 7. *Suppose that M is a metric space and*

$$T_n : BUC(M) \to BUC(M) \quad (n \in \mathbb{N})$$

is a sequence of positive linear operators such that $T_n(1) \to 1$ uniformly and

$$T_n(d(\cdot,t)^\alpha)(t) \to 0 \quad \textit{uniformly in } t, \tag{14}$$

for a positive real number α. Then $T_n(f) \to f$, uniformly for each $f \in BUC(M)$.

The usual technique of mollification for approximating the continuous functions by smooth functions can be derived as a consequence of Theorem 7. In the next theorem, a *mollifier* is meant as any nonnegative continuous function $\varphi : \mathbb{R}^N \to \mathbb{R}$ such that

$$\varphi(x) \le C(1 + \|x\|)^{-p} \text{ for some } C > 0 \text{ and } p > N$$

and

$$\int_{\mathbb{R}^N} \varphi(x)dx = 1.$$

The standard mollifier is the function $\varphi(x) = (2\pi)^{-N/2} e^{-\|x\|^2/2}$.

Theorem 8. *If $\varphi : \mathbb{R}^N \to \mathbb{R}$ is a mollifier and $f \in BUC(\mathbb{R}^N)$, then*

$$n^N \int_{\mathbb{R}^N} \varphi\left(n(y-x)\right) f(y)dy \to f(x)$$

uniformly on $\mathbb{R}^N$.

Proof. We apply Theorem 7 for $M = \mathbb{R}^N$, $\alpha \in (0, p-N)$ arbitrarily fixed, and the sequence of operators

$$T_n(f)(x) = n^N \int_{\mathbb{R}^N} \varphi\left(n(y-x)\right) f(y)ds.$$

In order to prove that the condition (14) is fulfilled we need the following estimate:

$$\|y-x\|^{\alpha}\varphi(n(y-x)) \leq \frac{n^N\|y-x\|^{\alpha}}{C(1+n\|y-x\|)^p} \leq \frac{n^{N-\alpha}}{C(1+n\|y-x\|)^{p-\alpha}}.$$

Then

$$0 \leq T_n(\|\cdot - x\|^{\alpha})(x) = n^N \int_{\mathbb{R}^N} \varphi(n(y-x))\|y-x\|^{\alpha}\,ds$$
$$\leq C'\frac{1}{n^{\alpha}}\int_{\mathbb{R}^N}\frac{ds}{(1+n\|y-x\|)^{p-\alpha}},$$

where C' is a constant and the integral in the right-hand side is convergent because $p-\alpha > N$. Consequently $T_n(\|\cdot - x\|^{\alpha})(x) \to 0$ uniformly, as $n \to \infty$, and the proof is complete. □

The technique of mollification works outside the framework of continuous functions. It would be interesting to enlarge the theory above to encompass some spaces of differentiable functions (for example, the Sobolev spaces). A nice account of the most significant developments in the Korovkin theory (including Bauer's approach [3] in terms of Choquet boundary) can be found in the monograph [1].

4. Absolute continuity and PDE

There are many instances when the concept of absolute continuity appears in PDE (see [8]) but we shall restrict here to the remarkable theorem of F. Rellich concerning the compact embedding of Sobolev spaces.

Theorem 9. *If Ω is a bounded open subset of $\mathbb{R}^N$ then the canonical injection*

$$i : \mathring{H}^{m+1}(\Omega) \to \mathring{H}^m(\Omega)$$

is compact.

Recall that $\mathring{H}^m(\Omega)$ is the closure of $C_c^{\infty}(\Omega)$ into $H^m(\Omega)$, the Sobolev space of all functions $f : \Omega \to \mathbb{R}$ that have weak derivatives $D^{\alpha} f \in L^2(\mathbb{R}^N)$ of all orders α with $|\alpha| \leq m$. The natural norm on $H^m(\Omega)$ (and thus on $\mathring{H}^m(\Omega)$) is

$$||f||_{H^m} = \left(\sum_{|\alpha|\leq m}\int_{\Omega}|D^{\alpha}f(x)|^2\,dx\right)^{1/2}.$$

Before to enter the details of Theorem 9, we shall discuss an easy (though important) application, related to a property of absolute continuity of compact operators.

Lemma 2. (*Ehrling's Lemma*). *Assume that E, F, G are Banach spaces. If $T \in L(E,F)$ is a compact linear operator and $S \in L(F,G)$ is an one-to-one bounded linear operator, then for every $\varepsilon > 0$ there is a $\delta(\varepsilon) > 0$ such that*

$$\|Tx\| \leq \varepsilon \|x\| + \delta(\varepsilon) \|S(Tx)\| \quad \textit{for all } x \in E.$$

The proof is similar to the proof of Lemma 1, and we shall omit the details.

By combining Ehrling's Lemma with Theorem 9 we get the estimate

$$||f||_{H^{m-1}} \leq \varepsilon ||f||_{H^m} + \delta(\varepsilon) ||f||_{L^2} \quad \text{for all } f \in \mathring{H}^m(\Omega),$$

which yields

$$||f||_{H^{m-1}} \leq \frac{1}{2} \left(\sum_{|\alpha|=m} \int_\Omega |D^\alpha f(x)|^2 \, dx \right)^{1/2} + \frac{1}{2} ||f||_{H^{m-1}} + \delta(1/2) ||f||_{L^2}$$

that is,

$$||f||_{H^{m-1}} \leq \left(\sum_{|\alpha|=m} \int_\Omega |D^\alpha f(x)|^2 \, dx \right)^{1/2} + 2\delta(1/2) ||f||_{L^2} .$$

Therefore the norm $||\cdot||_{H^m}$ is equivalent to the norm

$$|f|_{H^m} = \left(\sum_{|\alpha|=m} \int_\Omega |D^\alpha f(x)|^2 \, dx \right)^{1/2} + ||f||_{L^2} .$$

The above renorming argument is typical for many Banach spaces of differentiable functions. See [8].

The usual proof of Theorem 9 (and its generalization to the case of Sobolev spaces $\mathring{W}^{m,p}(\Omega)$) is obtained via the mollification technique described in Theorem 8. However it is possible to provide an alternative argument based on Fourier transform.

Indeed, $\mathring{H}^m(\Omega)$ can be viewed as a subspace of $\mathring{H}^m(\mathbb{R}^N)$. The later space has a very simple description in terms of Fourier transform:

$$\mathring{H}^m(\mathbb{R}^N) = \left\{ f \in L^2\left(\mathbb{R}^N\right) : \int_{\mathbb{R}^N} \left(1 + ||\xi||^2\right)^m \left|\widehat{f}(\xi)\right|^2 d\xi < \infty \right\}$$

Moreover, $||\cdot||_{H^m}$ on $\mathring{H}^m(\mathbb{R}^N)$ is equivalent to the norm $||\,|\cdot|\,||_{H^m}$, where

$$||\,|f|\,||_{H^m} = \left(\int_{\mathbb{R}^N} \left(1 + ||\xi||^2\right)^m \left|\widehat{f}(\xi)\right|^2 d\xi \right)^{1/2} .$$

This gives us a constant $C(m) > 0$ such that $||\cdot||_{H^m} \leq C(m) ||\,|\cdot|\,||_{H^m}$.

Let $\varepsilon > 0$. Then there is number $A > 0$ such that $1 + ||\xi||^2 \geq C(m-1)/\varepsilon$ for $||\xi|| \geq A$. Consequently, for every sequence $(f_k)_k$ of functions in the unit ball of $\overset{\circ}{H}{}^m(\mathbb{R}^N)$ we have

$$||f_j - f_k||^2_{H^{m-1}} \leq C(m-1) \int_{\mathbb{R}^N} \left(1 + ||\xi||^2\right)^{m-1} \left|\widehat{f}_j(\xi) - \widehat{f}_k(\xi)\right|^2 d\xi \tag{15}$$
$$\leq \varepsilon \int_{||\xi||>A} \left(1 + ||\xi||^2\right)^m \left|\widehat{f}_j(\xi) - \widehat{f}_k(\xi)\right|^2 d\xi$$
$$+ \delta(\varepsilon) \int_{||\xi|| \leq A} \left|\widehat{f}_j(\xi) - \widehat{f}_k(\xi)\right|^2 d\xi.$$

The Fourier transform of every function in $\overset{\circ}{H}{}^m(\Omega)$ is holomorphic on $\mathbb{C}^N$, and the Cauchy-Schwarz inequality shows that for every compact subset $K \subset \mathbb{C}^N$ there is a constant $M = M(K) > 0$ such that

$$\sup_{\xi \in K} \left|\widehat{f}(\xi)\right| \leq M \, ||f||_{H^m}$$

for all functions $f \in \overset{\circ}{H}{}^m(\Omega)$. Therefore the functions $(\widehat{f}_k)_k$ are uniformly bounded on the compact subsets of $\mathbb{C}^N$. Because they are holomorphic, a compactness principle due to P. Montel assures us that a subsequence should be uniformly convergent on each compact subset of $\mathbb{C}^N$. See [5], p. 209. Taking into account the estimate (15), that subsequence should also verify $\limsup_{j,k\to\infty} ||f_j - f_k||^2_{H^{m-1}} = 0$. The proof of Theorem 9 is done. □

Acknowledgement

This research was partially supported by the CNCSIS grant CEEX 05-D11-36.

References

[1] F. Altomare and M. Campiti, *Korovkin-Type Approximation Theory and Its Applications*, de Gruyter Studies in Mathematics **17**, de Gruyter, Berlin, 1994.

[2] A.V. Arkhangelskii and L.S. Pontryagin, *General Topology I*, Springer-Verlag, Berlin, 1990.

[3] H. Bauer, *Approximation and abstract boundaries*, Amer. Math. Monthly **85** (1978), 632–647.

[4] H. Brezis and E.H. Lieb, *A relation between pointwise convergence of functions and convergence of functionals*, Proc. Amer. Math. Soc. **88** (1983), 486–490.

[5] B. Chabat, *Introduction à l'analyse complexe*, Tome 1, Ed. Mir, Moscou, 1990.

[6] K.R. Davidson and A.P. Donsig, *Real Analysis with Real Applications*, Prentice Hall, Inc., Upper Saddle River, N.J., 2002.

[7] J. Diestel, H. Jarchow and A. Tonge, *Absolutely Summing Operators*, Cambridge University Press, 1995.

[8] L.C. Evans, *Partial Differential Equations*, American Mathematical Society, Providence, R.I., 3rd Printing, 2002.

[9] A. Grothendieck, *Sur les applications linéaires faiblement compactes d'espaces du type $C(K)$*, Canad. J. Math. **5** (1953), 129–173.

[10] E. Hewitt and K. Stromberg, *Real and Abstract Analysis*, Springer-Verlag, Berlin, 1965.

[11] H. Hueber, *On Uniform Continuity and Compactness in Metric Spaces*, Amer. Math. Monthly **88** (1981), 204–205.

[12] P.P. Korovkin, *On convergence of linear positive operators in the space of continuous functions*, Doklady Akad. Nauk. SSSR (NS) **90** (1953), 961–964. (Russian)

[13] H.E. Lomeli and C.L. Garcia, *Variations on a Theorem of Korovkin*, Amer. Math. Monthly **113** (2006), 744–750.

[14] C.P. Niculescu, *Opérateurs absolument continus*, Rev. Roum. Math. Pures et Appl. **19** (1974), 225–236.

[15] C.P. Niculescu, *Absolute Continuity and Weak Compactness*, Bull. Amer. Math. Soc. **81** (1975), 1064–1066.

[16] C.P. Niculescu, *Absolute Continuity in Banach Space Theory*, Rev. Roum. Math. Pures Appl. **24** (1979), 413–422.

[17] C.P. Niculescu, *Operators of type A and local absolute continuity*, J. Operator Theory, **13** (1985), 49–61.

[18] C.P. Niculescu and C. Buşe, *The Hardy-Landau-Littlewood inequalities with less smoothness*, Journal of Inequalities in Pure and Applied Mathematics (JIPAM) **4** (2003), Issue 3, article 51, p. 8.

[19] H.H. Schaefer, *Banach Lattices and Positive Operators*, Springer-Verlag, New York, 1974.

Constantin P. Niculescu
University of Craiova
Department of Mathematics
RO-200585 Craiova, Romania
e-mail: cniculescu47@yahoo.com

Part V

Generalizations of Convexity and Inequalities for Means

PHB'07

International Series of Numerical Mathematics, Vol. 157, 217–228

Normalized Jensen Functional, Superquadracity and Related Inequalities

Shoshana Abramovich and Silvestru S. Dragomir

Abstract. In this paper we generalize the inequality

$$MJ_n(f,\mathbf{x},\mathbf{q}) \geq J_n(f,\mathbf{x},\mathbf{p}) \geq mJ_n(f,\mathbf{x},\mathbf{q})$$

where

$$J_n(f,\mathbf{x},\mathbf{p}) = \sum_{i=1}^{n} p_i f(x_i) - f\left(\sum_{i=1}^{n} p_i x_i\right),$$

obtained by S.S. Dragomir for convex functions. We show that for the class of functions that we call superquadratic, strictly positive lower bounds of $J_n(f,\mathbf{x},\mathbf{p}) - mJ_n(f,\mathbf{x},\mathbf{q})$ and strictly negative upper bounds of $J_n(f,\mathbf{x},\mathbf{p}) - MJ_n(f,\mathbf{x},\mathbf{q})$ exist when the functions are also nonnegative. We also provide cases where we can improve the bounds m and M for convex functions and superquadratic functions. Finally, an inequality related to the Čebyšev functional and superquadracity is also given.

Mathematics Subject Classification (2000). 26D15.

Keywords. Convex functions, superquadratic functions, Jensen inequality, Jensen Steffensen inequality, Čebyšev inequality.

1. Introduction

In this paper we consider the normalized Jensen functional

$$J_n(f,\mathbf{x},\mathbf{p}) = \sum_{i=1}^{n} p_i f(x_i) - f\left(\sum_{i=1}^{n} p_i x_i\right), \tag{1.1}$$

where $\sum_{i=1}^{n} p_i = 1$, $f : I \longrightarrow \mathbb{R}$, and I is an interval in $\mathbb{R}$.

This type of functionals was considered by S.S. Dragomir in [7], where the following theorem was proved:

Theorem 1. *Consider the normalized Jensen functional* (1.1) *where* $f : C \longrightarrow \mathbb{R}$ *is a convex function on the convex set* C *in a real linear space, and* $\mathbf{x} = (x_1, \dots, x_n) \in C^n$, $\mathbf{p} = (p_1, \dots, p_n)$, $\mathbf{q} = (q_1, \dots, q_n)$ *are nonnegative* n*-tuples satisfying* $\sum_{i=1}^n p_i = 1$, $\sum_{i=1}^n q_i = 1$, $q_i > 0$, $i = 1, \dots, n$. *Then*

$$MJ_n(f, \mathbf{x}, \mathbf{q}) \geq J_n(f, \mathbf{x}, \mathbf{p}) \geq mJ_n(f, \mathbf{x}, \mathbf{q}), \tag{1.2}$$

provided

$$m = \min_{1 \leq i \leq n} \left(\frac{p_i}{q_i} \right), \quad M = \max_{1 \leq i \leq n} \left(\frac{p_i}{q_i} \right).$$

In the following section we show that for a class of functions we call superquadratic, defined below, strictly positive lower bounds of $J_n(f, \mathbf{x}, \mathbf{p}) - mJ_n(f, \mathbf{x}, \mathbf{q})$ and strictly negative upper bounds of $J_n(f, \mathbf{x}, \mathbf{p}) - MJ_n(f, \mathbf{x}, \mathbf{q})$ are obtained when the functions are also nonnegative. We also show when (1.2) holds for m^* larger than $\min_{1 \leq i \leq n} \left(\frac{p_i}{q_i} \right)$, and M^* smaller than $\max_{1 \leq i \leq n} \left(\frac{p_i}{q_i} \right)$. Although $\mathbf{x} = (x_1, \dots, x_n)$, $x_i \in I$, $i = 1, \dots, n$ is not necessarily a monotonic n-tuple, in order to get better bounds than m and M as defined in Theorem 1, both in the superquadratic case and in the convex case, we use Jensen-Steffensen's inequality that states that if $f : I \longrightarrow \mathbb{R}$ is convex, where I is an interval in $\mathbb{R}$, then

$$\sum_{i=1}^{n} a_i f(x_i) \geq A_n f(\overline{x}), \tag{1.3}$$

where $\overline{x} := \frac{\sum_{i=1}^n a_i x_i}{A_n}$, $\mathbf{x} = (x_1, \dots, x_n)$ is any monotone n-tuple in I^n, and $\mathbf{a} = (a_1, \dots, a_n)$ is a real n-tuple that satisfies the condition:

$$0 \leq A_i \leq A_n, \quad i = 1, \dots, n, \text{ where } A_i = \sum_{j=1}^{i} a_j, \text{ and } A_n > 0 \tag{1.4}$$

(see for instance [10, page 57]).

In order to get better bounds than m and M as defined in Theorem 1 for a superquadratic function, we use Theorem 2 below instead of the above Jensen-Steffensen inequality.

In addition, we get in the last section an inequality related to the Čebyšev's type functional and superquadracity.

Definition 1 ([3, Definition 1]). *A function* f *defined on an interval* $I = [0, a]$ *or* $[0, \infty)$ *is superquadratic, if for each* x *in* I *there exists a real number* $C(x)$ *such that*

$$f(y) - f(x) \geq f(|y - x|) + C(x)(y - x) \tag{1.5}$$

for all $y \in I$.

For example, the functions x^p, $p \geq 2$ and the functions $-x^p$, $0 \leq p \leq 2$ are superquadratic functions as well as the function $f(x) = x^2 \log x$, $x > 0$, $f(0) = 0$.

Remark 1. *The definition of superquadracity as stated here appeared first in* 2004 *in papers* [3] *and* [4] *and since then this terminology was used by several authors in many papers and journals.*

Unfortunately, the users of this definition were not aware that the term superquadracity was used in a different context since 1987 (*see* [11], [9] *and* [8]).

In [9] *it is stated: Let X be a real linear space and $\mathbb{R}$ be the set of all reals. Then every function $f : X \longrightarrow \mathbb{R}$ satisfying the inequality*

$$f(x+y) + f(x-y) \geq 2f(x) + 2f(y), \quad x, y \in X \tag{$*$}$$

is called superquadratic.

It is of interest to clarify the relations and the differences between the classes of functions satisfying these definitions. This is worth further investigation.

Definition 1 *and Definition* $(*)$ *are both well established in the mathematical literature. Although there is a conflict of terminology, we use in this paper our Definition* 1.

In Section 2 we use the following lemmas and theorem for superquadratic functions:

Lemma 1 ([3, Lemma 2.1]). *Let f be a superquadratic function with $C(x)$ as in* (1.5).

(i) *Then* $f(0) \leq 0$
(ii) *If* $f(0) = f'(0) = 0$, *then* $C(x) = f'(x)$ *wherever* f *is differentiable at* $x > 0$.
(iii) *If* $f \geq 0$, *then* f *is convex and* $f(0) = f'(0) = 0$.

Lemma 2 ([4, Lemma 2.3]). *Suppose that f is superquadratic. Let $x_i \geq 0$, $i = 1, \ldots, n$ and let $\overline{x} := \sum_{i=1}^{n} a_i x_i$, where $a_i \geq 0$, $i = 1, \ldots, n$ and $\sum_{i=1}^{n} a_i = 1$. Then*

$$\sum_{i=1}^{n} a_i f(x_i) - f(\overline{x}) \geq \sum_{i=1}^{n} a_i f(|x_i - \overline{x}|). \tag{1.6}$$

The following Theorem 2 was proved in [1, Theorem 1] for differentiable positive superquadratic functions f, but because of Lemma 1 (iii) it holds also when f is not always differentiable.

Theorem 2. *Let $f : I \longrightarrow \mathbb{R}$, where I is $[0, a]$ or $[0, \infty)$, be nonnegative superquadratic function. Let $\mathbf{x}$ be a monotone nonnegative n-tuple in I^n and $\mathbf{a}$ satisfies* (1.4). *Let*

$$\overline{x} := \frac{\sum_{i=1}^{n} a_i x_i}{A_n} \tag{1.7}$$

Then

$$\sum_{i=1}^{n} a_i f(x_i) - A_n f(\overline{x}) \geq (n-1) A_n f\left(\frac{\sum_{i=1}^{n} a_i |x_i - \overline{x}|}{(n-1) A_n}\right). \tag{1.8}$$

2. The main results

In Theorem 3 that deals with superquadratic functions we use the same techniques as used in [7] to prove Theorem 1 for convex functions.

Theorem 3. *Under the same conditions and definitions on* $\mathbf{p}$, $\mathbf{q}$, $\mathbf{x}$, m *and* M *as in Theorem* 1, *if* I *is* $[0,a)$ *or* $[0,\infty)$ *and* f *is a superquadratic function on* I, *then*

$$J_n(f,\mathbf{x},\mathbf{p}) - mJ_n(f,\mathbf{x},\mathbf{q}) \tag{2.1}$$
$$\geq mf\left(\left|\sum_{i=1}^{n}\left(q_i - p_i\right)x_i\right|\right) + \sum_{i=1}^{n}\left(p_i - mq_i\right)f\left(\left|x_i - \sum_{j=1}^{n}p_jx_j\right|\right)$$

and

$$J_n(f,\mathbf{x},\mathbf{p}) - MJ_n(f,\mathbf{x},\mathbf{q}) \tag{2.2}$$
$$\leq -\sum_{i=1}^{n}\left(Mq_i - p_i\right)f\left(\left|x_i - \sum_{j=1}^{n}q_jx_j\right|\right) - f\left(\left|\sum_{i=1}^{n}\left(p_i - q_i\right)x_i\right|\right).$$

Proof. To prove (2.1) we define $\mathbf{y}$ as

$$y_i = \begin{cases} x_i, & i = 1,\dots,n \\ \sum_{j=1}^{n} q_jx_j, & i = n+1 \end{cases},$$

and $\mathbf{d}$ as

$$d_i = \begin{cases} p_i - mq_i, & i = 1,\dots,n \\ m, & i = n+1 \end{cases}.$$

Then (1.6) for $\mathbf{y}$ and $\mathbf{d}$ is

$$\begin{aligned}
&\sum_{i=1}^{n}(p_i - mq_i)f(x_i) + mf\left(\sum_{i=1}^{n}q_ix_i\right) - f\left(\sum_{i=1}^{n}p_ix_i\right) \\
&\quad = \sum_{i=1}^{n+1}d_if\left(y_i\right) - f\left(\sum_{i=1}^{n+1}d_iy_i\right) \\
&\quad \geq \sum_{i=1}^{n+1}d_if\left(\left|y_i - \sum_{j=1}^{n+1}d_jy_j\right|\right) \\
&\quad = \sum_{i=1}^{n}(p_i - mq_i)f\left(\left|x_i - \sum_{j=1}^{n}p_jx_j\right|\right) + mf\left(\left|\sum_{i=1}^{n}\left(p_i - q_i\right)x_i\right|\right)
\end{aligned}$$

which is (2.1).

To get (2.2), we choose $\mathbf{z}$ and $\mathbf{r}$ as

$$z_i = \begin{cases} x_i, & i = 1,\dots,n \\ \sum_{j=1}^{n} p_jx_j, & i = n+1 \end{cases},$$

and

$$r_i = \begin{cases} q_i - \frac{p_i}{M}, & i = 1, \dots, n \\ \frac{1}{M}, & i = n+1 \end{cases} .$$

Then, as f is superquadratic and $\sum_{i=1}^{n+1} r_i = 1,\ r_i \geq 0$, we get that

$$\begin{aligned}
&\sum_{i=1}^{n} \left(q_i - \frac{p_i}{M} \right) f(x_i) + \frac{1}{M} f\left(\sum_{i=1}^{n} p_i x_i \right) - f\left(\sum_{i=1}^{n} q_i x_i \right) \\
&= \sum_{i=1}^{n+1} r_i f(z_i) - f\left(\sum_{i=1}^{n+1} r_i z_i \right) \\
&\geq \sum_{i=1}^{n+1} r_i f\left(\left| z_i - \sum_{i=1}^{n+1} r_i z_i \right| \right) \\
&= \sum_{i=1}^{n} \left(q_i - \frac{p_i}{M} \right) f\left(\left| x_i - \sum_{j=1}^{n} q_j x_j \right| \right) + \frac{1}{M} f\left(\left| \sum_{i=1}^{n} (p_i - q_i) x_i \right| \right)
\end{aligned}$$

which is equivalent to (2.2). □

Remark 2. *If the superquadratic function is also nonnegative and therefore according to Lemma* 1 *is convex, then* (2.1) *and* (2.2) *refine Theorem* 1.

In the sequel we use the following notations:

Let $\mathbf{x}_\uparrow = \left(x_{(1)}, \dots, x_{(n)} \right)$ be the *increasing rearrangement* of $\mathbf{x} = (x_1, \dots, x_n)$. Let π be the permutation that transfers $\mathbf{x}$ into $\mathbf{x}_\uparrow$ and let $(\overline{p}_1, \dots, \overline{p}_n)$ and $(\overline{q}_1, \dots, \overline{q}_n)$ be the n-tuples obtained by the same permutation π on $(p_1, \dots, p_n)$ and $(q_1, \dots, q_n)$ respectively. Then for an n-tuple $\mathbf{x} = (x_1, \dots, x_n)$, $x_i \in I$, $i = 1, \dots, n$ where I is an interval in $\mathbb{R}$ we get the following results:

Theorem 4. *Let* $\mathbf{p} = (p_1, \dots, p_n)$, *where* $0 \leq \sum_{j=1}^{i} \overline{p}_j \leq 1$, $i = 1, \dots, n$, $\sum_{i=1}^{n} p_i = 1$, *and* $\mathbf{q} = (q_1, \dots, q_n)$, $0 < \sum_{j=1}^{i} \overline{q}_j < 1$, $i = 1, \dots, n-1$, $\sum_{i=1}^{n} q_i = 1$, *and* $\mathbf{p} \neq \mathbf{q}$. *Denote*

$$m_i = \frac{\sum_{j=1}^{i} \overline{p}_j}{\sum_{j=1}^{i} \overline{q}_j}, \qquad \overline{m_i} = \frac{\sum_{j=i}^{n} \overline{p}_j}{\sum_{j=i}^{n} \overline{q}_j}, \qquad i = 1, \dots, n \tag{2.3}$$

where $(\overline{p}_1, \dots, \overline{p}_n)$ *and* $(\overline{q}_1, \dots, \overline{q}_n)$ *are as denoted above, and*

$$m^* = \min_{1 \leq i \leq n} \{m_i, \overline{m_i}\}, \qquad M^* = \max_{1 \leq i \leq n} \{m_i, \overline{m_i}\}. \tag{2.4}$$

If $\mathbf{x} = (x_1, \dots, x_n)$ *is any n-tuple in* I^n, *where* I *is an interval in* $\mathbb{R}$, *then*

$$M^* J_n(f, \mathbf{x}, \mathbf{q}) \geq J_n(f, \mathbf{x}, \mathbf{p}) \geq m^* J_n(f, \mathbf{x}, \mathbf{q}), \tag{2.5}$$

where $f : I \longrightarrow \mathbb{R}$ *is a convex function on the interval* I.

Proof. As $\mathbf{p} \neq \mathbf{q}$ it is clear that $m^* < 1$, and $M^* > 1$.

As $\sum_{i=1}^n q_i = 1$ and $1 \geq \sum_{i=1}^j q_i > 0$ it is easy to verify that there is an integer k, $2 \leq k \leq n$ such that $x_{(k-1)} \leq \sum_{i=1}^n q_i x_i \leq x_{(k)}$, see also [1].

We apply Jensen-Steffensen's inequality for the increasing $(n+1)$-tuple $\mathbf{y} = (y_1, \ldots, y_{n+1})$

$$y_i = \begin{cases} x_{(i)}, & i = 1, \ldots, k-1 \\ \sum_{j=1}^n q_j x_j, & i = k \\ x_{(i-1)}, & i = k+1, \ldots, n+1 \end{cases} \tag{2.6}$$

and

$$a_i = \begin{cases} \overline{p}_i - m^* \overline{q}_i, & i = 1, \ldots, k-1 \\ m^*, & i = k \\ \overline{p}_{i-1} - m^* \overline{q}_{i-1}, & i = k+1, \ldots, n+1 \end{cases} \tag{2.7}$$

where m^* is defined in (2.4).

It is clear that $\mathbf{a}$ satisfies (1.4). Therefore, (1.3) holds for the increasing $(n+1)$-tuple $\mathbf{y}$ and for a convex function f.

Hence

$$\begin{aligned} \sum_{i=1}^{n+1} a_i f(y_i) &= m^* f\left(\sum_{i=1}^n q_i x_i\right) + \sum_{i=1}^n (p_i - m^* q_i) f(x_i) \\ &\geq f\left(m^* \sum_{i=1}^n q_i x_i + \sum_{i=1}^n (p_i - m^* q_i) x_i\right) = f\left(\sum_{i=1}^n p_i x_i\right). \end{aligned}$$

In other words

$$\sum_{i=1}^n p_i f(x_i) - f\left(\sum_{i=1}^n p_i x_i\right) \geq m^* \left(\sum_{i=1}^n q_i f(x_i) - f\left(\sum_{i=1}^n q_i x_i\right)\right).$$

This completes the proof of the right side inequality in (2.5).

The proof of the left side of (2.5) is similar: We define an increasing $(n+1)$-tuple $\mathbf{z}$

$$z_i = \begin{cases} x_{(i)}, & i = 1, \ldots, s-1 \\ \sum_{j=1}^n p_j x_j, & i = s \\ x_{(i-1)}, & i = s+1, \ldots, n+1 \end{cases} \tag{2.8}$$

and

$$b_i = \begin{cases} \overline{q}_i - \frac{\overline{p}_i}{M^*}, & i = 1, \ldots, s-1 \\ \frac{1}{M^*}, & i = s \\ \overline{q}_{i-1} - \frac{\overline{p}_{i-1}}{M^*}, & i = s+1, \ldots, n+1, \end{cases} \tag{2.9}$$

where s satisfies $x_{(s-1)} \leq \sum_{j=1}^n p_j x_j \leq x_{(s)}$. As $\mathbf{b}$ satisfies (1.4) and $\sum_{i=1}^{n+1} b_i = 1$, by using Jensen-Steffensen's inequality, we get the left side of (2.5).

This completes the proof. $\square$

Remark 3. *If* $\min_{1\le i\le n}\left(\frac{p_i}{q_i}\right)=\frac{\overline{p}_k}{\overline{q}_k},\ k\neq 1,n$ *and* $\max_{1\le i\le n}\left(\frac{p_i}{q_i}\right)=\frac{\overline{p}_s}{\overline{q}_s},\ s\neq 1,n$ *then it is clear that for* $p_i\ge 0$, *and* $q_i>0$, *we get that* $m^*>m$ *and* $M^*<M$ *and in these cases* (2.5) *refines* (1.2).

The following result is proved for superquadratic functions using the same technique used in Theorem 4 for convex functions and by using Theorem 2, therefore, the proof is omitted.

Theorem 5. *Let* $f(x)$ *be a nonnegative superquadratic function on* I *where* I *is* $[0,a]$ *or* $[0,\infty)$. *Let* $\mathbf{x}$, $\mathbf{p}$, $\mathbf{q}$, m^*, M^* *be the same as in Theorem* 4. *Then*

$$J_n(f,\mathbf{x},\mathbf{p})-m^*J_n(f,\mathbf{x},\mathbf{q}) \ge nf\left(\frac{\sum_{i=1}^n(p_i-m^*q_i)\left|x_i-\sum_{j=1}^n p_jx_j\right|+m^*\left|\sum_{i=1}^n(p_i-q_i)x_i\right|}{n}\right), \tag{2.10}$$

and

$$J_n(f,\mathbf{x},\mathbf{p})-M^*J_n(f,\mathbf{x},\mathbf{q}) \le -M^*nf\left(\frac{\sum_{i=1}^n(M^*q_i-p_i)\left|x_i-\sum_{j=1}^n q_jx_j\right|+\left|\sum_{j=1}^n(q_j-p_j)x_j\right|}{M^*n}\right). \tag{2.11}$$

In the following Theorem 6 we state another generalisation of the Jensen inequality for superquadratic functions, then in Theorem 7 we extend Theorem 1 and in Theorem 8 we extend Theorem 3.

Theorem 6. *Assume that* $\mathbf{x}=(x_1,\dots,x_n)$ *with* $x_i\ge 0$ *for* $i\in\{1,\dots,n\}$, $\mathbf{p}=(p_1,\dots,p_n)$ *is a probability sequence and* $\mathbf{q}=(q_1,\dots,q_k)$ *is another probability sequence with* $n,k\ge 2$. *Then for any superquadratic function* $f:[0,\infty)\to\mathbb{R}$ *we have the inequality*

$$\sum_{i_1,\dots,i_k=1}^n p_{i_1}\dots p_{i_k}f\left(\sum_{j=1}^k q_jx_{i_j}\right) \ge f\left(\sum_{i=1}^n p_ix_i\right)+\sum_{i_1,\dots,i_k=1}^n p_{i_1}\dots p_{i_k}f\left(\left|\sum_{j=1}^k q_jx_{i_j}-\sum_{i=1}^n p_ix_i\right|\right). \tag{2.12}$$

Proof. By the definition of superquadratic functions, we have

$$f\left(\sum_{j=1}^k q_jx_{i_j}\right)\ge f\left(\sum_{i=1}^n p_ix_i\right)+C\left(\sum_{i=1}^n p_ix_i\right)\left(\sum_{j=1}^k q_jx_{i_j}-\sum_{i=1}^n p_ix_i\right) + f\left(\left|\sum_{j=1}^k q_jx_{i_j}-\sum_{i=1}^n p_ix_i\right|\right) \tag{2.13}$$

for any $x_{i_j}\ge 0, i_j\in\{1,\dots,n\}$.

Now, if we multiply (2.13) with $p_{i_1}\dots p_{i_k}\geq 0$, sum over $i_1,\dots,i_k$ from 1 to n and take into account that $\sum_{i_1,\dots,i_k=1}^{n} p_{i_1}\dots p_{i_k}=1$ we deduce

$$\begin{aligned}&\sum_{i_1,\dots,i_k=1}^{n} p_{i_1}\dots p_{i_k} f\Big(\sum_{j=1}^{k} q_j x_{i_j}\Big) \qquad (2.14)\\ &\geq f\Big(\sum_{i=1}^{n} p_i x_i\Big)+C\Big(\sum_{i=1}^{n} p_i x_i\Big)\sum_{i_1,\dots,i_k=1}^{n} p_{i_1}\dots p_{i_k}\Big(\sum_{j=1}^{k} q_j x_{i_j}-\sum_{i=1}^{n} p_i x_i\Big)\\ &\quad+\sum_{i_1,\dots,i_k=1}^{n} p_{i_1}\dots p_{i_k} f\Big(\Big|\sum_{j=1}^{k} q_j x_{i_j}-\sum_{i=1}^{n} p_i x_i\Big|\Big).\end{aligned}$$

However

$$I=\sum_{i_1,\dots,i_k=1}^{n} p_{i_1}\dots p_{i_k}\Big(\sum_{j=1}^{k} q_j x_{i_j}-\sum_{i=1}^{n} p_i x_i\Big)=\sum_{i_1,\dots,i_k=1}^{n} p_{i_1}\dots p_{i_k}\Big(\sum_{j=1}^{k} q_j x_{i_j}\Big)-\sum_{i=1}^{n} p_i x_i$$

and since

$$\begin{aligned}&\sum_{i_1,\dots,i_k=1}^{n} p_{i_1}\dots p_{i_k}\Big(\sum_{j=1}^{k} q_j x_{i_j}\Big)\\ &=q_1\sum_{i_1=1}^{n} p_{i_1}x_{i_1}\sum_{i_2,\dots,i_k=1}^{n} p_{i_2}\dots p_{i_k}+\dots+q_k\sum_{i_k=1}^{n} p_{i_k}x_{i_k}\sum_{i_1,\dots,i_{k-1}=1}^{n} p_{i_1}\dots p_{i_{k-1}}\\ &=q_1\sum_{i=1}^{n} p_i x_i+\dots+q_k\sum_{i=1}^{n} p_i x_i=\sum_{i=1}^{n} p_i x_i\end{aligned}$$

hence $I=0$ and by (2.14) we get the desired result (2.12). □

Theorem 7. *Assume that* $\mathbf{x}=(x_1,\dots,x_n)$ *with* $x_i\in I$, $i=1,\dots,n$, I *is an interval in* $\mathbb{R}$, $\mathbf{p}=(p_1,\dots,p_n)$, $\mathbf{r}=(r_1,\dots,r_n)$, $r_i>0$, $i=1,\dots,n$ *are probability sequences, and* $\mathbf{q}=(q_1,\dots,q_k)$, *another probability sequence with* $n,k\geq 2$. *Then, for any convex function* f *on* I *we have the inequality*

$$\begin{aligned}&M\Big(\sum_{i_1,\dots,i_k=1}^{n} r_{i_1}\dots r_{i_k} f\Big(\sum_{j=1}^{k} q_j x_{i_j}\Big)-f\Big(\sum_{i=1}^{n} r_i x_i\Big)\Big) \qquad (2.15)\\ &\geq\Big(\sum_{i_1,\dots,i_k=1}^{n} p_{i_1}\dots p_{i_k} f\Big(\sum_{j=1}^{k} q_j x_{i_j}\Big)-f\Big(\sum_{i=1}^{n} p_i x_i\Big)\Big)\\ &\geq m\Big(\sum_{i_1,\dots,i_k=1}^{n} r_{i_1}\dots r_{i_k} f\Big(\sum_{j=1}^{k} q_j x_{i_j}\Big)-f\Big(\sum_{i=1}^{n} r_i x_i\Big)\Big)\end{aligned}$$

where $m=\min\limits_{1\leq i_1,\dots,i_k\leq n}\left(\frac{p_{i_1}\dots p_{i_k}}{r_{i_1}\dots r_{i_k}}\right)$, $M=\max\limits_{1\leq i_1,\dots,i_k\leq n}\left(\frac{p_{i_1}\dots p_{i_k}}{r_{i_1}\dots r_{i_k}}\right)$.

Proof. The proof is similar to the proof of Theorem 1: We will prove the right side of the inequality. The left side of the inequality is similar. As

$$m\sum_{i=1}^{n} r_i x_i + \sum_{i_1,i_2,\dots,i_k=1}^{n} (p_{i_1}\dots p_{i_k} - m r_{i_1}\dots r_{i_k}) \sum_{j=1}^{k} q_j x_{i_j}$$
$$= \sum_{i_1,\dots,i_k}^{n} p_{i_1}\dots p_{i_k} \sum_{j=1}^{k} q_j x_{ij} = \sum_{i=1}^{n} p_i x_i,$$

$0 \le m \le 1, \quad 0 \le p_{i_1}\dots p_{i_k} - m r_{i_1}\dots r_{i_k} \le 1$ and

$$m + \sum_{i_1\dots i_k=1}^{n} (p_{i_1}\dots p_{i_k} - m r_{i_1}\dots r_k) = 1$$

we get as a result of the convexity of f that

$$mf\left(\sum_{i=1}^{n} r_i x_i\right) + \sum_{i_1,\dots,i_k=1}^{n} (p_{i_1}\dots p_{i_k} - m r_{i_1}\dots r_{i_k})\, f\left(\sum_{j=1}^{k} q_j x_{i_j}\right)$$
$$\ge f\left(m\sum_{i=1}^{n} r_i x_i + \sum_{i_1,\dots,i_k=1}^{n} (p_{i_1}\dots p_{i_k} - m r_{i_1}\dots r_{i_k})\, f\left(\sum_{j=1}^{k} q_j x_{i_j}\right)\right)$$
$$= f\left(\sum_{i=1}^{n} p_i x_i\right).$$

This completes the proof of the right inequality of (2.15). □

Below we state the analogue to Theorem 7 for superquadratic functions. The proof is similar to the proof of Theorem 3 and hence it is omitted.

Theorem 8. *Under the same conditions on* $\mathbf{p}$, $\mathbf{q}$, $\mathbf{r}$, m *and* M *as in Theorem 7, if* I *is* $[0,a)$ *or* $[0,\infty)$ *and* $f(x)$ *is a superquadratic function on* I*, then:*

$$\sum_{i_1,\dots,i_k=1}^{n} p_{i_1}\dots p_{i_k} f\left(\sum_{j=1}^{k} q_j x_{i_j}\right) - f\left(\sum_{i=1}^{n} p_i x_i\right)$$
$$- m\left(\sum r_{i_1}\dots r_{i_k} f\left(\sum_{j=1}^{k} q_j x_{i_j}\right) - f\left(\sum_{i=1}^{n} r_i x_i\right)\right)$$
$$\ge mf\left(\left|\sum_{i=1}^{n} (r_i - p_i)\, x_i\right|\right)$$
$$+ \sum_{i_1,\dots,i_k=1}^{n} (p_{i_1} p_{i_2}\dots p_{i_k} - m r_{i_1}\dots r_{i_k})\, f\left(\left|\sum_{j=1}^{k} q_j x_{i_j} - \sum_{s=1}^{n} p_s x_s\right|\right)$$

and

$$\sum_{i_1,\dots,i_k=1}^{n} p_{i_1}\dots p_{i_k} f\Big(\sum_{j=1}^{k} q_j x_{i_j}\Big) - f\Big(\sum_{i=1}^{n} p_i x_i\Big)$$

$$- M\left(\sum r_{i_1}\dots r_{i_k} f\Big(\sum_{j=1}^{k} q_j x_{i_j}\Big) - f\Big(\sum_{i=1}^{n} r_i x_i\Big)\right)$$

$$\le -f\left(\left|\sum_{i=1}^{n}(r_i - p_i)\, x_i\right|\right)$$

$$- \sum_{i_1,\dots,i_k=1}^{n} \left(M r_{i_1}\dots r_{i_k} - p_{i_1} p_{i_2}\dots p_{i_k}\right) f\left(\left|\sum_{j=1}^{k} q_j x_{i_j} - \sum_{s=1}^{n} r_s x_s\right|\right)$$

If f is also positive, then these inequalities refine (2.15).

3. Other inequalities

The definition of superquadratic functions and their properties draw our attention to the possibility of using the *Čebyšev functional* and its properties to get new type of reverse Jensen Inequality.

For a function $C : [0,\infty) \to \mathbb{R}$ we consider the Čebyšev type functional

$$T(C,\mathbf{x},\mathbf{p}) := \sum_{i=1}^{n} p_i x_i C(x_i) - \sum_{i=1}^{n} p_i x_i \sum_{i=1}^{n} p_i C(x_i).$$

It is well known that, if C is monotonic nondecreasing function on $[0,\infty)$ then the sequences $\mathbf{x}$ and $C(\mathbf{x}) := (C(x_1),\dots,C(x_n))$ are synchronous and for any probability sequence $\mathbf{p}$ we have the *Čebyšev inequality*

$$T(C,\mathbf{x},\mathbf{p}) \ge 0.$$

If certain bounds for the values of the function $C(x_i)$ are known, namely

$$-\infty < m \le C(x_i) \le M < \infty \qquad \text{for any } i \in \{1,\dots,n\} \tag{3.1}$$

then the following inequality due to Cerone & Dragomir [6] holds:

$$|T(C,\mathbf{x},\mathbf{p})| \le \frac{1}{2}(M-m)\sum_{i=1}^{n} p_i \left| x_i - \sum_{j=1}^{n} p_j x_j \right|. \tag{3.2}$$

The constant $\frac{1}{2}$ is best possible in the sense that it cannot be replaced by a smaller quantity.

We can state now the following reverse of the Jensen inequality for superquadratic functions:

Theorem 9. *Assume that* $\mathbf{x} = (x_1, \ldots, x_n)$ *with* $x_i \geq 0$ *for* $i \in \{1, \ldots, n\}$, *and* $\mathbf{p} = (p_1, \ldots, p_n)$ *is a probability sequence with* $n \geq 2$. *Then for any superquadratic function* $f : [0, \infty) \to \mathbb{R}$ *with* $C(x_i)$ *satisfying* (3.1), *where* $C(x)$ *is as in Definition 1 we have the inequality,*

$$\begin{aligned} &\frac{1}{2}(M-m)\sum_{i=1}^{n} p_j \left| x_j - \sum_{j=1}^{n} p_i x_i \right| - \sum_{j=1}^{n} p_j f\left(\left| \sum_{i=1}^{n} p_i x_i - x_j \right| \right) \qquad (3.3) \\ &\geq \sum_{j=1}^{n} p_j f(x_j) - f\left(\sum_{i=1}^{n} p_i x_i \right) \geq \sum_{j=1}^{n} p_j f\left(\left| \sum_{i=1}^{n} p_i x_i - x_j \right| \right). \end{aligned}$$

Proof. The right-hand side inequality of (3.3) is inequality (1.6).

Utilizing the definition of the superquadratic functions we have

$$f\left(\sum_{i=1}^{n} p_i x_i \right) \geq f(x_j) + C(x_j)\left(\sum_{i=1}^{n} p_i x_i - x_j \right) + f\left(\left| \sum_{i=1}^{n} p_i x_i - x_j \right| \right) \qquad (3.4)$$

for any $j \in \{1, \ldots, n\}$.

If we multiply (3.4) by $p_j \geq 0, j \in \{1, \ldots, n\}$, sum over j from 1 to n and take into account that $\sum_{j=1}^{n} p_j = 1$ we get

$$f\left(\sum_{i=1}^{n} p_i x_i \right) \geq \sum_{j=1}^{n} p_j f(x_j) + \sum_{j=1}^{n} p_j C(x_j)\left(\sum_{i=1}^{n} p_i x_i - x_j \right) + \sum_{j=1}^{n} p_j f\left(\left| \sum_{i=1}^{n} p_i x_i - x_j \right| \right). \qquad (3.5)$$

Hence by this inequality, by (3.2) and since

$$\sum_{j=1}^{n} p_j C(x_j)\left(\sum_{i=1}^{n} p_i x_i - x_j \right) = -T(C, \mathbf{x}, \mathbf{p})$$

we deduce the desired result (3.3). □

Remark 4. *As a "by-product" of* (3.5) *we get by using the right-hand side inequality of* (3.3) *that for superquadratic functions the following inequality*

$$\frac{1}{2} T(C, \mathbf{x}, \mathbf{p}) \geq \sum_{j=1}^{n} p_j f\left(\left| \sum_{i=1}^{n} p_i x_i - x_j \right| \right)$$

holds, while from (3.3) *we get*

$$\frac{1}{2}(M-m)\sum_{i=1}^{n} p_j \left| x_j - \sum_{j=1}^{n} p_i x_i \right| \geq \sum_{j=1}^{n} p_j f\left(\left| \sum_{i=1}^{n} p_i x_i - x_j \right| \right).$$

Remark 5. *During the conference of Inequalities and Application* 2007 *in Noszvaj Hungary, we realized that Theorem* 4 *in this paper* (*see also preprint* [2] *dated June* 26, 2007), *overlaps Theorem* 2 *and Corollary* 1 *in preprint* [5].

References

[1] S. Abramovich, S. Banić, M. Matić, and J. Pečarić, *Jensen-Steffensen's and related inequalities for superquadratic functions*, to appear in Math. Ineq. Appl. **11**.

[2] S. Abramovich and S.S. Dragomir, *Normalized Jensen functional, superquadracity and and related inequalities*, Preprint RGMIA Res. Rep. Coll., **10**(2007), Issue 3, Article 7, [Online http://rgmia.vu.edu.au/v10n3.html].

[3] S. Abramovich, G. Jameson and G. Sinnamon, *Refining Jensen's inequality*, Bull. Math. Soc. Math. Roum. **47**, (2004), 3–14.

[4] S. Abramovich, G. Jameson and G. Sinnamon, *Inequalities for averages of convex and superquadratic functions*, J. Inequal. Pure & Appl. Math. **5** (4), (2004), Article 91.

[5] J. Barić, M. Matić and J. Pečarić, *On the bounds for the Normalized functional and Jensen-Steffensen Inequality*, Preprint, August 4, 2007.

[6] P. Cerone and S.S. Dragomir, *A Refinement of the Grüss inequality and applications*, Tamkang J. Math. **38**(2007), 37–49. Preprint *RGMIA Res. Rep. Coll.*, **5**(2002), Article 14, [ONLINE http://rgmia.vu.edu.au/v5n2.html].

[7] S.S. Dragomir, *Bounds for the normalised Jensen functional*, Bull. Austral. Math. Soc. **74** (2006), 471–478.

[8] A. Gilányi, *Remark on subquadratic functions*, Presented at CIA07.

[9] Z. Kominek and K. Troczka, *Some remarks on subquadratic functions*, Demostratio Mathematica **39** (2006), 751–758

[10] J. Pečarić, F. Proschan, Y.L. Tong, *Convex Functions, Partial Orderings, and Statistical Applications*, Academic Press, Inc., 1992.

[11] W. Smajdor, *Subadditive and subquadratic set-valued functions*, Scientific Publications of the University of Silesia, **889**, Katowice, 1987.

Shoshana Abramovich
Department of Mathematics
University of Haifa
Haifa 31905, Israel
e-mail: abramos@math.haifa.ac.il

Silvestru S. Dragomir
School of Computer Science & Mathematics
Victoria University, PO Box 14428
Melbourne City, MC 8001, Australia
e-mail: Sever.Dragomir@vu.edu.au

International Series of Numerical Mathematics, Vol. 157, 229–232

Comparability of Certain Homogeneous Means

Pál Burai

Dedicated to professor Zoltán Daróczy on his 70th birthday

Abstract. We present some inequalities between two variables homogeneous means. Namely, we give necessary as well as sufficient condition on the comparability of Daróczy means.

Mathematics Subject Classification (2000). Primary 39B12; Secondary 39B22.

Keywords. Means, Daróczy means, homogeneous means, comparability of means.

1. Introduction

In [9] W. Janous introduced the class of generalized Heronian means as follows:

$$H_w(x,y) := \begin{cases} \dfrac{x + w\sqrt{xy} + y}{w+2}, & \text{if } 0 \le w < \infty\,, \\ \mathcal{G}(x,y) := \sqrt{xy}, & \text{if } w = \infty\,, \end{cases}$$

for all $x, y \in \mathbb{R}$. Here (and hereafter in this work) $\mathbb{R}_+$ denotes the positive real line.

In [6] Daróczy generalized this class (more precisely he defined the p-modification of the previous class (see also [14])):

$$\mathcal{D}_{\alpha,p}(x,y) := \begin{cases} \left(\dfrac{x^p + \alpha(\sqrt{xy})^p + y^p}{\alpha+2}\right)^{1/p} & \text{if } p \neq 0\,,\ -1 \le \alpha < \infty\,, \\ \sqrt{xy} & \text{if } p = 0 \text{ or } \alpha = \infty\,, \end{cases}$$

for all $x, y \in \mathbb{R}$.

In the sequel we call the members of this class Daróczy means. It is easy to prove that $\mathcal{D}_{\alpha,p}$ is a mean indeed. Namely, it is continuous as a two place function, and

$$\min\{x,y\} \le \mathcal{D}_{\alpha,p}(x,y) \le \max\{x,y\}\,, \qquad x, y \in \mathbb{R}_+\,.$$

Supported by the Hungarian Research Fund (OTKA) Grant No. NK 68040.

One can easily check that $\mathcal{D}_{\alpha,p}$ is homogeneous and symmetric. It is an obvious but important fact that the class of Hölder means is a subclass of Daróczy means ($\alpha = 0$). Furthermore the following limits hold: $\lim_{p\to 0}\mathcal{D}_{\alpha,p} = \mathcal{D}_{\alpha,0} = \mathcal{G}$ and $\lim_{\alpha\to\infty}\mathcal{D}_{\alpha,p} = \mathcal{D}_{\infty,p} = \mathcal{G}$.

In this paper we give necessary as well as sufficient conditions on the comparability of Daróczy means.

2. Comparability problem of Daróczy means

Our first lemma shows that the geometric mean separates the Daróczy means with special parameters.

Lemma 2.1. *Let $p \in \mathbb{R}_+$ and $\alpha \geq -1$ be real numbers, then*

$$\mathcal{D}_{\alpha,-p} \leq \mathcal{G} \leq \mathcal{D}_{\alpha,p}\,.$$

Proof. Easy calculation. □

In the following lemma we examine the comparability of Daróczy means when either the first or the second parameter is the same.

Lemma 2.2. *Let $p \in \mathbb{R}$ and $\alpha\,,\beta \geq -1$ and $p,q \in \mathbb{R}$. Then*

1. $\mathcal{D}_{\alpha,p} \leq \mathcal{D}_{\beta,p}$ *if and only if* $p\,(\alpha-\beta) \geq 0$;
2. $\mathcal{D}_{\alpha,p} \leq \mathcal{D}_{\alpha,q}$ *if and only if* $p \leq q$.

Proof. The first part is an easy calculation.

Proving the second part, we can assume that $pq > 0$, because of Lemma 1. We can apply [1, Theorem 5.], which says that in our case such comparability holds if and only if $\varepsilon_\psi \cdot \psi \circ \varphi^{-1}$ is convex, where $\psi(x) = x^q\,, \varphi(x) = x^p$ and $\varepsilon_\psi = 1$, when ψ is increasing and -1, when it is decreasing. An elementary calculation shows, that this is equivalent to our assertion. □

Remark 2.3. From the second part of the previous lemma we get the comparability theorem of Hölder means (see [8]) if $\alpha = 0$.

Lemma 2.4. *Let $p \in \mathbb{R}$ and $\alpha \geq -1$ and $p \in \mathbb{R}$. Then*

$$\lim_{x\to 1}\frac{\mathcal{D}_{\alpha,p}(x,1)-\frac{x+1}{2}}{(x-1)^2} = \frac{p}{4(\alpha+2)} - \frac{1}{4}.$$

Proof. Using the L'Hospital rule twice, we get our statement. Indeed

$$\lim_{x\to 1}\frac{\mathcal{D}_{\alpha,p}(x,1)-\frac{x+1}{2}}{(x-1)^2} = \lim_{x\to 1}\frac{\frac{\partial^2\mathcal{D}_{\alpha,p}(x,1)}{\partial x^2}}{2} = \frac{2p-2-\alpha}{8(\alpha+2)}\,,$$

where $\frac{\partial^2\mathcal{D}_{\alpha,p}(x,1)}{\partial x^2}$ denotes the second partial derivative of $\mathcal{D}_{\alpha,p}$ with respect to the first variable at the place $(x,1)$, and

$$\frac{\partial^2\mathcal{D}_{\alpha,p}(x,1)}{\partial x^2} = \left(\frac{x^p+\alpha x^{\frac{p}{2}}+1}{\alpha+2}\right)^{\frac{1}{p}}\frac{\left((x^{\frac{p}{2}}+x^{\frac{3p}{2}})(p-2)\alpha+4x^p(p-1)-\alpha^2x^p\right)}{4x^2(x^p+\alpha x^{\frac{p}{2}}+1)^2}\,.$$

□

Theorem 2.5 (Necessary condition). *If $\mathcal{D}_{\alpha,p} \le \mathcal{D}_{\beta,q}$, then $\frac{p}{\alpha+2} \le \frac{q}{\beta+2}$.*

Proof. Because of the symmetry and homogeneous property of Daróczy means $\mathcal{D}_{\alpha,p}(x,y) \le \mathcal{D}_{\beta,q}(x,y)$ if and only if $\frac{\mathcal{D}_{\alpha,p}(x,1)-\frac{x+1}{2}}{(x-1)^2} \le \frac{\mathcal{D}_{\beta,q}(x,1)-\frac{x+1}{2}}{(x-1)^2}$, $x>1$. As x tends to 1, we get from the previous lemma

$$\frac{p}{4(\alpha+2)} - \frac{1}{4} \le \frac{q}{4(\beta+2)} - \frac{1}{4}.$$

This inequality is equivalent to our assertion. □

Remark 2.6. The previous theorem with Lemma 2.1 give necessary condition for all $p\,,q$ and $\alpha\,,\beta \ge -1$.

Theorem 2.7 (Sufficient condition). *Let $p\,,q \in \mathbb{R}$ and $\alpha\,,\beta \ge 0$. If $0<2p\le q$ and $\frac{p}{\alpha+2} \le \frac{q}{\beta+2}$, then $\mathcal{D}_{\alpha,p} \le \mathcal{D}_{\beta,q}$.*

Proof. Similarly as in the proof of Lemma 2 we have to examine the positivity of function

$$A(s) := \frac{1}{r}\log(s^{2r}+\beta s^r+1) - \frac{1}{r}\log(\beta+2) - \log(s^2+\alpha s+1) + \log(\alpha+2)\,,$$

where $x^p = s^2$, $s \ge 1$ and $r = \frac{q}{p}$. It is clear that $A(1)=0$. Calculate the first derivative of A:

$$A'(s) = \frac{\left(s^{2r}-1\right)\alpha + s^{r-1}\left(1-s^2\right)\beta + 2s\left(s^{2(r-1)}-1\right)}{\left(s^{2\,r}+\beta\,s^r+1\right)\left(s^2+\alpha\,s+1\right)}.$$

Because the denominator is positive here, it is enough to analyze the numerator. According our assumption the numerator does not increase if we substitute β by $\alpha r+2r-2$. Therefore,

$$A'(s) \ge B(s) := \left(s^{r-1}r - s^{r+1}r + s^{2\,r} - 1\right)\alpha + 2\,s^{r+1} + 2\,s^{2\,r-1} + 2\,s^{r-1}r - 2\,s^{r+1}r - 2\,s - 2\,s^{r-1}\,,\ \text{where } s \ge 1\,.$$

We examine the coefficient of α and the other terms, respectively.

According to the comparability theorem of Stolarsky means (see [3], [13]) we get that $S_{0,0}(s^2,1) \le S_{r,1}(s^2,1)$, where $S_{a,b}(x,y)$ denotes the corresponding Stolarsky mean (see [3], [10], [11], [13]) and $s \ge 1\,,\ r>1$. In other terms

$$\sqrt{s^2\cdot 1} \le \left(\frac{s^{2r}-1}{r(s^2-1)}\right)^{\frac{1}{r-1}}\,, \qquad r>1\,,\ s\ge 1\,.$$

This means that the coefficient of α is nonnegative. It is an easy calculation that the coefficient of α is also nonnegative if $r=1\,,\ s\ge 1$. On the other hand, we get the other terms from the coefficient of α if we replace r by $r-1$. If $r\ge 2$ the nonnegativity remains valid on the other terms, so the proof is finished now. □

Remark 2.8. It remains an open problem to find a necessary and sufficient condition on all parameters such that $\mathcal{D}_{\alpha,p} \le \mathcal{D}_{\beta,q}$ be valid.

Acknowledgement
The author thanks for the valuable comments of the referees.

References

[1] M. Klaričič Bakula, Zs. Páles and J. Pečarič *On weighted L-conjugate means*, Commun. Math. Anal. **11** (2007), 95–110.

[2] P.S. Bullen, D.S. Mitrinović and P.M. Vasić, *Means and their inequalities*, D. Reidel Publishing Company, 1987.

[3] P. Czinder, *Inequalities on two variable Gini and Stolarsky means*, Ph.D. dissertation, University of Debrecen, 2005.

[4] P. Czinder and Zs. Páles, *An extension of the Hermite-Hadamard inequality and an application for Gini and Stolarsky means*, J. Inequal. Pure Appl. Math. **5** (2004), Article 42, p. 8.

[5] P. Czinder and Zs. Páles, *Some comparison inequalities for Gini and Stolarsky means*, Math. Inequal. Appl. **9** (2006), 607–616.

[6] Z. Daróczy, Oral communication

[7] Z. Daróczy and Zs. Páles, *On a class of means of several variables*, Math. Inequal. Appl. **4** (2001), 331–341.

[8] G.H. Hardy, J.E. Littlewood, and G. Pólya, *Inequalities*, Cambridge University Press, Cambridge, 1934 (first edition), 1952 (second edition).

[9] W. Janous, *A note on generalized Heronian means*, Math. Inequal. Appl. **4** (2001), 390–407.

[10] E.B. Leach and M.C. Sholander, *Multi-variable extended mean values*, J. Math. Anal. Appl. **104** (1984), 390–407.

[11] E. Neumann and J. Sándor, *Inequalities involving Stolarsky and Gini means*, Math. Pannonica **14** (2003), 29–44.

[12] Zs. Páles, Oral communication.

[13] Zs. Páles, *Inequalities for differences of powers*, J. Math. Anal. Appl. **131** (1988), 271–281.

[14] M.K. Vanamurthy and M. Vuorinen, *Inequalities for means*, J. Math. Anal. Appl. **183** (1994), 155–166.

Pál Burai
Institute of Mathematics
University of Debrecen
Pf. 12
H-4010 Debrecen, Hungary
e-mail: `buraip@inf.unideb.hu`

International Series of Numerical Mathematics, Vol. 157, 233–243

On Some General Inequalities Related to Jensen's Inequality

Milica Klaričić Bakula, Marko Matić and Josip Pečarić

Abstract. We present several general inequalities related to Jensen's inequality and the Jensen-Steffensen inequality. Some recently proved results are obtained as special cases of these general inequalities.

Mathematics Subject Classification (2000). 26D15.

Keywords. Convex functions, Jensen's inequality, Jensen-Steffensen inequality, Slater's inequality, Mercer's inequality.

1. Introduction

Let the real function φ be defined on some nonempty interval I of the real line $\mathbb{R}$. We say that φ is *convex* on I if

$$\varphi(\lambda x + (1-\lambda) y) \leq \lambda \varphi(x) + (1-\lambda)\varphi(y)$$

holds for all $x, y \in I$ and $\lambda \in [0,1]$.

An important property of convex functions is the existence of the left and the right derivative on the interior $\mathring{I}$ of I (see [11]). If $\varphi : I \to \mathbb{R}$ is convex then for any $x \in \mathring{I}$ the left derivative $\varphi'_{-}(x)$ and the right derivative $\varphi'_{+}(x)$ are increasing on $\mathring{I}$ and

$$\varphi'_{-}(x) \leq \varphi'_{+}(x) \quad \text{for all } x \in \mathring{I}.$$

It can be also proved that for any convex function $\varphi : I \to \mathbb{R}$ the inequalities

$$\varphi(z) + c(z)(y-z) \leq \varphi(y), \quad c(z) \in \left[\varphi'_{-}(z), \varphi'_{+}(z)\right] \tag{1.1}$$

$$\varphi(y) \leq \varphi(z) + c(y)(y-z), \quad c(y) \in \left[\varphi'_{-}(y), \varphi'_{+}(y)\right] \tag{1.2}$$

hold for all $y, z \in \mathring{I}$.

One consequence of (1.1) and (1.2) is that $\varphi : I \to \mathbb{R}$ is convex if and only if there is at least one line of support for φ at each $x_0 \in \mathring{I}$. Furthermore, φ is differentiable if and only if the line of support at $x_0 \in \mathring{I}$ is unique. In this case, the line of support is

$$A(x) = \varphi(x_0) + \varphi'(x_0)(x - x_0).$$

There are many known inequalities for convex functions, but surely the most important of them is Jensen's inequality. In its integral form it is stated as follows (see [10, p. 45]).

Theorem A. (*Jensen*) *Let* $(\Omega, \mathcal{A}, \mu)$ *be a measure space with* $0 < \mu(\Omega) < \infty$, *and let* $u : \Omega \to I$, $I \subset \mathbb{R}$, *be a function from* $L^1(\mu)$. *Then for any convex function* $\varphi : I \to \mathbb{R}$ *the inequality*

$$\varphi\left(\frac{1}{\mu(\Omega)}\int_\Omega u d\mu\right) \le \frac{1}{\mu(\Omega)}\int_\Omega (\varphi \circ u)\, d\mu \tag{1.3}$$

holds.

One of the inequalities which are strongly related to Jensen's inequality is the Jensen-Steffensen inequality for convex functions. An integral version was proved by Steffensen, but here we consider a variant given by R.P. Boas in [3].

Theorem B. (*Steffensen-Boas*) *Let* $f : [\alpha, \beta] \to (a, b)$ *be a continuous and monotonic function, where* $-\infty < \alpha < \beta < +\infty$ *and* $-\infty \le a < b \le +\infty$, *and let* $\varphi : (a, b) \to \mathbb{R}$ *be a convex function. If* $\lambda : [\alpha, \beta] \to \mathbb{R}$ *is either continuous or of bounded variation satisfying*

$$(\forall x \in [\alpha, \beta]) \quad \lambda(\alpha) \le \lambda(x) \le \lambda(\beta), \qquad \lambda(\beta) - \lambda(\alpha) > 0, \tag{1.4}$$

then

$$\varphi\left(\frac{\int_\alpha^\beta f(t)\, \mathrm{d}\lambda(t)}{\int_\alpha^\beta \mathrm{d}\lambda(t)}\right) \le \frac{\int_\alpha^\beta \varphi(f(t))\, \mathrm{d}\lambda(t)}{\int_\alpha^\beta \mathrm{d}\lambda(t)}. \tag{1.5}$$

In [7] a couple of companion inequalities to Jensen's inequality in its discrete and integral form were proved. The main result in its discrete form is stated as follows.

Theorem C. (*Matić, Pečarić*) *Let* $\varphi : C \to \mathbb{R}$ *be a convex function defined on an open convex subset* C *in a normed real linear space* X. *For the given vectors* $\boldsymbol{x}_i \in C$, $i = 1, 2, \ldots, n$, *and a nonnegative real* n*-tuple* $\mathbf{p}$ *such that* $P_n = \sum_{i=1}^n p_i > 0$ *let*

$$\overline{\boldsymbol{x}} = \frac{1}{P_n}\sum_{i=1}^n p_i \boldsymbol{x}_i, \qquad \overline{y} = \frac{1}{P_n}\sum_{i=1}^n p_i \varphi(\boldsymbol{x}_i).$$

If $\mathbf{c}, \mathbf{d} \in C$ *are arbitrarily chosen vectors, then*

$$\varphi(\mathbf{c}) + a^*(\mathbf{c}; \overline{\boldsymbol{x}} - \mathbf{c}) \le \overline{y} \le \varphi(\mathbf{d}) + \frac{1}{P_n}\sum_{i=1}^n p_i a^*(\boldsymbol{x}_i; \boldsymbol{x}_i - \mathbf{d}). \tag{1.6}$$

Also, when φ is strictly convex we have equality in the first inequality in (1.6) *if and only if $\boldsymbol{x}_i = \mathbf{c}$ for all indices i with $p_i > 0$, while equality holds in the second inequality in* (1.6) *if and only if $\boldsymbol{x}_i = \mathbf{d}$ for all indices i with $p_i > 0$.*

In the rest of the paper without any loss of generality for the convex function $\varphi : (a, b) \to \mathbb{R}$ we denote

$$\varphi'(x) := \varphi'_+(x), \quad x \in (a, b).$$

Theorem D. (*Klaričić, Matić, Pečarić*) *Let $\varphi : (a, b) \to \mathbb{R}, -\infty \le a < b \le +\infty$, be a convex function and $\boldsymbol{p} \in \mathbb{R}^n$ $(n \ge 2)$ such that*

$$0 \le P_k = \textstyle\sum_{i=1}^{k} p_i \le P_n, \ k = 1, \ldots, n, \quad P_n > 0. \tag{1.7}$$

Then for any $\boldsymbol{x} \in (a, b)^n$ such that

$$x_1 \le x_2 \le \cdots \le x_n \quad or \quad x_1 \ge x_2 \ge \cdots \ge x_n$$

the inequalities

$$\varphi(c) + \varphi'(c)(\overline{x} - c) \le \frac{1}{P_n} \sum_{i=1}^{n} p_i \varphi(x_i) \le \varphi(d) + \frac{1}{P_n} \sum_{i=1}^{n} p_i \varphi'(x_i)(x_i - d) \tag{1.8}$$

hold for all $c, d \in (a, b)$.

Under the stated assumptions on $\boldsymbol{x}$ and $\boldsymbol{p}$ the inequalities in (1.8) are valid for all $c, d \in (a, b)$, so in the first inequality in (1.8) we may choose $c = \overline{x}$ thus obtaining the discrete Jensen-Steffensen inequality. Moreover, the choice $c = \overline{x}$ is the best possible since

$$\varphi(c) + \varphi'(c)(\overline{x} - c) \le \varphi(\overline{x})$$

for all $c \in (a, b)$.

The integral version of Theorem D, stated in Theorem E, has been also proved in [6].

Theorem E. (*Klaričić, Matić, Pečarić*) *Suppose that f, φ and λ are as in Theorem B. Then $\overline{x}$ and $\overline{y}$ given by*

$$\overline{x} = \frac{1}{\lambda(\beta) - \lambda(\alpha)} \int_{\alpha}^{\beta} f(t)\, \mathrm{d}\lambda(t),$$

$$\overline{y} = \frac{1}{\lambda(\beta) - \lambda(\alpha)} \int_{\alpha}^{\beta} \varphi(f(t))\, \mathrm{d}\lambda(t)$$

are well defined and $\overline{x} \in (a, b)$. Furthermore, if $\varphi'(f)$ and λ have no common discontinuity points, then the inequalities

$$\begin{aligned} &\varphi(c) + \varphi'(c)(\overline{x} - c) \\ &\le \overline{y} \le \varphi(d) + \frac{1}{\lambda(\beta) - \lambda(\alpha)} \int_{\alpha}^{\beta} \varphi'(f(t))\, [f(t) - d]\, \mathrm{d}\lambda(t) \end{aligned} \tag{1.9}$$

hold for each $c, d \in (a, b)$.

In [9] the following theorem was proved.

Theorem F. (*Pečarić*) *Suppose that φ is convex on (a,b) and $a < x_1 \le \cdots \le x_n < b$. If $p_1,\dots,p_n$ are real numbers such that the conditions* (1.7) *hold and if*

$$\sum_{i=1}^{n} p_i\varphi'(x_i) \neq 0, \qquad \widetilde{x} = \frac{\sum_{i=1}^{n} p_i x_i \varphi'(x_i)}{\sum_{i=1}^{n} p_i\varphi'(x_i)} \in (a,b),$$

then

$$\frac{1}{P_n}\sum_{i=1}^{n} p_i\varphi(\boldsymbol{x}_i) \le \varphi(\widetilde{x}).$$

In paper [8] A. Mercer proved the following variant of Jensen's inequality:

$$\varphi\left(x_1 + x_n - \sum_{i=1}^{n} w_i x_i\right) \le \varphi(x_1) + \varphi(x_n) - \sum_{i=1}^{n} w_i\varphi(x_i), \tag{1.10}$$

which holds whenever φ is a convex function on an interval containing the n-tuple $\boldsymbol{x}$ such that $0 < x_1 \le x_2 \le \cdots \le x_n$ and where $\boldsymbol{w}$ is a positive n-tuple with $\sum_{i=1}^{n} w_i = 1$. His result was generalized for weights satisfying the conditions as in the Jensen-Steffensen inequality in [1], and two alternative proofs of (1.10) were given in [13] and [2].

2. The results

The goal of this paper is to obtain Mercer-type variants of Theorems C, D, E and F.

In the following with $(\Omega, \mathcal{A}, \mu)$ we denote a measure space with $0 < \mu(\Omega) < \infty$ and for $a, b, m, M \in \mathbb{R}$ we always assume $\infty \le a < m < M < b \le \infty$.

Theorem 1. *Let $\varphi : (a,b) \to \mathbb{R}$ be a convex function and let $u : \Omega \to [m,M]$ be a measurable function such that $\varphi' \circ u$ belongs to $L^1(\mu)$. Then the inequalities*

$$\begin{aligned} &\varphi(c) + \varphi'(c)\left(m + M - c - \frac{1}{\mu(\Omega)}\int_\Omega u\,d\mu\right) \\ &\le \varphi(m) + \varphi(M) - \frac{1}{\mu(\Omega)}\int_\Omega (\varphi\circ u)\,d\mu \\ &\le \varphi(d) + \varphi'(m)(m-d) + \varphi'(M)(M-d) - \frac{1}{\mu(\Omega)}\int_\Omega (u(t)-d)(\varphi'\circ u)\,d\mu \end{aligned} \tag{2.1}$$

hold for all $c, d \in [m,M]$.

Proof. We prove the first inequality in (2.1).

For all $u(t) \in [m,M]$, $t \in \Omega$, we can write

$$u(t) = \lambda_t m + (1-\lambda_t) M, \quad \lambda_t \in [0,1]$$

hence

$$(\varphi\circ u)(t) = \varphi(\lambda_t m + (1-\lambda_t)M) \le \lambda_t\varphi(m) + (1-\lambda_t)\varphi(M)$$

for all $t \in \Omega$. Also

$$\begin{aligned}\varphi(m+M-u(t)) = \varphi((1-\lambda_t)m+\lambda_t M) &\le (1-\lambda_t)\varphi(m)+\lambda_t\varphi(M)\\ &= \varphi(m)+\varphi(M)-[\lambda_t\varphi(m)+(1-\lambda_t)\varphi(M)]\\ &\le \varphi(m)+\varphi(M)-(\varphi\circ u)(t).\end{aligned}$$

If in (1.1) we choose $z=c$ and $y=m+M-u(t)$ we obtain

$$\begin{aligned}&\varphi(c)+\varphi'(c)(m+M-u(t)-c)\\ &\le \varphi(m+M-u(t)) \le \varphi(m)+\varphi(M)-(\varphi\circ u)(t).\end{aligned} \tag{2.2}$$

Integrating over Ω and dividing by $\mu(\Omega)$ we obtain

$$\begin{aligned}&\varphi(c)+\varphi'(c)\left(m+M-c-\frac{1}{\mu(\Omega)}\int_\Omega u\,d\mu\right)\\ &\le \frac{1}{\mu(\Omega)}\int_\Omega \varphi(m+M-u(t))\,d\mu \le \varphi(m)+\varphi(M)-\frac{1}{\mu(\Omega)}\int_\Omega(\varphi\circ u)\,d\mu.\end{aligned}$$

Now it remains to prove the second inequality in (2.1). Let $d, u(t) \in [m,M]$, $t \in \Omega$.

We consider two cases.

Case 1. $u(t) \ge d$. From (1.2) we have

$$\begin{aligned}\varphi(m)-\varphi(d) &\le \varphi'(m)(m-d),\\ \varphi(M)-(\varphi\circ u)(t) &\le \varphi'(M)(M-u(t)),\end{aligned}$$

hence

$$\begin{aligned}&\varphi(m)+\varphi(M)-(\varphi\circ u)(t)\\ &= \varphi(d)+\varphi(m)-\varphi(d)+\varphi(M)-(\varphi\circ u)(t)\\ &\le \varphi(d)+\varphi'(m)(m-d)+\varphi'(M)(M-u(t))\\ &= \varphi(d)+\varphi'(m)(m-d)+\varphi'(M)(M-d)-\varphi'(M)(u(t)-d).\end{aligned} \tag{2.3}$$

Since φ is convex the derivative φ' is nondecreasing and we know that from $u(t) \le M$ follows $(\varphi'\circ u)(t) \le \varphi'(M)$, hence (2.3) implies

$$\begin{aligned}&\varphi(m)+\varphi(M)-(\varphi\circ u)(t)\\ &\le \varphi(d)+\varphi'(m)(m-d)+\varphi'(M)(M-d)-(\varphi'\circ u)(t)(u(t)-d).\end{aligned} \tag{2.4}$$

Case 2. $u(t) < d$. Similarly as in the previous case we can write

$$\begin{aligned}&\varphi(m)+\varphi(M)-(\varphi\circ u)(t)\\ &= \varphi(d)+\varphi(m)-(\varphi\circ u)(t)+\varphi(M)-\varphi(d)\\ &\le \varphi(d)+\varphi'(m)(m-u(t))+\varphi'(M)(M-d)\\ &= \varphi(d)+\varphi'(m)(m-d)+\varphi'(M)(M-d)+\varphi'(m)(d-u(t)).\end{aligned}$$

From $m \leq u(t)$ we have $\varphi'(m) \leq (\varphi' \circ u)(t)$, hence

$$\begin{aligned}
&\varphi(m) + \varphi(M) - (\varphi \circ u)(t) \\
&\leq \varphi(d) + \varphi'(m)(m-d) + \varphi'(M)(M-d) + (\varphi' \circ u)(t)(d - u(t)) \\
&= \varphi(d) + \varphi'(m)(m-d) + \varphi'(M)(M-d) - (\varphi' \circ u)(t)(u(t) - d),
\end{aligned}$$

which is again (2.4).

In other words, for any $d, u(t) \in [m, M]$ the inequality in (2.4) holds. Integrating (2.4) over Ω and dividing by $\mu(\Omega)$ we obtain the second inequality in (2.1). The proof is complete. □

Corollary 1. *Let $\varphi : (a,b) \to \mathbb{R}$ be a convex function. If $\boldsymbol{p} \in \mathbb{R}_+^n$ and $\boldsymbol{x} \in [m, M]^n$ then the inequalities*

$$\begin{aligned}
&\varphi(c) + \varphi'(c)\left(m + M - c - \frac{1}{P_n}\sum_{i=1}^{n} p_i x_i\right) \\
&\leq \varphi(m) + \varphi(M) - \frac{1}{P_n}\sum_{i=1}^{n} p_i \varphi(x_i) \qquad (2.5) \\
&\leq \varphi(d) + \varphi'(m)(m-d) + \varphi'(M)(M-d) - \frac{1}{P_n}\sum_{i=1}^{n} p_i \varphi'(x_i)(x_i - d)
\end{aligned}$$

hold for all $c, d \in [m, M]$.

Proof. This is a straightforward consequence of Theorem 1. We simply choose

$$\begin{aligned}
\Omega &= \{1, 2, \ldots, n\}, \\
\mu(\{i\}) &= p_i, \quad i = 1, 2, \ldots, n, \\
u(i) &= x_i, \quad i = 1, 2, \ldots, n.
\end{aligned}$$
□

Corollary 2. *The following inequalities are valid under the assumptions of Corollary 1:*

$$\begin{aligned}
0 &\leq \varphi(m) + \varphi(M) - \frac{1}{P_n}\sum_{i=1}^{n} p_i \varphi(x_i) - \varphi(\overline{x}) \\
&\leq \varphi'(m)(m - \overline{x}) + \varphi'(M)(M - \overline{x}) - \frac{1}{P_n}\sum_{i=1}^{n} p_i \varphi'(x_i)(x_i - \overline{x}),
\end{aligned}$$

where $\overline{x} = m + M - 1/P_n \sum_{i=1}^{n} p_i x_i$.

Corollary 3. *Suppose that the conditions of Corollary* 1 *are satisfied and additionally assume*

$$\sum_{i=1}^{n} p_i\varphi'(x_i) \neq P_n\left[\varphi'(m) + \varphi'(M)\right],$$

$$\widetilde{x} = \frac{P_n\left[m\varphi'(m) + M\varphi'(M)\right] - \sum_{i=1}^{n} p_i x_i \varphi'(x_i)}{P_n\left[\varphi'(m) + \varphi'(M)\right] - \sum_{i=1}^{n} p_i\varphi'(x_i)} \in [m, M].$$

Then

$$\varphi(m) + \varphi(M) - \frac{1}{P_n}\sum_{i=1}^{n} p_i\varphi(x_i) \leq \varphi(\widetilde{x}).$$

The inequalities obtained in Corollary 2 and 3 are the Mercer-type variants of the corresponding inequalities given in [4] and [12].

Theorem 2. *Let* $\varphi : (a,b) \to \mathbb{R}$ *be a convex function and* $\boldsymbol{w} \in \mathbb{R}^l$ *such that*

$$0 \leq W_k = \sum_{i=1}^{k} w_i \leq W_l, \ k = 1, \ldots, l, \quad W_l > 0.$$

Let $\boldsymbol{\xi} \in [m, M]^l$ *be such that* $\xi_1 \leq \xi_2 \leq \cdots \leq \xi_l$ *or* $\xi_1 \geq \xi_2 \geq \cdots \geq \xi_l$. *Then the inequalities*

$$\begin{aligned}
&\varphi(c) + \varphi'(c)\left(m + M - c - \frac{1}{W_l}\sum_{i=1}^{l} w_i\xi_i\right) \\
&\leq \varphi(m) + \varphi(M) - \frac{1}{W_l}\sum_{i=1}^{l} w_i\varphi(\xi_i) \qquad (2.6) \\
&\leq \varphi(d) + \varphi'(m)(m-d) + \varphi'(M)(M-d) - \frac{1}{W_l}\sum_{i=1}^{l} w_i\varphi'(\xi_i)(\xi_i - d)
\end{aligned}$$

hold for all $c, d \in [m, M]$.

Proof. For $n = l + 2$ we define

$$\begin{array}{llllll}
x_1 = m, & x_2 = \xi_1, & x_3 = \xi_2, & \ldots & x_{n-1} = \xi_l, & x_n = M \\
p_1 = 1, & p_2 = -w_1/W_l, & p_2 = -w_2/W_l, & \ldots & p_{n-1} = -w_l/W_l, & p_n = 1
\end{array}.$$

It is obvious that $x_1 \leq x_2 \leq \cdots \leq x_n$ if $\xi_1 \leq \xi_2 \leq \cdots \leq \xi_l$ or $x_1 \geq x_2 \geq \cdots \geq x_n$ if $\xi_1 \geq \xi_2 \geq \cdots \geq \xi_l$ and that

$$0 \leq P_k = \sum_{i=1}^{k} p_i \leq P_n, \quad k = 1, 2, \ldots, n, \quad P_n = 1 > 0,$$

hence we can apply Theorem D on φ, $\boldsymbol{x}$ and $\boldsymbol{p}$ thus obtaining (2.6). □

Note that under the conditions of Theorem 2 we also have

$$\overline{\xi} = m + M - \frac{1}{W_l}\sum_{i=1}^{l} w_i \xi_i \in [m, M],$$

which means that in (2.6) we can choose $c = \overline{\xi}$ in which case the first inequality in (2.6) becomes the generalized Mercer inequality as it was stated in [1]. Mercer's inequality itself can be obtained in the same way as a special case of Corollary 1.

Corollary 4. *The following inequalities are valid under the assumptions of Theorem 2:*

$$0 \le \varphi(m) + \varphi(M) - \frac{1}{W_l}\sum_{i=1}^{l} w_i \varphi(\xi_i) - \varphi\left(\overline{\xi}\right)$$

$$\le \varphi'(m)\left(m - \overline{\xi}\right) + \varphi'(M)\left(M - \overline{\xi}\right) - \frac{1}{W_l}\sum_{i=1}^{l} w_i \varphi'(\xi_i)\left(\xi_i - \overline{\xi}\right),$$

where

$$\overline{\xi} = m + M - \frac{1}{W_l}\sum_{i=1}^{l} w_i \xi_i.$$

Corollary 5. *Suppose that the conditions of Theorem 2 are satisfied and additionally assume*

$$\sum\nolimits_{i=1}^{l} w_i \varphi'(\xi_i) \ne W_l\left[\varphi'(m) + \varphi'(M)\right],$$

$$\widetilde{\xi} = \frac{W_l\left[m\varphi'(m) + M\varphi'(M)\right] - \sum\nolimits_{i=1}^{l} w_i \xi_i \varphi'(\xi_i)}{W_l\left[\varphi'(m) + \varphi'(M)\right] - \sum\nolimits_{i=1}^{l} w_i \varphi'(\xi_i)} \in (m, M).$$

Then

$$\varphi(m) + \varphi(M) - \frac{1}{W_l}\sum_{i=1}^{l} w_i \varphi(\xi_i) \le \varphi\left(\widetilde{\xi}\right).$$

The inequalities given in Corollary 4 are the Mercer type variants of a result from [5] and the inequality given in Corollary 5 is the Mercer type variant of Theorem F.

Now we prove the integral case of Theorem 2.

Theorem 3. *Suppose that $f : [\alpha, \beta] \to [m, M], \varphi, \lambda, \overline{x}$ and $\overline{y}$ are all as in Theorem E and additionally assume that φ is continuously differentiable. Then the inequalities*

$$\begin{aligned}
&\varphi(c) + \varphi'(c)(m + M - c - \overline{x}) \le \varphi(m) + \varphi(M) - \overline{y} \\
&\le \varphi(d) + \varphi'(m)(m - d) + \varphi'(M)(M - d) \\
&- \frac{1}{\lambda(\beta) - \lambda(\alpha)}\int_{\alpha}^{\beta} \varphi'(f(t))\left[f(t) - d\right] \mathrm{d}\lambda(t)
\end{aligned} \tag{2.7}$$

hold for each $c, d \in [m, M]$.

Proof. Suppose that f is nondecreasing (for f nonincreasing the proof is analogous). For arbitrary $\tilde{\alpha}, \tilde{\beta} \in \mathbb{R}$ such that $\tilde{\alpha} < \alpha$ and $\tilde{\beta} > \beta$ we define a new function $\tilde{f} : [\tilde{\alpha}, \tilde{\beta}] \to [m, M]$ by

$$\tilde{f}(t) = \begin{cases} m + \frac{f(\alpha)-m}{\alpha-\tilde{\alpha}}\,(t-\tilde{\alpha})\,, & t \in [\tilde{\alpha}, \alpha], \\ f\,(t)\,, & t \in [\alpha, \beta], \\ M + \frac{M-f(\beta)}{\tilde{\beta}-\beta}\,(t-\tilde{\beta}), & t \in [\beta, \tilde{\beta}]. \end{cases}$$

It can be easily seen that the function $\tilde{f}$ is continuous and nondecreasing.

Next we define two new functions $\tilde{\lambda}_s : [\tilde{\alpha}, \tilde{\beta}] \to \mathbb{R}$ and $\tilde{\lambda}_c : [\tilde{\alpha}, \tilde{\beta}] \to \mathbb{R}$ by

$$\tilde{\lambda}_s\,(t) = \begin{cases} 1, & t = \tilde{\alpha}, \\ 0, & t \in (\tilde{\alpha}, \tilde{\beta}), \\ -1, & t = \tilde{\beta}, \end{cases}$$

and

$$\tilde{\lambda}_c\,(t) = \begin{cases} 1, & t \in [\tilde{\alpha}, \alpha], \\ \frac{\lambda(\beta)-\lambda(t)}{\lambda(\beta)-\lambda(\alpha)}, & t \in [\alpha, \beta], \\ 0, & t \in [\beta, \tilde{\beta}]. \end{cases}$$

Note that for any function $g : [\tilde{\alpha}, \tilde{\beta}] \to \mathbb{R}$ continuous at the points $\tilde{\alpha}$ and $\tilde{\beta}$ we have

$$\begin{aligned} \int_{\tilde{\alpha}}^{\tilde{\beta}} g\,(t)\,\mathrm{d}\tilde{\lambda}_s\,(t) &= g\,(\tilde{\alpha})\,[\tilde{\lambda}_s\,(\tilde{\alpha}+0) - \tilde{\lambda}_s\,(\tilde{\alpha})] + g(\tilde{\beta})[\tilde{\lambda}_s(\tilde{\beta}) - \tilde{\lambda}_s(\tilde{\beta}-0)] \\ &= -g\,(\tilde{\alpha}) - g(\tilde{\beta}). \end{aligned} \tag{2.8}$$

Also, if λ is continuous on $[\alpha, \beta]$ then $\tilde{\lambda}_c$ is continuous on $[\tilde{\alpha}, \tilde{\beta}]$, and if λ is of bounded variation on $[\alpha, \beta]$ then $\tilde{\lambda}_c$ is of bounded variation on $[\tilde{\alpha}, \tilde{\beta}]$. This means that for any continuous and piecewise monotonic function $g : [\tilde{\alpha}, \tilde{\beta}] \to \mathbb{R}$ the integral $\int_{\tilde{\alpha}}^{\tilde{\beta}} g(t)\mathrm{d}\tilde{\lambda}_c(t)$ is well defined and

$$\begin{aligned} \int_{\tilde{\alpha}}^{\tilde{\beta}} g\,(t)\,\mathrm{d}\tilde{\lambda}_c\,(t) &= \int_{\tilde{\alpha}}^{\alpha} g\,(t)\,\mathrm{d}\tilde{\lambda}_c\,(t) + \int_{\alpha}^{\beta} g\,(t)\,\mathrm{d}\tilde{\lambda}_c\,(t) + \int_{\beta}^{\tilde{\beta}} g\,(t)\,\mathrm{d}\tilde{\lambda}_c\,(t) \\ &= \int_{\alpha}^{\beta} g\,(t)\,\mathrm{d}\tilde{\lambda}_c\,(t) = \int_{\alpha}^{\beta} g\,(t)\,\mathrm{d}\left[\frac{\lambda\,(\beta)-\lambda\,(t)}{\lambda\,(\beta)-\lambda\,(\alpha)}\right] \\ &= -\frac{1}{\lambda\,(\beta)-\lambda\,(\alpha)}\int_{\alpha}^{\beta} g\,(t)\,\mathrm{d}\lambda\,(t)\,. \end{aligned} \tag{2.9}$$

Now we define $\tilde{\lambda} : [\tilde{\alpha}, \tilde{\beta}] \to \mathbb{R}$ by

$$\tilde{\lambda}\,(t) = \tilde{\lambda}_c\,(t) - \tilde{\lambda}_s\,(t)\,, \quad t \in [\tilde{\alpha}, \tilde{\beta}].$$

From (2.8) and (2.9) we conclude that the integral $\int_{\tilde{\alpha}}^{\tilde{\beta}} g(t)\mathrm{d}\tilde{\lambda}(t)$ is well defined for any continuous and piecewise monotonic function $g : [\tilde{\alpha}, \tilde{\beta}] \to \mathbb{R}$ and

$$\begin{aligned}\int_{\tilde{\alpha}}^{\tilde{\beta}} g(t)\,\mathrm{d}\tilde{\lambda}(t) &= \int_{\tilde{\alpha}}^{\tilde{\beta}} g(t)\,\mathrm{d}\tilde{\lambda}_c(t) - \int_{\tilde{\alpha}}^{\tilde{\beta}} g(t)\,\mathrm{d}\tilde{\lambda}_s(t)\\ &= -\frac{1}{\lambda(\beta)-\lambda(\alpha)}\int_{\alpha}^{\beta} g(t)\,\mathrm{d}\lambda(t) + g(\tilde{\alpha}) + g(\tilde{\beta}). \qquad (2.10)\end{aligned}$$

We also have

$$\tilde{\lambda}(\tilde{\beta}) - \tilde{\lambda}(\tilde{\alpha}) = \tilde{\lambda}_c(\tilde{\beta}) - \tilde{\lambda}_c(\tilde{\alpha}) - \tilde{\lambda}_s(\tilde{\beta}) + \tilde{\lambda}_s(\tilde{\alpha}) = 0 - 1 + 1 + 1 = 1.$$

If we apply Theorem E on the functions $\tilde{f}, \varphi$ and $\tilde{\lambda}$ (we can do that even if the function $\tilde{\lambda}$ is neither continuous nor of bounded variation since all the integrals are well defined) we obtain

$$\begin{aligned}&\varphi(c) + \varphi'(c)(\tilde{x} - c)\\ &\le \tilde{y} \le \varphi(d) + \frac{1}{\tilde{\lambda}(\beta) - \tilde{\lambda}(\alpha)}\int_{\tilde{\alpha}}^{\tilde{\beta}} \varphi'\left(\tilde{f}(t)\right)\left[\tilde{f}(t) - d\right]\mathrm{d}\tilde{\lambda}(t)\end{aligned}$$

where

$$\begin{aligned}\tilde{x} &= \frac{1}{\tilde{\lambda}\left(\tilde{\beta}\right) - \tilde{\lambda}(\tilde{\alpha})}\int_{\tilde{\alpha}}^{\tilde{\beta}} \tilde{f}(t)\,\mathrm{d}\tilde{\lambda}(t) = \int_{\tilde{\alpha}}^{\tilde{\beta}} \tilde{f}(t)\,\mathrm{d}\tilde{\lambda}(t)\\ &= -\frac{1}{\lambda(\beta)-\lambda(\alpha)}\int_{\alpha}^{\beta} f(t)\,\mathrm{d}\lambda(t) + \tilde{f}(\tilde{\alpha}) + \tilde{f}(\tilde{\beta})\\ &= m + M - \overline{x}\end{aligned}$$

and

$$\begin{aligned}\tilde{y} &= \frac{1}{\tilde{\lambda}\left(\tilde{\beta}\right) - \tilde{\lambda}(\tilde{\alpha})}\int_{\tilde{\alpha}}^{\tilde{\beta}} \varphi\left(\tilde{f}(t)\right)\mathrm{d}\tilde{\lambda}(t) = \int_{\tilde{\alpha}}^{\tilde{\beta}} \varphi\left(\tilde{f}(t)\right)\mathrm{d}\tilde{\lambda}(t)\\ &= \varphi(m) + \varphi(M) - \overline{y}.\end{aligned}$$

Now we have

$$\begin{aligned}&\varphi(c) + \varphi'(c)(m + M - c - \overline{x}) \le \varphi(m) + \varphi(M) - \overline{y}\\ &\le \varphi(d) + \int_{\tilde{\alpha}}^{\tilde{\beta}} \varphi'\left(\tilde{f}(t)\right)\left[\tilde{f}(t) - d\right]\mathrm{d}\tilde{\lambda}(t), \qquad (2.11)\end{aligned}$$

and if in the second inequality in (2.11) we apply (2.10) for the function $g : [\tilde{\alpha}, \tilde{\beta}] \to \mathbb{R}$ defined by

$$g(t) = \varphi'\left(\tilde{f}(t)\right)\left[\tilde{f}(t) - d\right]$$

we obtain (2.7). The proof is complete. □

References

[1] S. Abramovich, M. Klaričić Bakula, M. Matić, J. Pečarić, *A variant of Jensen-Steffensen's inequality and quasi-arithmetic means*, J. Math. Anal. Appl. **307**(3) (2005), 370–376.

[2] S. Abramovich, M. Klaričić Bakula, S. Banić, *A variant of Jensen-Steffensen's inequality for convex and superquadratic functions*, JIPAM **7** (2) (2006), Article 70.

[3] R.P. Boas, *The Jensen-Steffensen inequality*, Univ. Beograd. Publ. Elektrotehn. Fak. Ser. Mat. Fiz. **302-319** (1970), 1–8.

[4] S.S. Dragomir, C.J. Goh, *A counterpart of Jensen's inequality*, Math. Comput. Modelling **24**(2) (1996), 1–11.

[5] N. Elezović, J. Pečarić, *A counterpart to Jensen-Steffensen's discrete inequality for differentiable convex mappings and applications in information theory*, Rad HAZU **481**, Mat. znan. 14 (2003), 25–28.

[6] M. Klaričić Bakula, M. Matić, J. Pečarić, *Generalizations of the Jensen-Steffensen and related inequalities*, submitted for publication.

[7] M. Matić, J. Pečarić, *Some companion inequalities to Jensen's inequality*, Math. Ineq. Appl. **3**(3) (2000), 355–368.

[8] A. McD. Mercer, *A variant of Jensen's inequality*, JIPAM **4** (4) (2003), Article 73.

[9] J. Pečarić, *A companion to Jensen-Steffensen's inequality*, J. Approx. Theory **44**(3) (1985), 289–291.

[10] J.E. Pečarić, F. Proschan and Y.L. Tong, *Convex Functions, Partial Orderings, and Statistical Applications*, Academic Press, Inc., 1992.

[11] A.W. Roberts, D.E. Varberg, *Convex Functions*, Academic Press, 1973.

[12] M.L. Slater, *A companion inequality to Jensen's inequality*, J. Aprox. Theory **32**(2) (1981), 160–166.

[13] A. Witkowski, *A new Proof of the Monotonicity Property of Power Means*, JIPAM **5**(3) (2004), Article 73.

Milica Klaričić Bakula and Marko Matić
Department of Mathematics
Faculty of Natural Sciences, Mathematics and Education
University of Split
Teslina 12
21000 Split, Croatia
e-mail: milica@pmfst.hr
e-mail: mmatic@fesb.hr

Josip Pečarić
Faculty of Textile Technology
University of Zagreb
Prilaz Baruna Filipovića 30
10000 Zagreb, Croatia
e-mail: pecaric@hazu.hr

International Series of Numerical Mathematics, Vol. 157, 245–250

Schur-Convexity, Gamma Functions, and Moments

Albert W. Marshall and Ingram Olkin

Abstract. The gamma function is a central function that arises in many contexts. A wide class of inequalities is obtained by showing that certain gamma functions are Schur-convex coupled with majorization of two vectors.

Mathematics Subject Classification (2000). 26A51, 26D07, 26D20, 33B15, 39B62.

Keywords. Majorization, constrained majorization.

1. Background and preliminaries

A function defined on a subset of $\mathbb{R}_n$ is Schur-convex (concave) if it preserves (reverses) the ordering of majorization. Upper and lower bounds for such a function can often be found by identifying vectors that, within the domain of the function, are external in the sense of majorization. This procedure is employed here to obtain bounds on several functions, all defined in terms of gamma functions. The required Schur-convexity is verified by making use of inequalities that result from connections with moments. For a more detailed discussion of majorization and Schur-convexity, see [8] or [2].

The gamma function introduced by Euler in 1729 is a central function that arises in many contexts, and is discussed in many books on special functions or applied mathematics. The book by [3] deals solely with the gamma function, and provides a detailed discussion of its properties. Two historical accounts are provided by [5] and [10]. A brief description that may suffice for general use is given by [8, Chapter 23]. The compendium by [1] lists formulas and asymptotic behavior of the gamma function.

The concern here is with Schur-convexity of gamma functions, which in turn leads to bounds for the function. For definitions of Schur-convexity and majorization, see [8] or [2]. In particular, let $x = (x_1, \dots, x_n)$ and $y = (y_1, \dots, y_n)$; if $\varphi(x)$ is a Schur-convex function, and $x \succ y$ (x majorizes y), then $\varphi(x) \geq \varphi(y)$. Because

$$(x_1, \dots, x_n) \succ (\bar{x}, \dots, \bar{x}), \tag{1}$$

where $\bar{x} = \sum x_i/n$ for all vectors x, $\varphi(x) \geq \varphi(\bar{x}, \ldots, \bar{x})$. However, an upper bound for gamma functions may be elusive if x is restricted to be positive because the majorization

$$\left(\sum x_i, 0, \ldots, 0\right) \succ (x_1, \ldots, x_n) \tag{2}$$

does not provide a bound. An upper bound can be obtained if the x vector is constrained, as in the following discussion.

The notation $\mathcal{R} = (0, \infty), \mathcal{R}_+ = [0, \infty), \mathcal{R}_{++} = (0, \infty)$ is used throughout.

2. Majorization under constraints

Here a variety of constrained majorization results are stated, each of which will yield a bound. For further discussion and proofs of constrained majorization, see [8, page 132].

Fact 1. [6] Suppose that $m \leq x_i \leq M$, $i = 1, \ldots, n$. Then there exist a unique $\theta \in [m, M]$, and a unique integer L such that

$$\sum x_i = (n - L - 1)m + \theta + LM. \tag{3}$$

With L and θ so determined,

$$x \prec (\underbrace{M, \ldots, M}_{L},\ \theta,\ \underbrace{m, \ldots, m}_{n-L-1}) \equiv v. \tag{4}$$

Fact 2. If $x_i \geq m, i = 1, \ldots, n$, and $\sum x_i = t$, then

$$x \prec (m, \ldots, m, t - (n-1)m); \tag{5}$$

if $x_i \leq M, i = 1, \ldots, n$, and $\sum x_i = t$, then

$$x \prec \left(\frac{t-M}{n-1}, \ldots, \frac{t-M}{n-1}, M\right). \tag{6}$$

Fact 3. If $x_{[1]} \geq \cdots \geq x_{[n]}$ and $x_{[n]} \leq c x_{[n-1]}$, then

$$x \prec \left(x_{[n]}, \frac{x_{[n]}}{c}, \ldots, \frac{x_{[n]}}{c}, M\right), \tag{7}$$

where M is determined by the equation $\sum x_i = x_{[n]} + (n-2)x_{[n]}/c + M$.

Fact 4. If $x_{[n]} \leq x_{[n-1]} + d$, then

$$x \prec (x_{[n]}, x_{[n]} + d, \ldots, x_{[n]} + d, M), \tag{8}$$

where M is determined by the equation $\sum x_i = x_{[n]} + (n-2)(x_{[n]} + d) + M$.

Fact 5. If $\min x_i = m, \max x_i = M, \sum x_i = t$, then

$$x \succ \left(m, \frac{t-m-M}{n-2}, \ldots, \frac{t-m-M}{n-2}, M\right). \tag{9}$$

As a consequence of the majorizations 5–9, if $\varphi(x)$ is Schur-convex and if, for example, $x_i \geq m, \sum x_i = t$, then an upper bound is given by

$$\varphi(x) \leq \varphi(m, \dots, m, t - (n-1)m).$$

Each of the constraints noted in Facts 1–5 leads to bounds.

3. Gamma functions and Schur-convexity

Each of the following functions will be shown to be Schur-convex (or concave):

$$\varphi(x) = \prod_1^n \Gamma(x_i + a), \qquad x_i + a > 0,\ a \geq 0; \tag{10}$$

$$\varphi(x) = \prod_1^n \frac{\Gamma(x_i + a)}{\Gamma(x_i + a + b)}, \qquad x_i + a > 0,\ a, b > 0; \tag{11}$$

$$\varphi(x) = \prod_1^n \frac{\Gamma(mx_i + a)}{\Gamma^s(x_i + a)}, \qquad mx_i + a > 0,\ a \geq 1,\ m \geq 2,\ s \leq m; \tag{12}$$

$$\varphi(x) = \prod_1^n \frac{x_i^{x_i+1}}{\Gamma(x_i + 1)}, \qquad x_i \geq 0; \tag{13}$$

The key to the proofs of 10–13 is the fact that a product of moments of nonnegative random variables is Schur-convex.

Fact 6. Let g be a continuous positive function defined on an interval $I \subset \mathbb{R}$. Then

$$\varphi(x) = \prod_1^n g(x_i), \quad x \in I^n, \tag{14}$$

is Schur-convex on I^n if and only if $\log g$ is convex. Moreover, φ is strictly Schur-convex on I^n if and only if $\log g$ is strictly convex on I.

A particularly useful sufficient condition for the convexity of $\log g$ is the following.

Fact 7. If ν is a measure on $[0, \infty)$ such that $g(x) = \int_0^\infty z^x d\nu(z)$ exists for all x in an interval I, then $\log g$ is convex on I. Unless ν concentrates its mass on a set of the form $\{0, z_0\}$, $\log g$ is strictly convex on I.

The following is a consequence: If μ_r is the rth moment of a non-negative variable, that is

$$\mu_r = \int_0^\infty z^r d\nu(z) \tag{15}$$

for some probability measure ν, and if μ_r exists for all r in the interval $I \subset \mathbb{R}$, then

$$\varphi(r_1, \dots, r_n) = \prod_1^n \mu_{r_i} \tag{16}$$

is Schur-convex in $r = (r_1, \dots, r_n) \in I^n$.

The application of Fact 7 is central to showing that 10–13 are Schur-convex.

Proof of the Schur-convexity of 10. With $d\nu(z) = z^{a-1}e^{-z}dz$, $0 \le z \le \infty$, and $I = (-a, \infty)$, $\mu_r = \Gamma(r+a)$, $r > -a$. □

Proof of the Schur-convexity of 11. With $d\nu(z) = z^{a-1}(1-z)^{b-1}dz$, $0 \le z \le 1$, and $I = (-a, \infty)$, $\mu_r = \Gamma(r+a)/\Gamma(r+a+b)$. □

Proof of the Schur-convexity of 12. The Legendre multiplication formula permits the expansion

$$\frac{\Gamma(mz+a)}{\Gamma^s(z+a)} = \frac{m^{mz+a-1/2}}{(2\pi)^{(m-1)/2}} \prod_{j=1}^{s} \left(\frac{\Gamma\left(z + \frac{a+j-1}{m}\right)}{\Gamma(z+a)} \right) \prod_{j=s+1}^{m} \Gamma\left(z + \frac{a+j-1}{m}\right). \quad (17)$$

Define $\psi_j(z_i) = \Gamma\left(z_i + \frac{a+j-1}{m}\right) \Big/ \Gamma(z_i + a)$ for $j = 1, \ldots, s$, and $\psi_j(z_i) = \Gamma(z_i + \frac{a+j-1}{m})$ for $j = s+1, \ldots, m$. Further, note that $\frac{a+j-1}{m} \le a$ for $a \ge 1$. Then

$$\varphi(z) = c \prod_{i=1}^{n} \prod_{j=1}^{m} \psi_j(z_i), \quad (18)$$

where c depends on m, n, a, and $\sum z_i$, which does not affect Schur-convexity. From the Schur-convexity of 10 and 11 it follows that the product 18 is Schur-convex. □

Proof of the Schur-convexity of 13. The function $\varphi(x)$ in 13 is Schur-concave on R^n_+. Here note that

$$\frac{\Gamma(r+1)}{r^{r+1}} = \int_0^\infty (te^{-t})^r dt = \int_0^\infty z^r d\nu(z),$$

where

$$\nu[-y, z] = \int_{te^{-t} \le z} te^{-t} dt.$$

Consequently, $\log(\Gamma(x+1)/x^{x+1})$ is convex in $x > 0$ and $\log(x^{x+1}/\Gamma(x+1))$ is concave in $x > 0$. □

Other results can be obtained in a similar manner. In particular, if g is a Laplace transform, i.e.,

$$g(s) = \int_0^\infty e^{-sz} d\mu(z),$$

then $\varphi(x) = \Pi g(x_i)$ is Schur-convex on $\mathbb{R}^n_+$.

This opens up many possible results on Schur-convexity.

4. Bounds from constrained majorizations

Each of the constrained majorizations 4 to 9 coupled with the Schur-convexity results of Section 3 leads to new bounds. The following examples are illustrative of these bounds.

4.1 If $x_i \leq M$, $x_i + a > 0$, $i = 1, \dots, n$, $\sum x_i = t$, $\sum x_i/n = \bar{x}$, then

$$[\Gamma(\bar{x}+a)]^n \leq \prod_1^n \Gamma(x_i + a) \leq \Big[\Gamma\Big(\frac{t-M}{n-1} + a\Big)\Big]^{n-1} \Gamma(M+a). \tag{19}$$

The upper bound 19 follows from the Schur-convexity of 10 and the constrained majorization 6; the lower bound follows from $x \succ (\bar{x}, \dots, \bar{x})$.

4.2 If $x_i \geq m$, $x_i + a > 0$, $i = 1, \dots, n$, $b > 0$, $\sum x_i = t$, $\sum x_i/n = \bar{x}$, then

$$\begin{aligned}\Big[\frac{\Gamma(\bar{x}+a)}{\Gamma(\bar{x}+a+b)}\Big]^n &\leq \prod_1^n \frac{\Gamma(x_i+a)}{\Gamma(x_i+a+b)} \\ &\leq \Big[\frac{\Gamma(m+a)}{\Gamma(m+a+b)}\Big]^{n-1} \frac{\Gamma(t-(n-1)m+a)}{\Gamma(t-(n-1)m+a+b)}.\end{aligned} \tag{20}$$

The upper bound 20 follows from the Schur-convexity of 11 and the constrained majorization 5; the lower bound follows from $x \succ (\bar{x}, \dots, \bar{x})$.

These examples show how the pairing of a constrained majorization with a Schur-convex function leads to a bound.

References

[1] M. Abramowitz and I.A. Stegun (Eds.) *Handbook of Mathematical Functions.* National Bureau of Standards, Applied Mathematics Series **55**, U.S. Government Printing Office, Washington DC.,1964. (Reprinted by Dover Publications, 1970).

[2] T. Ando, *Majorization, doubly stochastic matrices, and comparison of eigenvalues,* Linear Algebra Appl. **118** (1989), 163–248.

[3] E. Artin, *Einführung in die Theorie der Gammafunktion.* Hamburger Mathematische Einzelschriften, Heft II. (Transl. by M. Butler, *The Gamma Function* (1964)) Holt, Rinehart & Winston, New York, 1931.

[4] C.P. Chen, *Complete monotonicity properties for a ratio of gamma functions,* Univ. Beograd. Publ. Elektrotehn. Fac. Ser. Mat. **16** (2005), 26–28.

[5] P.J. Davis, *Leonard Euler's integral: A historical profile of the gamma function,* Amer. Math. Monthly **66** (1959), 849–869.

[6] J.H.B. Kemperman, *Moment problems for sampling without replacement. I, II, III.* Nederl. Akad. Wetensch. Proc. Ser. A 76 Indag. Math. **35** (1973) 149–164, 165–180 and 181–188.

[7] A.J. Li, W.Z. Zhao& C.P. Chen, *Logarithmically complete monotonicity and Schur-convexity for some ratios of gamma functions,* Univ. Beograd. Publ. Elektrotehn. Fac. Ser. Mat. **17** (2006), 88–92.

[8] A.W. Marshall & I. Olkin, *Inequalities: Theory of majorization and its applications*, Academic Press, New York, 1979.

[9] M. Merkle, *On log-convexity of a ratio of gamma functions*, Univ. Beograd. Publ. Elektrotehn. Fac. Ser. Mat. **8** (1997), 114–119.

[10] G.K. Srinivasan, *The gamma function: An eclectic tour*, Amer. Math. Monthly **114** (2007), 297–315.

Albert W. Marshall
Statistics Department
University of British Columbia
Vancouver BG, V6T 1Z2, Canada

Ingram Olkin
Department of Statistics
Stanford University
Sequoia Hall
390 Serra Mall
Stanford, CA 94305-4065, USA
e-mail: iolkin@stat.stanford.edu

International Series of Numerical Mathematics, Vol. 157, 251–260

A Characterization of Nonconvexity and Its Applications in the Theory of Quasi-arithmetic Means

Zoltán Daróczy and Zsolt Páles

Abstract. In this paper, we give necessary and sufficient conditions for the comparison, equality and homogeneity problems of two-variable means of the form

$$M\big(A_{\varphi,w_1}(x,y),\ldots,A_{\varphi,w_n}(x,y)\big) \qquad (x,y\in I)$$

where M is an n-variable mean on the open interval I and A_{φ,w_i} denotes the weighted quasi-arithmetic mean generated by a strictly increasing continuous function $\varphi : I \to \mathbb{R}$ and by a weight function $w_i : I^2 \to]0,1[$. The approach is based on a characterization of lower semicontinuous nonconvex function.

Mathematics Subject Classification (2000). Primary 26D10.

Keywords. Nonconvexity; comparison, equality, and homogeneity of means; quasi-arithmetic mean.

1. Introduction

Throughout this paper let I denote a nonempty open interval of real numbers and let $\mathcal{CM}(I)$ stand for the class of continuous strictly monotone functions $f : I \to \mathbb{R}$.

For a function $\varphi \in \mathcal{CM}(I)$, define the (two-variable symmetric) quasi-arithmetic mean A_φ by

$$A_\varphi(x,y) := \varphi^{-1}\left(\frac{\varphi(x)+\varphi(y)}{2}\right) \qquad (x,y\in I).$$

When $I = \mathbb{R}_+$ and there exists a real parameter $p \in \mathbb{R}$ such that, for $x > 0$,

$$\varphi(x) = \sigma_p(x) := \begin{cases} x^p & \text{if } p \neq 0, \\ \ln(x) & \text{if } p = 0, \end{cases}$$

This research has been supported by the Hungarian Scientific Research Fund (OKTA) Grants K-62316, NK-68040.

then the two-variable symmetric quasi-arithmetic mean A_φ becomes the so-called two-variable symmetric power or Hölder mean H_p defined by

$$H_p(x,y) := \begin{cases} \left(\dfrac{x^p + y^p}{2}\right)^{\frac{1}{p}} & \text{if } p \neq 0 \\ \sqrt{xy} & \text{if } p = 0 \end{cases} \qquad (x, y > 0).$$

Obviously, in the particular cases $p = 1, p = 0$, and $p = -1$, the Hölder means reduce to the arithmetic, geometric, and harmonic means, respectively.

Concerning the comparison, equality and homogeneity problems of quasi-arithmetic means, we have the following three classical results presented in the book [8] of Hardy, Littlewood, and Pólya (cf. also [6], [11]).

Theorem A. *Let $\varphi, \psi \in \mathcal{CM}(I)$. Then the comparison inequality*

$$A_\varphi(x,y) \leq A_\psi(x,y) \qquad (x, y \in I)$$

holds if and only if $\psi \circ \varphi^{-1}$ is convex (concave) on $\varphi(I)$ provided that ψ is increasing (decreasing).

Theorem B. *Let $\varphi, \psi \in \mathcal{CM}(I)$. Then the identity*

$$A_\varphi(x,y) = A_\psi(x,y) \qquad (x, y \in I)$$

holds if and only if there exist real constants $a \neq 0$ and b such that $\psi = a\varphi + b$.

In this case, we say that the generating functions φ and ψ are equivalent and we write $\varphi \sim \psi$.

Theorem C. *Let $\varphi \in \mathcal{CM}(\mathbb{R}_+)$. Then the quasi-arithmetic mean A_φ is homogeneous, i.e.,*

$$A_\varphi(tx, ty) = tA_\varphi(x,y) \qquad (x, y, t > 0)$$

if and only if there exists a real parameter p such that $\varphi \sim \sigma_p$, i.e., the quasi-arithmetic mean A_φ is equal to the Hölder mean H_p.

In order to define quasi-arithmetic means and Hölder means weighted by a weight function, denote by $\mathcal{W}(I)$ the class of functions $w : I^2 \to]0,1[$. For $\varphi \in \mathcal{CM}(I)$ and for a weight function $w \in \mathcal{W}(I)$, define the weighted quasi-arithmetic mean $A_{\varphi,w} : I^2 \to I$ by

$$A_{\varphi,w}(x,y) := \varphi^{-1}\big(w(x,y)\varphi(x) + (1 - w(x,y))\varphi(y)\big) \qquad (x, y \in I).$$

For a parameter $p \in \mathbb{R}$ and for a weight function $w \in \mathcal{W}(\mathbb{R}_+)$, define the weighted Hölder mean $H_{p,w} : \mathbb{R}_+^2 \to \mathbb{R}_+$ by

$$H_{p,w}(x,y) := \begin{cases} \big(w(x,y)x^p + (1 - w(x,y))y^p\big)^{\frac{1}{p}} & \text{if } p \neq 0 \\ x^{w(x,y)} y^{1-w(x,y)} & \text{if } p = 0 \end{cases} \qquad (x, y > 0).$$

Clearly, if the weight function w is equal to the constant $\frac{1}{2}$, then we have that $A_\varphi = A_{\varphi,\frac{1}{2}}$ and $H_p = H_{p,\frac{1}{2}}$.

Given a generating function $\varphi \in \mathcal{CM}(I)$, an n-variable mean $M : I^n \to I$, and an n-tuple of weight functions $w_1, \ldots, w_n \in \mathcal{W}(I)$, we can define a two-variable mean by the following formula

$$M\big(A_{\varphi,w_1}(x,y), \ldots, A_{\varphi,w_n}(x,y)\big) \qquad (x, y \in I). \tag{1.1}$$

The main problem investigated in this paper is to find necessary and sufficient conditions for the comparison, equality and homogeneity problems of means of the form (1.1). Our approach is based on a characterization of the nonconvexity of lower semicontinuous real-valued functions. In the last section examples are provided to demonstrate the applicability of our results and also some open questions are discussed.

2. Nonconvex functions

Let X be a real linear space and $D \subseteq X$ be a nonvoid convex subset. By the standard definition of convexity, a function $f : D \to \mathbb{R}$ is convex if, for all $x, y \in D$ and $t \in [0, 1]$,

$$f\big(tx + (1-t)y\big) \le tf(x) + (1-t)f(y). \tag{2.1}$$

Thus, the nonconvexity of f yields the existence of elements $x \neq y$ in D and $t \in]0, 1[$ such that

$$f\big(tx + (1-t)y\big) > tf(x) + (1-t)f(y). \tag{2.2}$$

It seems to be a natural problem if, for some fixed $x \neq y$, one could get further properties of the values t satisfying (2.2).

Let $X = \mathbb{R}$ and consider the function $f = \chi_{\{0\}} : \mathbb{R} \to \mathbb{R}$, the characteristic function of the singleton $\{0\}$. Obviously, f is nonconvex. On the other hand, (2.2) can hold for this function only if $f\big(tx + (1-t)y\big) = 1$, i.e., when $tx + (1-t)y = 0$. This is satisfied only if either $x < 0 < y$ or $y < 0 < x$ and $t = \frac{y}{y-x}$. Therefore, if $x < 0 < y$ or $y < 0 < x$, then there exists exactly one $t \in]0, 1[$ such that (2.2) holds and for $x, y \le 0$ and $0 \le x, y$ there is no $t \in]0, 1[$ satisfying (2.2). Thus, in order to assure a larger set of the values t satisfying (2.2) for some fixed $x \neq y$, we need to assume additional regularity properties on the function f.

Since the restriction of a convex functions to any open segment of D is automatically continuous, the continuity of f in this sense seems to be a relevant regularity property. In fact, we shall see that lower semicontinuity is sufficient for our purposes. We say that $f : D \to \mathbb{R}$ is *lower semicontinuous along the segments of* D if, for all $u, v \in D$, the real function $t \mapsto f\big(tu + (1-t)v\big)$ is lower semicontinuous on $[0, 1]$ in the standard sense.

Theorem 2.1. *Let $f : D \to \mathbb{R}$ be a nonconvex function which is lower semicontinuous along the segments of D. Then there exist $x \neq y$ in D such that* (2.2) *holds for all* $0 < t < 1$.

Proof. By the nonconvexity of f, we have that there exist elements $u \neq v$ in D and $0 < \tau < 1$ such that

$$f\big(\tau u + (1-\tau)v\big) > \tau f(u) + (1-\tau)f(v).$$

Thus, the set

$$T := \big\{t \in]0,1[\colon f\big(tu + (1-t)v\big) - tf(u) - (1-t)f(v) > 0\big\}$$

is nonempty. On the other hand, by the lower semicontinuity property of f, the set T is also open. Let $S =]r,s[$ be a maximal open subinterval of T. Then r, s cannot belong to T, hence

$$\begin{aligned} f\big(ru + (1-r)v\big) &\le rf(u) + (1-r)f(v),\\ f\big(su + (1-s)v\big) &\le sf(u) + (1-s)f(v), \end{aligned} \tag{2.3}$$

and

$$f\big(\tau u + (1-\tau)v\big) > \tau f(u) + (1-\tau)f(v) \qquad (\tau \in]r,s[). \tag{2.4}$$

Denote

$$x := su + (1-s)v \qquad \text{and} \qquad y := ru + (1-r)v. \tag{2.5}$$

Then $y - x = (s-r)(v-u) \neq 0$, hence $x \neq y$. To show that (2.2) holds, let $t \in]0,1[$ be arbitrary. Then, using the definitions in (2.5), the inequality (2.4) with $\tau := ts + (1-t)r$ and finally the two inequalities of (2.3), we get

$$\begin{aligned} f\big(tx + (1-t)y\big) &= f\Big(t\big(su + (1-s)v\big) + (1-t)\big(ru + (1-r)v\big)\Big)\\ &= f\Big(\big(ts + (1-t)r\big)u + \big(t(1-s) + (1-t)(1-r)\big)v\Big)\\ &> \big(ts + (1-t)r\big)f(u) + \big(t(1-s) + (1-t)(1-r)\big)f(v)\\ &= t\big(sf(u) + (1-s)f(v)\big) + (1-t)\big(rf(u) + (1-r)f(v)\big)\\ &\ge tf\big(su + (1-s)v\big) + (1-t)f\big(ru + (1-r)v\big)\\ &= tf(x) + (1-t)f(y). \end{aligned}$$

Therefore (2.2) holds for all $t \in]0,1[$. □

Remark 2.2. A characterization of nonconvexity for upper semicontinuous functions was found by the second author in [13]. This characterization was used to obtain sandwich-type theorems for the separation of quasi-arithmetic means by Hölder means.

The following result, which is an obvious consequence of Theorem 2.1, offers a new characterization of convexity.

Corollary 2.3. *Let $f : D \to \mathbb{R}$ be lower semicontinuous function along the segments of D. Let $w : D^2 \to]0,1[$ and assume that f is w-convex, i.e., for all $x, y \in D$,*

$$f\big(w(x,y)x + (1 - w(x,y))y\big) \le w(x,y)f(x) + \big(1 - w(x,y)\big)f(y). \tag{2.6}$$

Then f is convex on D.

Remark 2.4. In the case when, instead of lower semicontinuity, the local upper boundedness is assumed for the function f and the function w is continuous in both variables, the second author proved in [12, Remark 2] that inequality (2.6) also characterizes the convexity of f. Using the results of Adamek [1, 2], one can see that, for this statement, the local upper boundedness of f at a single point of D is sufficient. These results generalize the so-called Bernstein–Doetsch Theorem (cf. [5], [9]). On the other hand, by the example considered at the beginning of this section, one can see that the lower semicontinuity of f at a single point of D and the validity of (2.6) does not imply the convexity of f in general.

3. Main results

Our basic result presents an alternative for the comparison of quasi-arithmetic means.

Theorem 3.1. *Let $\varphi, \psi \in \mathcal{CM}(I)$. Then*

(i) *either, for all elements $x, y \in I$ and for all constant weights $w \in]0,1[$,*

$$A_{\varphi,w}(x,y) \leq A_{\psi,w}(x,y), \tag{3.1}$$

(ii) *or there exist two distinct elements $x, y \in I$ such that, for all constant weights $w \in]0,1[$,*

$$A_{\varphi,w}(x,y) > A_{\psi,w}(x,y). \tag{3.2}$$

Proof. Obviously, the two alternatives cannot hold simultaneously. Without loss of generality, we may assume that ψ is increasing.

Suppose that the first alternative (i) is not valid: Then there exist elements $x, y \in I$ and $w \in]0,1[$, such that (3.1) is not satisfied, i.e., (3.2) holds. By the increasingness of ψ, (3.2) can be rewritten as

$$\psi \circ \varphi^{-1}\big(w\varphi(x) + (1-w)\varphi(y)\big) > w\psi(x) + (1-w)\psi(y). \tag{3.3}$$

Now, set $u := \varphi(x)$ and $v := \varphi(y)$. Then $u, v \in \varphi(I)$, $x = \varphi^{-1}(u)$, $y = \varphi^{-1}(v)$, and (3.3) yields

$$\psi \circ \varphi^{-1}\big(wu + (1-w)v\big) > w\psi \circ \varphi^{-1}(u) + (1-w)\psi \circ \varphi^{-1}(v),$$

showing that the continuous function $f := \psi \circ \varphi^{-1}$ is not convex on the interval $\varphi(I)$. By our characterization of nonconvexity, it follows that there exist elements $u \neq v$ in $\varphi(I)$ such that, for all $w \in]0,1[$,

$$f\big(wu + (1-w)v\big) > wf(u) + (1-w)f(v).$$

Define $x := \varphi^{-1}(u)$ and $y := \varphi^{-1}(v)$. Then $u = \varphi(x)$ and $v = \varphi(y)$ and the above inequality yields that (3.3) holds for all $w \in]0,1[$. By the increasingness of ψ again, we get that the second alternative (ii) must be valid. $\square$

Our main result below gives a general comparison theorem for means of the form (1.1). It is a generalization of Theorem A. In the formulation of the result, an n-variable mean $M : I^n \to I$ is called *increasing* if, for all elements $x_1, \dots, x_n$, $y_1, \dots, y_n$ in I with $x_1 \le y_1, \dots, x_n \le y_n$, we have the inequality

$$M(x_1, \dots, x_n) \le M(y_1, \dots, y_n)$$

The mean M is called *strictly increasing* if it is increasing and the strict inequality

$$M(x_1, \dots, x_n) < M(y_1, \dots, y_n)$$

is valid whenever $x_1 < y_1, \dots, x_n < y_n$.

Theorem 3.2. *Let $\varphi, \psi \in \mathcal{CM}(I)$, let $M : I^n \to I$ be a strictly increasing n-variable mean, and let $w_1, \dots, w_n \in \mathcal{W}(I)$. Then*

$$M\big(A_{\varphi,w_1}(x,y), \dots, A_{\varphi,w_n}(x,y)\big) \le M\big(A_{\psi,w_1}(x,y), \dots, A_{\psi,w_n}(x,y)\big) \quad (x, y \in I) \tag{3.4}$$

holds if and only if $\psi \circ \varphi^{-1}$ is convex (concave) on $\varphi(I)$ if ψ is increasing (decreasing).

Proof. Throughout the proof, we may assume that ψ is increasing.

If $\psi \circ \varphi^{-1}$ is convex on $\varphi(I)$ then, by Theorem A, the comparison inequality $A_\varphi \le A_\psi$ holds on I^2, i.e., for all $x, y \in I$,

$$A_{\varphi,\frac{1}{2}}(x,y) \le A_{\psi,\frac{1}{2}}(x,y).$$

Therefore, the second alternative (ii) of Theorem 3.1 cannot be valid. Thus, the first alternative (i) of this theorem must hold, i.e., for all $x, y \in I$ and for all constant weights $w \in]0,1[$, we have

$$A_{\varphi,w}(x,y) \le A_{\psi,w}(x,y).$$

In particular, with $w = w_i(x,y)$, we get that

$$A_{\varphi,w_i}(x,y) \le A_{\psi,w_i}(x,y) \qquad (x, y \in I,\ i = 1, \dots, n).$$

Hence, by the increasingness of the mean M, (3.4) follows.

Now assume that $\psi \circ \varphi^{-1}$ is nonconvex on $\varphi(I)$. Then, by Theorem A, the comparison inequality $A_\varphi \le A_\psi$ is not satisfied on I^2. Therefore, the first alternative (i) of Theorem 3.1 cannot hold. Thus, the second alternative (ii) of this theorem must be valid, i.e., there exist two elements $x, y \in I$ such that, for all constant weights $w \in]0,1[$,

$$A_{\varphi,w}(x,y) > A_{\psi,w}(x,y).$$

In particular, with $w = w_i(x,y)$, we get that

$$A_{\varphi,w_i}(x,y) > A_{\psi,w_i}(x,y) \qquad (i = 1, \dots, n).$$

Now, using the strict increasingness of the mean M, it follows from these strict inequalities that (3.4) cannot be valid. □

Theorem 3.3. *Let $\varphi, \psi \in \mathcal{CM}(I)$, let $M : I^n \to I$ be a strictly increasing n-variable mean, and let $w_1, \dots, w_n \in \mathcal{W}(I)$. Then*

$$M\big(A_{\varphi,w_1}(x,y),\dots,A_{\varphi,w_n}(x,y)\big) = M\big(A_{\psi,w_1}(x,y),\dots,A_{\psi,w_n}(x,y)\big) \quad (x,y \in I) \tag{3.5}$$

holds if and only if there exist real constants $a \neq 0$ and b such that $\psi = a\varphi + b$.

Proof. Without loss of generality, we may assume that φ and ψ are increasing functions. Then the identity (3.5) holds if and only if, instead of the equality, the two inequalities $\leq$ and $\geq$ hold in (3.5) simultaneously. By Theorem 3.2, this is valid if and only if $\psi \circ \varphi^{-1}$ and $\varphi \circ \psi^{-1}$ are convex functions on $\varphi(I)$ and $\psi(I)$, respectively. Using Theorem A, it follows that the two inequalities $A_\varphi \leq A_\psi$ and $A_\psi \leq A_\varphi$ hold on I^2 simultaneously, i.e., we have the identity $A_\varphi = A_\psi$. By Theorem B, this is satisfied if and only if $\varphi \sim \psi$, i.e., there exist real constants $a \neq 0$ and b such that $\psi = a\varphi + b$. □

Theorem 3.4. *Let $\varphi, \psi \in \mathcal{CM}(\mathbb{R}_+)$, let $M : \mathbb{R}_+^n \to \mathbb{R}_+$ be a strictly increasing n-variable mean which is homogeneous, i.e.,*

$$M(tx_1,\dots,tx_n) = tM(x_1,\dots,x_n) \qquad (t, x_1,\dots,x_n > 0).$$

Let $w_1,\dots,w_n \in \mathcal{W}(\mathbb{R}_+)$ be nullhomogeneous weight functions, i.e.,

$$w_i(tx,ty) = w_i(x,y) \qquad (t,x,y > 0,\ i = 1,\dots,n)$$

Then the homogeneity property

$$M\big(A_{\varphi,w_1}(tx,ty),\dots,A_{\varphi,w_n}(tx,ty)\big) = tM\big(A_{\varphi,w_1}(x,y),\dots,A_{\varphi,w_n}(x,y)\big) \tag{3.6}$$

holds for all $t,x,y > 0$ if and only if there exists a real parameter p such that $\varphi \sim \sigma_p$, i.e., for all $x,y > 0$,

$$M\big(A_{\varphi,w_1}(x,y),\dots,A_{\varphi,w_n}(x,y)\big) = M\big(H_{p,w_1}(x,y),\dots,H_{p,w_n}(x,y)\big). \tag{3.7}$$

Proof. Assume first that (3.6) holds. For a fixed $t > 0$, define the function $\varphi_t : \mathbb{R}_+ \to \mathbb{R}_+$ by $\varphi_t(x) := \varphi(tx)$. Then, by the nullhomogeneity of w_i, for all $x,y > 0$, we have

$$\begin{aligned} A_{\varphi,w_i}(tx,ty) &= \varphi^{-1}\big(w_i(tx,ty)\varphi(tx) + (1 - w_i(tx,ty))\varphi(ty)\big) \\ &= \varphi^{-1}\big(w_i(x,y)\varphi_t(x) + (1 - w_i(x,y))\varphi_t(y)\big) \\ &= t\varphi_t^{-1}\big(w_i(x,y)\varphi_t(x) + (1 - w_i(x,y))\varphi_t(y)\big) = tA_{\varphi_t,w_i}(x,y). \end{aligned}$$

Thus, using the homogeneity of M, (3.6) yields

$$M\big(A_{\varphi_t,w_1}(x,y),\dots,A_{\varphi_t,w_n}(x,y)\big) = M\big(A_{\varphi,w_1}(x,y),\dots,A_{\varphi,w_n}(x,y)\big).$$

Applying Theorem 3.3, it follows that $\varphi_t \sim \varphi$ for all $t > 0$. Hence, by Theorem B, we get that $A_{\varphi_t} = A_\varphi$ for all $t > 0$, i.e.,

$$A_\varphi(x,y) = A_{\varphi_t}(x,y) = \frac{1}{t}\varphi^{-1}\Big(\frac{\varphi(tx) + \varphi(ty)}{2}\Big) = \frac{1}{t}A_\varphi(tx,ty).$$

This shows that A_φ is a homogeneous mean. In view of Theorem C, we get that $\varphi \sim \sigma_p$, and hence $A_{\varphi,w_i} = H_{p,w_i}$ $(i = 1,\dots,n)$ are valid for some $p \in \mathbb{R}$. Therefore, (3.7) holds with this value of $p \in \mathbb{R}$.

Conversely, assume that $\varphi \sim \sigma_p$ holds for some $p \in \mathbb{R}$. By the nullhomogeneity of w_i, we have that

$$\begin{aligned} H_{p,w_i}(tx,ty) &= \big(w_i(tx,ty)(tx)^p + (1 - w_i(tx,ty))(ty)^p\big)^{\frac{1}{p}} \\ &= \big(t^p\big(w_i(x,y)x^p + (1 - w_i(x,y))y^p\big)\big)^{\frac{1}{p}} \\ &= t\big(w_i(x,y)x^p + (1 - w_i(x,y))y^p\big)^{\frac{1}{p}} = tH_{p,w_i}(x,y) \end{aligned}$$

if $p \neq 0$, and the same is also valid if $p = 0$. In other words, H_{p,w_i} is homogeneous for all $i = 1,\dots,n$. Using $\varphi \sim \sigma_p$, the homogeneity of the means H_{p,w_i} and M, we get

$$\begin{aligned} &M\big(A_{\varphi,w_1}(tx,ty),\dots,A_{\varphi,w_n}(tx,ty)\big) = M\big(H_{p,w_1}(tx,ty),\dots,H_{p,w_n}(tx,ty)\big) \\ &= M\big(tH_{p,w_1}(x,y),\dots,tH_{p,w_n}(x,y)\big) = tM\big(H_{p,w_1}(x,y),\dots,H_{p,w_n}(x,y)\big) \\ &= tM\big(A_{\varphi,w_1}(x,y),\dots,A_{\varphi,w_n}(x,y)\big), \end{aligned}$$

which proves the homogeneity property (3.6). □

4. Examples

In the subsequent examples we demonstrate the applicability of our main results.

Example 1. For a generating function $\varphi \in \mathcal{CM}(I)$ and for a constant weight $w \in]0,1[$, consider the mean

$$\widetilde{A}_{\varphi,w}(x,y) := \frac{\varphi^{-1}\big(w\varphi(x) + (1-w)\varphi(y)\big) + \varphi^{-1}\big((1-w)\varphi(x) + w\varphi(y)\big)}{2},$$

which is called a *symmetrized weighted quasi-arithmetic mean*. Then $\widetilde{A}_{\varphi,w}$ is of the form (1.1), where $n = 2$ and

$$M(x,y) = \frac{x+y}{2}, \qquad w_1(x,y) = w, \qquad w_2(x,y) = 1 - w \qquad (x,y \in I).$$

Observe that M is a strictly increasing and homogeneous mean, and the weight functions w_1 and w_2 are trivially nullhomogeneous. Thus, each of Theorem 3.2, Theorem 3.3, and Theorem 3.4 can be applied. By these theorems, we have that:

1. For $\varphi,\psi \in \mathcal{CM}(I)$ and $w \in]0,1[$, the inequality $\widetilde{A}_{\varphi,w} \le \widetilde{A}_{\psi,w}$ holds on I^2 if and only if $\psi\circ\varphi^{-1}$ is convex (concave) on $\varphi(I)$ if ψ is increasing (decreasing).
2. For $\varphi,\psi \in \mathcal{CM}(I)$ and $w \in]0,1[$, the identity $\widetilde{A}_{\varphi,w} = \widetilde{A}_{\psi,w}$ holds on I^2 if and only if $\varphi \sim \psi$.
3. For $\varphi \in \mathcal{CM}(\mathbb{R}_+)$ and $w \in]0,1[$, the mean $\widetilde{A}_{\varphi,w}$ is homogeneous on $\mathbb{R}_+^2$ if and only if $\varphi \sim \sigma_p$ for some $p \in \mathbb{R}$.

The results of this paper have grown out from the investigation of means of the form $\widetilde{A}_{\varphi,w}$. Originally, we wanted to describe the comparison, equality and homogeneity problems of these means. The more general comparison problem $\widetilde{A}_{\varphi,w_1} \leq \widetilde{A}_{\psi,w_2}$ and the related equality problem $\widetilde{A}_{\varphi,w_1} = \widetilde{A}_{\psi,w_2}$ (where $\varphi, \psi \in \mathcal{CM}(I)$ and $w_1, w_2 \in]0,1[$), i.e., when the constant weights w_1 and w_2 are possibly different seem to be much more difficult, and they have not been solved yet.

Example 2. For a generating function $\varphi \in \mathcal{CM}(I)$ and for a positive function $p : I \to \mathbb{R}_+$, consider the mean

$$M_{\varphi,p}(x,y) := \varphi^{-1}\left(\frac{p(x)\varphi(x) + p(y)\varphi(y)}{p(x) + p(y)}\right) \qquad (x, y \in I), \tag{4.1}$$

which is called a *quasi-arithmetic mean weighted by the weight function* p. These means were considered first by Bajraktarević [3, 4]. It is easy to see that $M_{\varphi,p}$ is also of the form (1.1), where $n = 1$ and

$$M(x) = x, \qquad w_1(x,y) = \frac{p(x)}{p(x) + p(y)} \qquad (x, y \in I).$$

Then M is a strictly increasing and homogeneous mean. The weight functions w_1 is not nullhomogeneous in general except if p is a power function. Thus, again Theorem 3.2, Theorem 3.3, and Theorem 3.4 can be applied. By these theorems, we have that:

1. For $\varphi, \psi \in \mathcal{CM}(I)$ and $p : I \to \mathbb{R}_+$, the inequality $M_{\varphi,p} \leq M_{\psi,p}$ holds on I^2 if and only if $\psi \circ \varphi^{-1}$ is convex (concave) on $\varphi(I)$ if ψ is increasing (decreasing).
2. For $\varphi, \psi \in \mathcal{CM}(I)$ and $p : I \to \mathbb{R}_+$, the identity $M_{\varphi,p} = M_{\psi,p}$ holds on I^2 if and only if $\varphi \sim \psi$.
3. For $\varphi \in \mathcal{CM}(\mathbb{R}_+)$ and $p(x) = x^q$ $(x > 0)$, the mean $M_{\varphi,p}$ is homogeneous on $\mathbb{R}_+^2$ if and only if $\varphi \sim \sigma_r$ for some $r \in \mathbb{R}$, i.e., if $M_{\varphi,p}$ is a two-variable Gini mean.

The more general equality problem $M_{\varphi,p} = M_{\psi,q}$ with four unknown functions $\varphi, \psi \in \mathcal{CM}(I)$, $p, q : I \to \mathbb{R}_+$ was solved under six-times continuous differentiability assumptions by Losonczi [10]. The general comparison problem $M_{\varphi,p} \leq M_{\psi,q}$ (again with four unknown functions) has not been solved yet. The solution for the homogeneity problem is also unknown. If the definition (4.1) is extended to more than two-variable means, the comparison problem was solved in [7].

Example 3. For a generating function $\varphi \in \mathcal{CM}(\mathbb{R}_+)$, consider the mean

$$N_\varphi(x,y) := \sqrt{\varphi^{-1}\left(\frac{\varphi(x)+\varphi(y)}{2}\right) \cdot \varphi^{-1}\left(\cos^2\left(\frac{x}{x+y}\right)\varphi(x) + \sin^2\left(\frac{x}{x+y}\right)\varphi(y)\right)}.$$

Then N_φ is of the form (1.1), where $n = 2$ and

$$M(x,y) = \sqrt{xy}, \qquad w_1(x,y) = \frac{1}{2}, \qquad w_2(x,y) = \cos^2\left(\frac{x}{x+y}\right) \qquad (x, y > 0).$$

Observe that M is a strictly increasing and homogeneous mean, and the weight functions w_1 and w_2 are nullhomogeneous. Thus, each of Theorem 3.2, Theorem 3.3, and Theorem 3.4 can be applied. By these theorems, we have that:

1. For $\varphi, \psi \in \mathcal{CM}(\mathbb{R}_+)$, the inequality $N_\varphi \leq N_\psi$ holds on $\mathbb{R}_+^2$ if and only if $\psi \circ \varphi^{-1}$ is convex (concave) on $\varphi(\mathbb{R}_+)$ if ψ is increasing (decreasing).
2. For $\varphi, \psi \in \mathcal{CM}(\mathbb{R}_+)$, the identity $N_\varphi = N_\psi$ holds on $\mathbb{R}_+^2$ if and only if $\varphi \sim \psi$.
3. For $\varphi \in \mathcal{CM}(\mathbb{R}_+)$, the mean N_φ is homogeneous if and only if $\varphi \sim \sigma_p$ for some $p \in \mathbb{R}$.

References

[1] M. Adamek, *On λ-quasiconvex and λ-convex functions*, Rad. Mat. **11** (2002/03), 171–181.

[2] M. Adamek, *A characterization of λ-convex functions*, JIPAM. J. Inequal. Pure Appl. Math. **5** (2004), Article 71, p. 5 (electronic).

[3] M. Bajraktarević, *Sur une équation fonctionnelle aux valeurs moyennes*, Glasnik Mat.-Fiz. Astronom. Društvo Mat. Fiz. Hrvatske. Ser. II **13** (1958), 243–248.

[4] M. Bajraktarević, *Über die Vergleichbarkeit der mit Gewichtsfunktionen gebildeten Mittelwerte*, Studia Sci. Math. Hungar. **4** (1969), 3–8.

[5] F. Bernstein and G. Doetsch, *Zur Theorie der konvexen Funktionen*, Math. Ann. **76** (1915), 514–526.

[6] P.S. Bullen, D.S. Mitrinović, and P.M. Vasić, *Means and Their Inequalities*, D. Reidel Publishing Co., Dordrecht, 1988.

[7] Z. Daróczy and L. Losonczi, *Über den Vergleich von Mittelwerten*, Publ. Math. Debrecen **17** (1970), 289–297.

[8] G.H. Hardy, J.E. Littlewood, and G. Pólya, *Inequalities*, Cambridge University Press, Cambridge, 1934, (first edition), 1952 (second edition).

[9] M. Kuczma, *An Introduction to the Theory of Functional Equations and Inequalities*, PWN – Uniwersytet Śląski, Warszawa – Kraków – Katowice, 1985.

[10] L. Losonczi, *Equality of two variable weighted means: reduction to differential equations*, Aequationes Math. **58** (1999), 223–241.

[11] D.S. Mitrinović, J.E. Pečarić, and A.M. Fink, *Classical and New Inequalities in Analysis*, Kluwer Academic Publishers, Dordrecht, 1993.

[12] Zs. Páles, *Bernstein–Doetsch-type results for general functional inequalities*, Rocznik Nauk.-Dydakt. Prace Mat. **17** (2000), 197–206.

[13] Zs. Páles, *Nonconvex functions and separation by power means*, Math. Inequal. Appl. **3** (2000), 169–176.

Zoltán Daróczy and Zsolt Páles
Institute of Mathematics, University of Debrecen
H-4010 Debrecen, Pf. 12, Hungary
e-mail: `daroczy@math.klte.hu`
e-mail: `pales@math.klte.hu`

International Series of Numerical Mathematics, Vol. 157, 261–265

Approximately Midconvex Functions

Jacek Mrowiec, Jacek Tabor and Józef Tabor

Abstract. Let X be a vector space and let $D \subset X$ be a nonempty convex set. We say that a function f is δ-midconvex if

$$f\left(\frac{x+y}{2}\right) \leq \frac{f(x)+f(y)}{2} + \delta \quad \text{for } x, y \in D.$$

We find the smallest function $C : [0,1] \cap \mathbb{Q} \to \mathbb{R}$ such that for every δ-midconvex function $f : D \to \mathbb{R}$ the following estimate holds

$$f(qx + (1-q)y) \leq qf(x) + (1-q)f(y) + C(q)\delta$$

for $x, y \in D, q \in [0,1] \cap \mathbb{Q}$.

Mathematics Subject Classification (2000). 26A51, 39B82.

Keywords. Stability, convexity, Jensen convex functions.

Let X be a real vector space and let D be a convex subset of X. A function $f : D \to \mathbb{R}$ is convex if

$$f(tx + (1-t)y) \leq tf(x) + (1-t)f(y) \quad \text{for } x, y \in D, t \in [0,1].$$

We say that f is midconvex (or Jensen convex) if it satisfies

$$f\left(\frac{x+y}{2}\right) \leq \frac{f(x)+f(y)}{2} \quad \text{for } x, y \in D.$$

D. Hyers and S. Ulam introduced in [4] the notion of approximate convexity. Function f is said to be δ-*convex* if

$$f(tx + (1-t)y) \leq tf(x) + (1-t)f(y) + \delta \quad \text{for } x, y \in D, t \in [0,1],$$

and f is δ-*midconvex* if

$$f\left(\frac{x+y}{2}\right) \leq \frac{f(x)+f(y)}{2} + \delta \quad \text{for } x, y \in D.$$

It is clear that every discontinuous additive function midconvex, but is is not δ-convex with any δ. C.T. Ng and K. Nikodem proved in [8] (for the discussion of related results see also [3]) that every locally bounded above at a point δ-midconvex function is automatically 2δ-convex:

Theorem NN. *Let X be a real topological vector space. Let $D \subset X$ be an open and convex set, and let $f : D \to \mathbb{R}$ be δ-midconvex. If f is bounded from above at a point of D, then f is 2δ-convex.*

On the other hand, as it is well known [5], a midconvex function is $\mathbb{Q}$-convex, that is

$$f(qx + (1-q)y) \le qf(x) + (1-q)f(y) \quad \text{for } x, y \in D, q \in [0,1]_{\mathbb{Q}},$$

where

$$[0,1]_{\mathbb{Q}} := [0,1] \cap \mathbb{Q}.$$

Thus it seemed natural to investigate whether a version of the result of C.T. Ng and K. Nikodem holds for approximately midconvex functions, but without the assumption of local boundedness. J. Mrowiec showed in [6] (see also [7]) that this is the case. Namely, he proved that if $f : D \to \mathbb{R}$ is a δ-midconvex function, then

$$f(qx + (1-q)y) \le qf(x) + (1-q)f(y) + 2\delta \quad \text{for } x, y \in D, q \in [0,1]_{\mathbb{Q}}.$$

Our aim in this paper is to improve the result of J. Mrowiec and provide the optimal function C such that the following estimation holds

$$f(qx + (1-q)y) \le qf(x) + (1-q)f(q) + C(q)\delta.$$

for every δ-midconvex function $f : D \to \mathbb{R}$.

At the end of the introduction let us mention that in our results we apply the modified (and simplified) version of the method used by A. Házy and Zs. Páles [1].

Let us now proceed to our results. From now on X stands for a real vector space, $D \subset X$ for a nonempty $\mathbb{Q}$-convex subset of X and δ for a nonnegative number.

Following the idea of C.T. Ng and K. Nikodem, see [8, Lemma 1], we define the function $C : [0,1] \to \mathbb{R}$ by the formula

$$C(t) = \begin{cases} 0 \text{ for } t = 0 \text{ or } t = 1; \\ 2 - 2^{1-n} \text{ for } t = \frac{k}{2^n},\ k \text{ odd}; \\ 2 \text{ otherwise.} \end{cases}$$

Clearly C is a nonnegative function which is bounded from above by 2. As we will show later in Remark 1 in general C gives us optimal estimation.

In the following lemma we investigate the properties of the function C.

Lemma 1. *The function C satisfies the following conditions*

(i) $C(t) = C(2t)/2 + 1$ *for* $t \in (0, 1/2]$;

(ii) $C(t) = C(2t-1)/2 + 1$ *for* $t \in [1/2, 1)$;

(iii) C *is* 1-*midconvex, that is*

$$C\left(\frac{s+t}{2}\right) \le \frac{C(s) + C(t)}{2} + 1 \tag{1}$$

for all $s, t \in [0,1]$.

Before proceeding to the proof let us point out that the functional equations (i) and (ii) are to similar to the equations investigated in [1, Section 2].

Proof. Since conditions (i) and (ii) are obvious, we consider only (iii).

Let $s, t \in [0,1]$ be arbitrary. If either $C(s) + C(t) \geq 2$ or $s = 0 = t$ (or $s = 1 = t$) then (1) trivially holds. It remains to consider the case when $s = k/2^n$ (k odd) and $t = 0$ or $t = 1$. If $t = 0$ then by (ii) we have

$$C(\tfrac{s+0}{2}) = C(s/2) = C(s)/2 + 1 = \tfrac{C(s)+C(0)}{2} + 1.$$

Similarly, if $t = 1$ applying (ii) we get

$$C(\tfrac{s+1}{2}) = C(s)/2 + 1 = \tfrac{C(s)+C(1)}{2} + 1. \qquad \square$$

The following proposition will play a crucial role in the proof of our main result.

Proposition 1. *Let $n \in \mathbb{N}$ be fixed and let $I_n = \{0, \frac{1}{n}, \ldots, \frac{n-1}{n}, 1\}$. Let $g : I_n \to \mathbb{R}$ be a function such that $g(0) = 0 = g(1)$ and that*

$$g(q) \leq g(2q)/2 + \delta \qquad \text{for } q \in I_n, q \leq 1/2, \tag{2}$$

$$g(q) \leq g(2q-1)/2 + \delta \quad \text{for } q \in I_n, q \geq 1/2. \tag{3}$$

Then

$$g(q) \leq C(q)\delta \quad \text{for } q \in I_n. \tag{4}$$

Proof. Let

$$M := \max\{g(q) - C(q)\delta \,:\, q \in I_n\},$$

and let $r \in I_n$ be chosen such that

$$g(r) - C(r)\delta = M.$$

If $r = 0$ then $M = g(0) - C(0)\delta = 0$. Analogously we obtain that $M = 0$ if $r = 1$.

Now let us consider the case when $r \in (0, \frac{1}{2}]$. Then by (2) and property (i) from Lemma 1 we obtain

$$M \geq g(2r) - C(2r)\delta \geq (2g(r) - 2\delta) - (2C(r) - 2)\delta = 2M,$$

which means that $M \leq 0$.

If $r \in [\frac{1}{2}, 1)$ then by similar argumentation using (3) and (ii) from Lemma 1 we again obtain that $M \leq 0$. $\square$

Now we are ready to prove the main result of the paper.

Theorem 1. *Let D be a nonempty $\mathbb{Q}$-convex subset of a vector space X and let $f : D \to \mathbb{R}$ be a δ-midconvex function. Then*

$$f(qx + (1-q)y) \leq qf(x) + (1-q)f(y) + C(q)\delta \quad \text{for } x, y \in D, q \in [0,1]_{\mathbb{Q}}.$$

Proof. Let us fix arbitrarily $x, y \in D$ and $n \in \mathbb{N}$. We define the function $g : I_n \to \mathbb{R}$ by the formula

$$g(q) = f(qx + (1-q)y) - (qf(x) + (1-q)f(y)) \quad \text{for } q \in I_n,$$

where as before $I_n = \{0, \frac{1}{n}, \ldots, \frac{n-1}{n}, 1\}$. One can easily check that g satisfies all the assumptions of Proposition 1. Consequently we get

$$g(q) \le C(q)\delta \quad \text{for } q \in I_n,$$

which means that

$$f(\tfrac{k}{n}x + (1-\tfrac{k}{n})y) \le \tfrac{k}{n}f(x) + (1-\tfrac{k}{n})f(y) + C(\tfrac{k}{n})\delta \quad \text{for } k \in \{0, 1, \ldots, n\}.$$

Since $n \in N$ was arbitrarily chosen, the proof is complete. □

Since the maximum of the function C is 2, directly from Theorem 1 we obtain that every δ midconvex function is 2δ $\mathbb{Q}$-midconvex.

Remark 1. Let us mention that the function $C|_{[0,1]_{\mathbb{Q}}}$ is the best possible in the class of functions for which Theorem 1 holds.

To observe this, suppose that $\tilde{C} : [0,1]_{\mathbb{Q}} \to \mathbb{R}$ is such a function that condition

$$f(qx + (1-q)y) \le qf(x) + (1-q)f(y) + \tilde{C}(q)\delta \quad \text{for } x, y \in D, q \in [0,1]_{\mathbb{Q}}. \tag{5}$$

holds for for every δ-midconvex function $f : D \to \mathbb{R}$. By Lemma 1 (iii) we know that C is 1-midconvex. Taking in (5) $D = [0,1]$, $f = C$, $\delta = 1$ and $x = 0$, $y = 1$ we get

$$C(q) \le \tilde{C}(q) \quad \text{for } q \in [0,1]_{\mathbb{Q}}.$$

At the end let us show some applications of our theorem. The first corollary gives an improvement of the result of C.T. Ng and Nikodem [8].

Corollary 1. *Let D be a nonempty open convex subset of a topological vector space X and let $f : D \to \mathbb{R}$ be a δ-midconvex function which is bounded from above in a neighbourhood of a point of D. Then*

$$f(tx + (1-t)y) \le tf(x) + (1-t)f(y) + C(t)\delta \quad \textit{for } t \in [0,1]. \tag{6}$$

Proof. If t is not rational, then $C(t) = 2$ and the above estimation follows from the result of C.T. Ng and K. Nikodem [8]. If t is rational, (6) follows directly from Theorem 1. □

Now we show that we can obtain something even in the case when the bound on the "Jensen difference" is not constant (see also [2] for different results in the similar spirit).

Corollary 2. *Let D be a nonempty $\mathbb{Q}$-convex subset of a normed space X, and let $\Psi : \mathbb{R}_+ \to \mathbb{R}_+$ be a non-decreasing function. Let $f : D \to \mathbb{R}$ satisfy the inequality*

$$f\left(\frac{x+y}{2}\right) \le \frac{f(x)+f(y)}{2} + \Psi(\|x-y\|) \quad \textit{for } x, y \in D.$$

Then

$$f(qx + (1-q)y) \le qf(x) + (1-q)f(y) + C(q)\Psi(\|x-y\|) \quad \textit{for } q \in [0,1]_{\mathbb{Q}}.$$

Proof. We fix arbitrarily $x, y \in D$ and put $\delta := \Psi(\|x-y\|)$,

$$D_{[x,y]} := \{qx + (1-q)y \,:\, q \in [0,1]_{\mathbb{Q}}\}.$$

Then one can easily notice that since Ψ is non-decreasing $f|_{D_{[x,y]}}$ is δ-midconvex. Therefore applying Theorem 1 we obtain that

$$f(qx+(1-q)y) \le qf(x)+(1-q)f(y)+C(q)\delta = qf(x)+(1-q)f(y)+C(q)\Psi(\|x-y\|)$$

for $q \in [0,1]_{\mathbb{Q}}$. □

References

[1] A. Házy and Zs. Páles, *On approximately midconvex functions,* Bulletin London Math. Soc. **36**(3) (2004), 339–350.

[2] A. Házy and Zs. Páles, *On approximately t-convex functions,* Publ. Math. Debrecen **66**(3-4) (2005), 489–501.

[3] D.H. Hyers, G. Isac and Th. M. Rassias, *Stability of Functional Equations in Several Variables*, Birkhäuser, Basel, 1998.

[4] D.H. Hyers and S.M. Ulam, *Approximately convex functions*, Proc. Amer. Math. Soc. **3** (1952), 821–828.

[5] M. Kuczma, *An Introduction to the Theory of Functional Equations and Inequalities*, PWN – Uniwersytet Śląski, Warszawa – Kraków – Katowice, 1985.

[6] J. Mrowiec, *Remark on approximately Jensen-convex functions,* C.R. Math. Acad. Sci. Soc. R. Canada **23** (2001), 16–21.

[7] J. Mrowiec, *Stabilność nierówności opisującej funkcje wypukłe w sensie Wrighta i pewnych nierówności pokrewnych*, PhD thesis, Katowice, 2004.

[8] C.T. Ng and K. Nikodem, *On approximately convex functions,* Proc. Amer. Math. Soc. **118** (1993), 103–108.

Jacek Mrowiec
Department of Mathematics, University of Bielsko-Biała
Willowa 2, 43-309 Bielsko-Biała, Poland
e-mail: `jmrowiec@ath.bielsko.pl`

Jacek Tabor
Institute of Mathematics, Jagiellonian University
Reymonta 4, 30-059 Kraków, Poland
e-mail: `jacek.tabor@im.uj.edu.pl`

Józef Tabor
Institute of Mathematics, University of Rzeszów
Rejtana 16A, Rzeszów 35-310, Poland
e-mail: `tabor@univ.rzeszow.pl`

Part VI

Inequalities, Stability, and Functional Equations

PHB
'07

International Series of Numerical Mathematics, Vol. 157, 269–281

Sandwich Theorems for Orthogonally Additive Functions

Włodzimierz Fechner and Justyna Sikorska

Abstract. Let p be an orthogonally subadditive mapping, q an orthogonally superadditive mapping such that $p \leq q$ or $q \leq p$. We prove that under some additional assumptions there exists a unique orthogonally additive mapping f such that $p \leq f \leq q$ or $q \leq f \leq p$, respectively.

Mathematics Subject Classification (2000). 39B55, 39B62, 39B82.

Keywords. Orthogonally additive functions, orthogonally subadditive functions, sandwich theorem.

1. Introduction

Throughout the paper $\mathbb{R}$ denotes the set of real numbers, $\mathbb{R}_+ = \{t \in \mathbb{R} : t \geq 0\}$ and $\mathbb{N} = \{1, 2, \ldots\}$.

Let $(X, \langle\cdot|\cdot\rangle)$ be a real inner product space of dimension at least 2 and let $(Y, +)$ be an abelian group. We say that a mapping $f\colon X \to Y$ is *orthogonally additive* iff it satisfies the following conditional functional equation:

$$x \perp y \implies f(x+y) = f(x) + f(y), \quad x, y \in X,$$

where $\perp \subset X \times X$ denotes the standard orthogonality relation on X, i.e., $x \perp y$ iff $\langle x|y\rangle = 0$.

Each additive mapping $b\colon X \to Y$, i.e., the solution of the Cauchy functional equation:

$$b(x+y) = b(x) + b(y), \quad x, y \in X$$

is an example of orthogonally additive mapping. Moreover, one may calculate that the map

$$X \ni x \mapsto \langle x|x\rangle = \|x\|^2 \in \mathbb{R}$$

is orthogonally additive (the Pythagoras Theorem for inner product spaces). In 1995 K. Baron and J. Rätz [1] proved that if $(Y, +)$ is an abelian group and

$f\colon X \to Y$ is orthogonally additive, then there exists exactly one pair of additive mappings $a\colon \mathbb{R} \to Y$, $b\colon X \to Y$ such that

$$f(x) = a(\|x\|^2) + b(x), \quad x \in X.$$

A map $g\colon X \to Y$ is called *quadratic* if it satisfies the equation

$$g(x+y) + g(x-y) = 2g(x) + 2g(y), \quad x, y \in X.$$

If $a\colon \mathbb{R} \to Y$ is an additive function, then the map $X \ni x \mapsto a(\|x\|^2) \in Y$ is an example of a quadratic mapping which is orthogonally additive. On the other hand, a quadratic map needs not to be orthogonally additive.

Orthogonality may be introduced on more general structures than the inner product space. W. Blaschke [4], G. Birkhoff [3], R.C. James [9] and Gy. Szabó [19] investigated an orthogonality relation, called *Birkhoff-James orthogonality*, defined in a real normed linear space X as follows:

$$x \perp_{BJ} y \iff \forall_{\lambda \in \mathbb{R}} \left(\|x + \lambda y\| \geq \|x\|\right).$$

One may check that if X is an inner product space, then $\perp_{BJ}$ coincides with the standard orthogonality $\perp$.

In 1974 S. Gudder and D. Strawther [8] proposed an axiomatic framework for the orthogonal additivity so that the case of standard orthogonality and the case of Birkhoff-James orthogonality is covered. The idea has been developed in the 1980's by J. Rätz and Gy. Szabó [12, 13, 14, 15]. In particular, they introduced the *orthogonality space*, also called the *Rätz space*. An ordered pair $(X, \perp)$ is called an orthogonality space whenever X is a vector space over a field K such that $1+1 \neq 0$ in K, $\dim X \geq 2$ and $\perp$ is a binary relation on X such that

(i) $x \perp 0$ and $0 \perp x$ for all $x \in X$;
(ii) if $x, y \in X \setminus \{0\}$ and $x \perp y$, then x and y are linearly independent;
(iii) if $x, y \in X$ and $x \perp y$, then for all $\xi, \eta \in K$ we have $\xi x \perp \eta y$;
(iv) for any two-dimensional subspace P of X and for every $x \in P$, there exists a $y \in P$ such that $x \perp y$ and $x + y \perp x - y$.

In 1998 K. Baron and P. Volkmann [2], developing the above-mentioned earlier results of J. Rätz and Gy. Szabó, obtained the representation of orthogonally additive mappings defined on an orthogonality space as a sum of an additive mapping and a quadratic mapping without additional assumptions on the range of investigated functions. On the other hand, Gy. Szabó [20] and D. Yang [21] proved that if an orthogonality space $(X, \perp)$ admits an orthogonally additive mapping which is not additive, then X is an inner product space and $\perp$ coincides with the standard orthogonality.

In 2006 the first author [5] made an attempt to describe orthogonally subadditive mappings. In particular, the following result has been obtained.

Theorem A. *Assume that* $(X, \langle\cdot|\cdot\rangle)$ *is a real inner product space with* $\dim X \geq 3$. *If* $f\colon X \to \mathbb{R}$ *is orthogonally subadditive, i.e., it satisfies*

$$x \perp y \Longrightarrow f(x+y) \leq f(x) + f(y), \tag{1}$$

and

$$f(2x) \geq 4f(x), \quad x \in X, \tag{2}$$

then there exists a nonnegative sublinear map $P\colon \mathbb{R}_+ \to \mathbb{R}$ *such that*

$$f(x) = P(\|x\|^2), \quad x \in X. \tag{3}$$

Conversely, for each sublinear map $P\colon \mathbb{R}_+ \to \mathbb{R}$ *a function* $f\colon X \to \mathbb{R}$ *given by* (3) *fulfills* (1) *and* (2).

By sublinear map we understand a subadditive and n-homogeneous map for each $n \in \mathbb{N}$. For more details see M. Kuczma [10, pp. 414–417].

In [5] we have provided some conditions weaker than (2) which jointly with (1) allow to describe f in terms of solutions of unconditional functional equations and inequalities. However, we were not able to get rid of such conditions completely. Therefore, our state of knowledge connected with orthogonally subadditive mappings is much less satisfactory than in case of (unconditionally) subadditive mappings.

In 1995 R. Ger and the second author [7] investigated the Hyers-Ulam stability of the equation of the orthogonal additivity. Further results in this direction have been obtained by M. Moslehian [11] and by the second author [16, 17, 18]. Theorem 1 below is a joint generalization of [18, Theorem 4.2] and [18, Theorem 6.1].

Let $(X, \perp)$ be an orthogonality space. Consider function $\varphi\colon X \times X \to [0, \infty)$ such that

(a) one of the series $\sum_{n=1}^{\infty} 2^{-n}\varphi(2^{n-1}x, 2^{n-1}x)$ and $\sum_{n=1}^{\infty} 2^{n-1}\varphi(2^{-n}x, 2^{-n}x)$ is convergent; denote such a sum by $\Phi(x)$;

(b) one of the series $\sum_{n=0}^{\infty} 4^{1-n}\varphi(2^{n-1}x, 2^{n-1}x)$ and $\sum_{n=1}^{\infty} 4^{n}\varphi(2^{-n}x, 2^{-n}x)$ is convergent; denote such a sum by $\Psi(x)$;

(c) for all $x, y \in X$ such that $x \perp y$ we have

$$\lim_{n\to\infty} 2^{-n}\varphi(2^n x, 2^n y) = 0 \quad \text{or} \quad \lim_{n\to\infty} 2^{n}\varphi(2^{-n} x, 2^{-n} y) = 0$$

for respective cases from (a);

(d) for all $x, y \in X$ such that $x \perp y$ we have

$$\lim_{n\to\infty} 4^{-n}\varphi(2^n x, 2^n y) = 0 \quad \text{or} \quad \lim_{n\to\infty} 4^{n}\varphi(2^{-n} x, 2^{-n} y) = 0$$

for respective cases from (b);

(e) there exists an $M > 0$ such that for all $x, y \in X$ if $x \perp y$ and $x + y \perp x - y$ then

$$\max\{\varphi(x,y), \varphi(x,-y), \varphi(x+y, x-y), \varphi(x+y, y-x)\} \le M\varphi(x,x).$$

Theorem 1. *Let $(X, \perp)$ be an orthogonality space, $(Y, \|\cdot\|)$ be a real Banach space and $\psi\colon X \times X \to [0,\infty)$. If $F, G, H\colon X \to Y$ satisfy the condition*

$$x \perp y \implies \|F(x+y) - G(x) - H(y)\| \le \psi(x,y),$$

and function $\varphi\colon X \times X \to [0,\infty)$ given by

$$\varphi(x,y) := \psi(x,y) + \psi(x,0) + \psi(0,y), \quad x, y \in X,$$

satisfies conditions (a)–(e), *then there exists a unique orthogonally additive function $f\colon X \to Y$ such that*

$$\|F(x) - G(0) - H(0) - f(x)\| \le M(3\overline{\Phi}(x) + \overline{\Psi}(x)), \quad x \in X;$$
$$\|G(x) - G(0) - f(x)\| \le M(3\overline{\Phi}(x) + \overline{\Psi}(x)) + \psi(x,0), \quad x \in X;$$
$$\|H(x) - H(0) - f(x)\| \le M(3\overline{\Phi}(x) + \overline{\Psi}(x)) + \psi(0,x), \quad x \in X,$$

where $\overline{\Phi}(x) := \frac{1}{2}[\Phi(x) + \Phi(-x)]$, $x \in X$, and $\overline{\Psi}(x) := \frac{1}{2}[\Psi(x) + \Psi(-x)]$, $x \in X$.

In particular case, if $F = G = H$, the approximation

$$\|F(x) - f(x)\| \le M(3\overline{\Phi}(x) + \overline{\Psi}(x)), \quad x \in X$$

is valid with $\varphi := \psi$ satisfying (a)–(e).

Proof. This statement can be proved by a modification of the proof of Theorem 4.2 in [18]. One needs to replace the value $\varepsilon(\|x\|^p + \|y\|^p)$ from that proof by $\psi(x,y)$ and observe that thanks to our assumptions upon ψ all the calculations can be repeated. □

2. Main results

From now on we will be assuming that $(X, \perp)$ is an orthogonality space.

We are going to show that if $p\colon X \to \mathbb{R}$ and $q\colon X \to \mathbb{R}$ satisfy

$$x \perp y \implies p(x+y) \le p(x) + p(y), \tag{4}$$
$$x \perp y \implies q(x+y) \ge q(x) + q(y), \tag{5}$$

jointly with some auxiliary technical conditions, then p and q can be separated by an orthogonally additive mapping.

Clearly, each subadditive map p and superadditive map q satisfy (4) and (5), respectively, and each orthogonally additive mapping satisfies both inequalities. Moreover, if X is an inner product space and $P\colon \mathbb{R}_+ \to \mathbb{R}$ is a sublinear map, then $p := P(\|\cdot\|^2)$ and $q := -P(\|\cdot\|^2)$ also fulfill these inequalities. One may check that for all above-mentioned examples of solutions of (4) and (5) the following estimations are satisfied:

$$p(2x) \le 3p(x) + p(-x), \quad x \in X; \tag{6}$$

$$q(2x) \ge 3q(x) + q(-x), \quad x \in X. \tag{7}$$

However, we do not know whether they hold true for arbitrary solutions of (4) and (5). In the sequel, we will provide some conditions sufficient for orthogonally subadditive or superadditive mapping to satisfy (6) or (7), respectively. Note that for an orthogonally subadditive and even map p the inequality (6) is the opposite to (2).

Now, consider the following properties of functions $\alpha\colon X \to [0, \infty)$ and $\upsilon\colon X \to X$:

1° either (i) $\sum_{n=1}^{\infty} 2^{-n}\alpha(2^n x)$ is convergent or (ii) $\sum_{n=0}^{\infty} 2^{n}\alpha(2^{-n} x)$ is convergent; denote such a sum by $A(x)$;

2° either (i) $\sum_{n=0}^{\infty} 4^{1-n}\alpha(2^n x)$ is convergent or (ii) $\sum_{n=0}^{\infty} 4^{n+1}\alpha(2^{-n} x)$ is convergent; denote such a sum by $B(x)$;

3° there exists an $M > 0$ such that for all $x, y \in X$ if $x \perp y$ and $x + y \perp x - y$ then:
 (i) $\max\{\alpha(x+y), \alpha(x-y), \alpha(2y)\} \le M\alpha(2x)$;
 (ii) $\max\{S(x,y), S(x,-y), S(x+y, x-y), S(x+y, y-x)\} \le M\,S(x,x)$, where $S(x,y) := \alpha(x+y) + \alpha(x) + \alpha(y)$;
 (iii) $\max\{\alpha(y), \alpha(-y), \alpha(x+y) + \alpha(x-y), \alpha(x+y) + \alpha(y-x)\} \le M\alpha(x)$;
 (iv) $\max\{\alpha(y), \alpha(x+y), \alpha(x-y)\} \le M(\alpha(x) + \alpha(0))$;

4° $\upsilon^2 = \upsilon \circ \upsilon = \mathrm{id}_X$ and $\alpha(x) = \alpha(\upsilon(x))$ for all $x \in X$.

Proposition 1. *Assume that $p\colon X \to \mathbb{R}$ and $q\colon X \to \mathbb{R}$ satisfy (4), (5) and*

$$p(x) \le q(x), \quad x \in X. \tag{8}$$

If there exists an $\alpha\colon X \to [0, \infty)$ such that 1°, 2°, 3°(i) hold true and

$$|q(x) - p(x)| \le \alpha(x), \quad x \in X, \tag{9}$$

then there exists a unique orthogonally additive mapping $f\colon X \to \mathbb{R}$ such that

$$p(x) - \mu(x) \le f(x) \le q(x) + \mu(x), \quad x \in X, \tag{10}$$

where $\mu(x) = M[3A(x) + B(x)]$ for all $x \in X$.

Proof. Observe that for arbitrarily fixed vectors $x, y \in X$ such that $x \perp y$ we have

$$p(x+y) - p(x) - p(y) \ge q(x+y) - \alpha(x+y) - q(x) - q(y) \ge -\alpha(x+y),$$

and, obviously,

$$p(x+y) - p(x) - p(y) \le 0 \le \alpha(x+y).$$

Therefore, we get the estimation

$$x \perp y \Longrightarrow |p(x+y) - p(x) - p(y)| \leq \alpha(x+y).$$

We are at the point to apply Theorem 1 with $\psi(x,y) = \varphi(x,y) := \alpha(x+y)$. We infer that there exists a unique orthogonally additive mapping $f\colon X \to Y$ such that:

$$|p(x) - f(x)| \leq \mu(x), \quad x \in X. \tag{11}$$

Moreover, one may easily calculate that $\mu(x) = M[3A(x) + B(x)]$ for all $x \in X$.

From this it follows that

$$p(x) - \mu(x) \leq f(x) \leq p(x) + \mu(x) \leq q(x) + \mu(x), \quad x \in X.$$

Uniqueness of f as a solution of (10) can be derived by use of standard tools, repeating argumentation from the proof of Theorem 2.4 in [18]. This completes the proof. □

In the next proposition condition (9) is replaced by a more flexible one, but we need to strengthen assumption upon α and we obtain a less sharp estimation.

Proposition 2. *Assume that $p\colon X \to \mathbb{R}$ and $q\colon X \to \mathbb{R}$ satisfy (4), (5) and (8). If there exist mappings $\alpha\colon X \to [0,\infty)$ and $\upsilon\colon X \to X$ such that 1°, 2°, 3°(ii), 4° hold true, and*

$$|q(x) - p(\upsilon(x))| \leq \alpha(x), \quad x \in X, \tag{12}$$

then there exists a unique orthogonally additive mapping $f\colon X \to \mathbb{R}$ such that the estimation (10) is satisfied with $\mu(x) = M\left[3A(x) + 6A\left(\frac{x}{2}\right) + B(x) + 2B\left(\frac{x}{2}\right)\right] + \alpha(x)$ for all $x \in X$.

Proof. One can check that for each $x, y \in X$ such that $x \perp y$ the following estimations are valid:

$$p(\upsilon(x+y)) - p(x) - p(y) \geq q(x+y) - \alpha(x+y) - q(x) - q(y) \geq -\alpha(x+y)$$

and

$$\begin{aligned}
p(\upsilon(x+y)) &- p(x) - p(y) \\
&= p(x+y) - p(x) - p(y) + p(\upsilon(x+y)) - p(x+y) \\
&\leq p(\upsilon(x+y)) - p(x+y) \leq q(\upsilon(x+y)) - p(x+y) \leq \alpha(\upsilon(x+y)).
\end{aligned}$$

Thus, we have shown that

$$x \perp y \Longrightarrow |p(\upsilon(x+y)) - p(x) - p(y)| \leq \alpha(x+y).$$

From (4) and (5) it follows that $p(0) \geq 0 \geq q(0)$ which jointly with (8) means that $p(0) = q(0) = 0$. From Theorem 1 applied for $F := p \circ \upsilon$, $G = H := p$ and $\psi(x,y) := \alpha(x+y)$ we derive the existence of a unique orthogonally additive mapping $f\colon X \to Y$ such that

$$\begin{aligned}
|p(x) - f(x)| &= |G(x) - G(0) - f(x)| \\
&\leq M\left[3A(x) + 6A\left(\frac{x}{2}\right) + B(x) + 2B\left(\frac{x}{2}\right)\right] + \alpha(x)
\end{aligned}$$

for all $x \in X$. From this it follows that

$$p(x) - \mu(x) \leq f(x) \leq q(x) + \mu(x) \leq p(x) + \mu(x), \quad x \in X.$$

Uniqueness of f as a solution of (10) is shown in a standard way. This completes the proof. □

In the next two statements we assume the converse inequality to (8) and we obtain analogous results.

Proposition 3. *Assume that* $p\colon X \to \mathbb{R}$ *and* $q\colon X \to \mathbb{R}$ *satisfy* (4), (5) *and*

$$q(x) \leq p(x), \quad x \in X. \tag{13}$$

If there exists an $\alpha\colon X \to [0,\infty)$ *such that* 1°, 2°, 3°(iii) *and* (9) *hold true, then there exists a unique orthogonally additive mapping* $f\colon X \to \mathbb{R}$ *such that*

$$q(x) - \mu(x) \leq f(x) \leq p(x) + \mu(x), \quad x \in X, \tag{14}$$

where $\mu(x) = M\left[6A\left(\frac{x}{2}\right) + 2B\left(\frac{x}{2}\right)\right]$ *for all* $x \in X$.

Proof. It is enough to check that

$$p(x+y) - p(x) - p(y) \geq q(x+y) - q(x) - \alpha(x) - q(y) - \alpha(y) \geq -\alpha(x) - \alpha(y),$$

whenever $x \perp y$, and apply Theorem 1 for $\psi(x,y) = \varphi(x,y) := \alpha(x) + \alpha(y)$. The rest of the proof is analogous to the proof of Proposition 1. □

Proposition 4. *Assume that* $p\colon X \to \mathbb{R}$ *and* $q\colon X \to \mathbb{R}$ *satisfy* (4), (5) *and* (13). *If there exist a mapping* $\alpha\colon X \to [0,\infty)$ *and an orthogonally additive mapping* $v\colon X \to X$ *such that* 1°, 2°, 3°(iv), 4° *and* (12) *hold true, then there exists a unique orthogonally additive mapping* $f\colon X \to \mathbb{R}$ *such that the estimation* (14) *is satisfied with* $\mu(x) = 4M\left[3A\left(\frac{x}{2}\right) + B\left(\frac{x}{2}\right)\right] + \alpha(x) + \left(\frac{50}{3}M + 3\right)\alpha(0)$ *for all* $x \in X$.

Proof. For $x \perp y$ we have

$$\begin{aligned}
&p(v(x) + v(y)) - p(x) - p(y) \\
&\quad \geq q(v(x) + v(y)) - q(v(x)) - \alpha(v(x)) - q(v(y)) - \alpha(v(y)) \\
&\quad \geq -\alpha(v(x)) - \alpha(v(y)),
\end{aligned}$$

and

$$\begin{aligned}
&p(v(x) + v(y)) - p(x) - p(y) \\
&\quad = p(v(x) + v(y)) - p(v(x)) - p(v(y)) + p(v(x)) + p(v(y)) - p(x) - p(y) \\
&\quad \leq p(v(x)) - q(x) + p(v(y)) - q(y) \leq \alpha(x) + \alpha(y),
\end{aligned}$$

so,

$$|p(v(x+y)) - p(x) - p(y)| = |p(v(x) + v(y)) - p(x) - p(y)| \leq \alpha(x) + \alpha(y). \tag{15}$$

Now, it is enough to apply Theorem 1 with $F := p \circ v$, $G = H := p$ and $\psi(x,y) := \alpha(x) + \alpha(y)$. From (15) and additivity of v we have $|p(0)| \leq 2\alpha(0)$. Moreover, it is easy to observe that cases 1°(ii) and 2°(ii) force $\alpha(0) = 0$. Calculating the exact formula of μ completes the proof. □

Now, we will prove our main result.

Theorem 2. *Assume that* $\alpha\colon X \to [0,\infty)$ *fulfills* 2°(ii), $p\colon X \to \mathbb{R}$ *and* $q\colon X \to \mathbb{R}$ *satisfy* (4), (5), (6), (7) *and* (8). *If at least one set of following conditions:* (9) *and* 3°(i) *or* (12), 3°(ii), 4° *with some* $v\colon X \to X$ *is valid, then there exists a unique orthogonally additive mapping* $f\colon X \to \mathbb{R}$ *such that*

$$p(x) \le f(x) \le q(x), \quad x \in X. \tag{16}$$

Proof. Define operator $\Lambda\colon \mathbb{R}^X \to \mathbb{R}^X$ by the formula

$$\Lambda(f)(x) := 3f\left(\frac{x}{2}\right) + f\left(-\frac{x}{2}\right), \quad f \in \mathbb{R}^X, x \in X.$$

Observe that Λ is a linear operator which is monotonic, i.e., if $g \le h$, then $\Lambda(g) \le \Lambda(h)$. Moreover, from (6) and (7) it follows that $p \le \Lambda(p)$ and $\Lambda(q) \le q$. Further, if $f\colon X \to \mathbb{R}$ is an orthogonally additive mapping, then $\Lambda(f) = f$.

Now, we join Proposition 1 or Proposition 2 with the already established properties of Λ to obtain the existence of an orthogonally additive mapping f satisfying

$$\begin{aligned} p - \Lambda^n(\mu) \le \Lambda^n(p) - \Lambda^n(\mu) \le \Lambda^n(f) \\ \le \Lambda^n(q) + \Lambda^n(\mu) \le q + \Lambda^n(\mu), \quad n \in \mathbb{N}, \end{aligned}$$

where μ is defined in the respective proposition.

Our assumptions upon α imply that

$$\lim_{n\to+\infty} \Lambda^n(\mu(x)) = 0, \quad x \in X.$$

Therefore, letting n tend to $+\infty$ in the foregoing estimation we arrive at $p \le f \le q$. The estimation (16) is stronger than (10) and thus f is unique. This completes the proof. □

Now, we derive two corollaries from Proposition 3 and Proposition 4 which provide conditions sufficient for the separation of p and q by an additive and by a quadratic mapping in case $q \le p$. Assumptions (6) and (7) are replaced by other ones, which force the map f to be odd or even, respectively.

Theorem 3. *Assume that* $\alpha\colon X \to [0,\infty)$ *fulfills* 1°(i), $p\colon X \to \mathbb{R}$ *and* $q\colon X \to \mathbb{R}$ *satisfy* (4), (5), (13) *and*

$$p(2x) \le 2p(x), \quad x \in X; \tag{17}$$

$$q(2x) \ge 2q(x), \quad x \in X. \tag{18}$$

If at least one set of conditions (9) *and* 3°(iii) *or* (12), 3°(iv), 4° *with some orthogonally additive* $v\colon X \to X$ *is valid, then there exists a unique additive mapping* $f\colon X \to \mathbb{R}$ *such that*

$$q(x) \le f(x) \le p(x), \quad x \in X. \tag{19}$$

Proof. Let $f\colon X \to \mathbb{R}$ be a unique orthogonally additive mapping postulated by Proposition 3 or Proposition 4. Assume that f has decomposition $f = f_o + f_e$, where f_o is the odd part of f and f_e is the even one. Thus f_o is additive, whereas f_e is quadratic. In particular,

$$f_o(2x) = 2f_o(x), \quad f_e(2x) = 4f_e(x), \quad x \in X.$$

On the other hand, from (14) jointly with (17) and (18) we infer that

$$\begin{aligned} 2^n q(x) - \mu(2^n x) &\le q(2^n x) - \mu(2^n x) \\ &\le f(2^n x) \\ &= 2^n f_o(x) + 4^n f_e(x) \\ &\le p(2^n x) + \mu(2^n x) \\ &\le 2^n p(x) + \mu(2^n x), \quad x \in X, n \in \mathbb{N}. \end{aligned}$$

Divide this estimations side by side by 2^n and let n tend to $+\infty$. Condition 1°(i) implies 2°(i), so from the form of μ and the properties of α we derive that necessarily $f_e = 0$ and

$$q(x) \le f_o(x) = f(x) \le p(x), \quad x \in X,$$

which was to be proved. □

Theorem 4. *Assume that* $\alpha\colon X \to [0,\infty)$ *fulfills* 1°, 2°(i), $p\colon X \to \mathbb{R}$ *and* $q\colon X \to \mathbb{R}$ *satisfy* (4), (5), (13) *and*

$$p(2x) \le 4p(x), \quad x \in X; \tag{20}$$

$$q(2x) \ge 4q(x), \quad x \in X. \tag{21}$$

If at least one set of conditions (9) *and* 3°(iii) *or* (12), 3°(iv) *and* 4° *with some additive* $v\colon X \to X$ *is valid, then there exists a unique quadratic orthogonally additive mapping* $f\colon X \to \mathbb{R}$ *such that* (19) *holds true.*

Proof. Preserving the notations of the foregoing proof, for all $x \in X$, $n \in \mathbb{N}$ we obtain

$$\begin{aligned} 4^n q(x) - \mu(2^n x) \le q(2^n x) - \mu(2^n x) \le f(2^n x) &= 2^n f_o(x) + 4^n f_e(x) \\ \le p(2^n x) + \mu(2^n x) &\le 4^n p(x) + \mu(2^n x). \end{aligned}$$

Dividing the above inequality side by side by 4^n, letting n tend to $+\infty$, using the form of μ and the properties of α we arrive at

$$q(x) \le f_e(x) \le p(x), \quad x \in X,$$

which was to be proved. □

Remark 1. A simple observation shows that the above results can be easily applied to the situation, where $(X, \|\cdot\|)$ is a real normed linear space with the Birkhoff-James orthogonality, $\alpha := \|\cdot\|^r$ and $v := -\mathrm{id}_X$. One may check using the properties of the Birkhoff-James orthogonality that for various possibilities $r < 1$, $r > 1$, $r < 2$, $r > 2$ the corresponding cases 1°(i), 1°(ii), 2°(i), 2°(ii), and 3° holds true.

We will terminate the paper by providing some conditions sufficient for (6) and (7).

Theorem 5. *Assume that* $p\colon X \to \mathbb{R}$ *and* $q\colon X \to \mathbb{R}$ *satisfy* (4), (5) *and*

$$\sum_{n=1}^{+\infty} 2^n \left[p\left(\frac{x}{2^n}\right) + p\left(-\frac{x}{2^n}\right)\right] \le p(x) + p(-x), \quad x \in X; \tag{22}$$

$$\sum_{n=1}^{+\infty} 2^n \left[q\left(\frac{x}{2^n}\right) + q\left(-\frac{x}{2^n}\right)\right] \ge q(x) + q(-x), \quad x \in X. \tag{23}$$

Then (6) *and* (7) *hold true.*

Proof. We will prove that (4) jointly with (22) implies (6). Fix arbitrarily $x \in X$ and choose $y \in X$ such that $x \perp y$ and $x + y \perp x - y$. Using the properties of the orthogonality relation $\perp$, by the multiple use of (4) we obtain

$$\begin{aligned} p(2x) &\le p(x+y) + p(x-y) \\ &\le 2p(x) + p(y) + p(-y) \\ &\le 2p(x) + \left[2p\left(\frac{y}{2}\right) + p\left(\frac{x}{2}\right) + p\left(-\frac{x}{2}\right)\right] + \left[2p\left(-\frac{y}{2}\right) + p\left(-\frac{x}{2}\right) + p\left(\frac{x}{2}\right)\right] \\ &= 2p(x) + 2p\left(\frac{x}{2}\right) + 2p\left(-\frac{x}{2}\right) + 2p\left(\frac{y}{2}\right) + 2p\left(-\frac{y}{2}\right) \\ &\le 2p(x) + \sum_{n=1}^{N} 2^n \left[p\left(\frac{x}{2^n}\right) + p\left(-\frac{x}{2^n}\right)\right] + 2^N \left[p\left(\frac{y}{2^N}\right) + p\left(-\frac{y}{2^N}\right)\right], \end{aligned}$$

for each $N \in \mathbb{N}$. Now, apply (22) to deduce that

$$\lim_{N \to +\infty} 2^N \left[p\left(\frac{y}{2^N}\right) + p\left(-\frac{y}{2^N}\right)\right] = 0$$

and

$$\begin{aligned} p(2x) &\le 2p(x) + \sum_{n=1}^{+\infty} 2^n \left[p\left(\frac{x}{2^n}\right) + p\left(-\frac{x}{2^n}\right)\right] \\ &\le 3p(x) + p(-x). \end{aligned}$$

To prove that (5) and (23) imply (7) put $p := -q$ and apply the already proved statement for p. This completes the proof. □

Remark 2. Under assumptions of the previous theorem, if additionally X is an inner product space with $\dim X \ge 3$, then there exist sublinear functions $P, Q\colon \mathbb{R}_+ \to \mathbb{R}$ such that

$$\sum_{n=1}^{+\infty} 2^n \left[p\left(\frac{x}{2^n}\right) + p\left(-\frac{x}{2^n}\right)\right] = P(\|x\|^2), \quad x \in X;$$

$$\sum_{n=1}^{+\infty} 2^n \left[q\left(\frac{x}{2^n}\right) + q\left(-\frac{x}{2^n}\right)\right] = -Q(\|x\|^2), \quad x \in X.$$

Indeed, denote $f(x) := \sum_{n=1}^{+\infty} 2^n \left[p\left(\frac{x}{2^n}\right) + p\left(-\frac{x}{2^n}\right)\right]$ for $x \in X$. It is clear that f satisfies (1) and

$$\begin{aligned} f(2x) &= 2p(x) + 2p(-x) + 2f(x) \\ &\geq 4f(x), \quad x \in X. \end{aligned}$$

Therefore, from Theorem A it follows that f has the desired representation. The argument for q is analogous.

Further, by (22) we have

$$P(\|x\|^2) \leq p(x) + p(-x), \quad x \in X,$$

and, consequently, from this and by the 2-homogeneity of P we get

$$\begin{aligned} \sum_{n=1}^{+\infty} 2^n \left[p\left(\frac{x}{2^n}\right) + p\left(-\frac{x}{2^n}\right)\right] &\geq \sum_{n=1}^{+\infty} 2^n \left[P\left(\left\|\frac{x}{2^n}\right\|^2\right)\right] \\ &= P(\|x\|^2), \quad x \in X. \end{aligned}$$

This means that the following equality holds true:

$$\begin{aligned} p(x) + p(-x) &= \sum_{n=1}^{+\infty} 2^n \left[p\left(\frac{x}{2^n}\right) + p\left(-\frac{x}{2^n}\right)\right] \\ &= P(\|x\|^2), \quad x \in X. \end{aligned}$$

Similarly,

$$q(x) + q(-x) = -Q(\|x\|^2), \quad x \in X.$$

Now, if p and q are Borel-measurable or p is bounded from above at a point and q is bounded from below at a point, then P and Q are continuous (see M. Kuczma [10, pp. 414–417]) and thus there exist constants $c_p, c_q \in \mathbb{R}$ such that $P(t) = c_p t$ and $Q(t) = c_q t$ for $t \in \mathbb{R}_+$. Therefore, if we define

$$\begin{aligned} p_o(x) &:= p(x) - \frac{1}{2}c_p\|x\|^2, \quad x \in X; \\ q_o(x) &:= q(x) - \frac{1}{2}c_q\|x\|^2, \quad x \in X, \end{aligned}$$

then p_o and q_o are odd mappings and one may calculate that p_o is orthogonally subadditive, whereas q_o is orthogonally superadditive. Thus, both mappings, as odd maps, are orthogonally additive and consequently, additive and thus linear, since p and q enjoy a regularity property. In particular, p and q are orthogonally additive.

Remark 3. In a number of sandwich theorems assumptions that the functions to be separated are increasing or decreasing in a sense appears (see, e.g., Z. Gajda [6]). For orthogonally subadditive mapping p it seems natural to assume that p is *orthogonally increasing*, i.e., $p(x + y) \geq p(x)$ whenever $x \perp y$. S. Gudder and D. Strawther proved in [8] that orthogonally increasing functions are of the form $X \ni x \mapsto g(\|x\|) \in \mathbb{R}$ with a nondecreasing map $g\colon \mathbb{R}_+ \to \mathbb{R}$. In fact, orthogonally subadditive and orthogonally increasing function p satisfies (6) and (20). Indeed, fix $x \in X$ arbitrarily and choose $y \in X$ such that $x \perp y$ and $\|x\| = \|y\|$. From the result of S. Gudder and D. Strawther we infer that in particular $p(\pm x) = p(\pm y)$. Now, we have

$$\begin{aligned} p(2x) &\leq p(x+y) + p(x-y) \\ &\leq 2p(x) + p(y) + p(-y) \\ &= 4p(x) \\ &= 3p(x) + p(-x). \end{aligned}$$

Similarly, an orthogonally subadditive and orthogonally decreasing function q satisfies (7) and (21).

References

[1] K. Baron, J. Rätz, *On orthogonally additive mappings on inner product spaces*, Bull. Polish Acad. Sci. Math. **43**(3) (1995), 187–189.

[2] K. Baron, P. Volkmann, *On orthogonally additive functions*, Publ. Math. Debrecen **52**(3-4) (1998), 291–297.

[3] G. Birkhoff, *Orthogonality in linear metric spaces*, Duke Math. J. **1**(2) (1935), 169–172.

[4] W. Blaschke, *Räumliche Variationsprobleme mit symmetrischer Transversalitätsbedingung*, Ber. Verh. Königl. Sächs. Ges. Wiss. Leipzig, Math.-phys. Kl. **68** (1916), 50–55.

[5] W. Fechner, *On functions with the Cauchy difference bounded by a functional. III*, Abh. Math. Sem. Univ. Hamburg **76** (2006), 57–62.

[6] Z. Gajda, *Sandwich theorems and amenable semigroups of transformations*, Selected topics in functional equations and iteration theory, Graz, 1991, 43–58, Grazer Math. Ber. **316**, Karl-Franzens-Univ. Graz, Graz (1992).

[7] R. Ger, J. Sikorska, *Stability of the orthogonal additivity*, Bull. Polish Acad. Sci. Math. **43**(2) (1995), 143–151.

[8] S. Gudder, D. Strawther, *Orthogonally additive and orthogonally increasing functions on vector spaces*, Pacific J. Math. **58**(2) (1975), 427–436.

[9] R.C. James, *Orthogonality and linear functionals in normed linear spaces*, Trans. Amer. Math. Soc. **61** (1947), 265–292.

[10] M. Kuczma, *An introduction to the theory of functional equations and inequalities*, Państwowe Wydawnictwo Naukowe & Uniwersytet Śląski, Katowice, 1985.

[11] M. Moslehian, *On the stability of the orthogonal Pexiderized Cauchy equation*, J. Math. Anal. Appl. **318**(1) (2006), 211–223.

[12] J. Rätz, *On orthogonally additive mappings*, Aequationes Math. **281–2** (1985), 35–49.

[13] J. Rätz, *On orthogonally additive mappings II*, Publ. Math. Debrecen **35/3-4** (1988), 241–249.

[14] J. Rätz, *On orthogonally additive mappings III*, Abh. Math. Sem. Univ. Hamburg **59** (1989), 23–33.

[15] J. Rätz, Gy. Szabó, *On orthogonally additive mappings IV*, Aequationes Math. **38/1** (1989), 73–85.

[16] J. Sikorska, *Orthogonal stability of the Cauchy equation on balls*, Demonstratio Math. **33**(3) (2000), 527–546.

[17] J. Sikorska, *Orthogonal stability of the Cauchy functional equation on balls in normed linear spaces*, Demonstratio Math. **37**(3) (2004), 579–596.

[18] J. Sikorska, *Generalized orthogonal stability of some functional equations*, J. Inequal. Appl.(2006), Art. ID 12404, 23.

[19] Gy. Szabó, *On mappings orthogonally additive in the Birkhoff-James sense*, Aequationes Math. **30**(1) (1986), 93–105.

[20] Gy. Szabó, *On orthogonality spaces admitting nontrivial even orthogonally additive mappings*, Acta Math. Hungar. **56** (1990), 177–187.

[21] D. Yang, *Orthogonality spaces and orthogonally additive mappings*, Acta Math. Hungar. **113** (2006), 269–280.

W. Fechner and J. Sikorska
Institute of Mathematics
Silesian University
Bankowa 14
PL-40-007 Katowice, Poland

e-mail: fechner@math.us.edu.pl
e-mail: sikorska@math.us.edu.pl

International Series of Numerical Mathematics, Vol. 157, 283–290

On Vector Pexider Differences Controlled by Scalar Ones

Roman Ger

Abstract. We deal with a functional inequality

$$\|F(x+y) - G(x) - H(y)\| \leq g(x) + h(y) - f(x+y)$$

where F, G, H map a given commutative (semi)group $(S, +)$ into a Banach space and $f, g, h : S \to \mathbb{R}$ are given scalar functions. This is a pexiderized version of the stability problem

$$\|F(x+y) - F(x) - F(y)\| \leq f(x) + f(y) - f(x+y)$$

examined in connection with the singular case ($p = 1$) in

$$\|F(x+y) - F(x) - F(y)\| \leq \varepsilon\,(\|x\|^p + \|y\|^p)$$

We show, among others, that the maps F, G and H have to be, in a sense, close to an additive map provided that the function $g + h - 2f$ is upper bounded.

Mathematics Subject Classification (2000). 39B82, 39B62, 39B52.

Keywords. Pexider difference, stability, functional inequality, vector-valued solutions.

1. Introduction

Let $(X, \|\cdot\|)$ and $(Y, \|\cdot\|)$ be two Banach spaces and let $F : X \longrightarrow Y$ and $\varphi : X^2 \longrightarrow \mathbb{R}$ be two functions such that

$$\| F(x+y) - F(x) - F(y) \| \leq \varphi(x, y), \qquad x, y \in X. \tag{1}$$

Z.Gajda [3] has proved that in the case where

$$\varphi(x, y) = \varepsilon\,(\| x \|^p + \| y \|^p), \quad (x, y) \in X^2,$$

with given $\varepsilon \geq 0$ and $p \in \mathbb{R}\backslash\{1\}$, there exists a unique additive mapping $A : X \longrightarrow Y$ such that

$$\| F(x) - A(x) \| \leq \frac{2\varepsilon \operatorname{sgn}(p-1)}{2^p - 2} \| x \|^p, \quad x \in X.$$

For $p = 0$ the result was known already in 1941 by D.H. Hyers [7] and for $p \in [0, 1)$ it was proved in 1950 by T. Aoki [1] and rediscovered in 1978 by Th.M. Rassias [10]. The value $p = 1$ seems to be critical and it actually is; in [3] (see also Th.M. Rassias and P. Šemrl [11]) Z. Gajda presents an example of a continuous function $F : \mathbb{R} \longrightarrow \mathbb{R}$ such that $\mid F(x + y) - F(x) - F(y) \mid \leq \mid x \mid + \mid y \mid$ for all $x, y \in \mathbb{R}$, but there exists no additive mapping $A : \mathbb{R} \longrightarrow \mathbb{R}$ for which the function

$$\mathbb{R}\backslash\{0\} \ni x \longrightarrow \frac{1}{\mid x \mid} \mid F(x) - A(x) \mid$$

were bounded. In other words, the inequality

$$\| F(x + y) - F(x) - F(y) \| \leq \| x \| + \| y \|, \quad x, y \in X, \tag{2}$$

is too weak for F to enforce its asymptotic additivity. In [4] it is shown, among others, that diminishing the right-hand side of (2) by taking

$$\|F(x + y) - F(x) - F(y)\| \leq \|x\| + \|y\| - \|x + y\|, \quad x, y \in X, \tag{3}$$

does the job, i.e., then there exists an additive mapping $A : X \longrightarrow Y$ such that the function $x \longrightarrow \frac{1}{\|x\|} \| F(x) - A(x) \|$ is bounded (i.e., F is asymptotically additive, as required). However, the target space $(Y, \| \cdot \|)$ is assumed to be finite dimensional in [4]. The latter assumption has then been replaced in [5] by the requirement that the space $(Y, \| \cdot \|)$ is either reflexive or has the Hahn-Banach extension property or forms a boundedly complete Banach lattice with a strong unit. Moreover, inequality (3) was there generalized to

$$\|F(x + y) - F(x) - F(y)\| \leq f(x) + f(y) - f(x + y), \quad x, y \in X, \tag{4}$$

whereas the domain X was replaced by an amenable group.

Kil-Woung Jun, Dong-Soo Shin & Byung-Do Kim [8] and a year later Yang-Hi Lee and Kil-Woung Jun [9] described the stability behaviour of the following pexiderization of inequality (1):

$$\|F(x + y) - G(x) - H(y)\| \leq \varphi(x, y) .$$

In both papers the dominating function φ is assumed to satisfy a convergence condition corresponding to the so-called direct method – the technique of Hyers sequences applied by him already in [7]. That standard approach is useless while dealing with the most delicate (singular) cases we have spoken of. Facing the lack of stability we then try to diminish the dominating function to get the desired result. To cover such cases in the pexiderized case, in the present paper we deal with a Pexider analogue of inequality (4), namely

$$\|F(x + y) - G(x) - H(y)\| \leq g(x) + h(y) - f(x + y) \tag{5}$$

where F, G, H map a given commutative (semi)group $(S, +)$ into a Banach space and $f, g, h : S \to \mathbb{R}$ are given scalar functions. However, we have to remark that without any additional assumptions upon the scalar functions occurring here such a setting is too general; actually, observe that even in the case where $G = H$ and $g = h$, quite *arbitrary* mappings F and G solve the inequality

$$\|F(x+y) - G(x) - G(y)\| \le \|G(x)\| + \|G(y)\| - (-\|F(x+y)\|)\,.$$

Therefore some shrinking conditions upon the dominating difference in (5) are indispensable to avoid trivialities.

2. Reducing inequality (5) to four unknown functions

We begin with a rearrangement of inequality (5) to reduce it to a slightly simpler form. Recall that a monoid is a semigroup admitting a neutral element.

Theorem 1. *Let $(S, +)$ be an Abelian semigroup and let $(X, \|\cdot\|)$ stand for a normed linear space. Given mappings $F, G, H : S \longrightarrow X$ and functions $f, g, h : S \longrightarrow \mathbb{R}$ satisfying inequality* (5) *for all $x, y \in S$, we have*

$$\|F_1(x+y) - G_1(x) - G_1(y)\| \le g_1(x) + g_1(y) - f_1(x+y)\,, \quad x, y \in S,$$

where $F_1 := 2F$, $G_1 := G + H$, $f_1 := 2f$ and $g_1 := g + h$.

Proof. Inequality (5) jointly with the commutativity of the semigroup $(S, +)$ imply that

$$\|F(x+y) - G(y) - H(x)\| \le g(y) + h(x) - f(x+y)\,, \quad x, y \in S.$$

Summing this inequality with (5) side by side and applying the triangle inequality of the norm we derive the assertion.

Theorem 2. *Let $(S, +, 0)$ be a monoid and let $(X, \|\cdot\|)$ stand for a normed linear space. Given mappings $F, G : S \longrightarrow X$ and functions $f, g : S \longrightarrow \mathbb{R}$ such that*

$$\|F(x+y) - G(x) - G(y)\| \le g(x) + g(y) - f(x+y)\,, \quad x, y \in S, \tag{6}$$

we have

$$\|F(x+y) - F(x) - F(y)\| \le \tilde{g}(x) + \tilde{g}(y) - f(x+y)\,, \quad x, y \in S, \tag{7}$$

and

$$\|G(x+y) - G(x) - G(y)\| \le g(x) + g(y) - \tilde{f}(x+y)\,, \quad x, y \in S, \tag{8}$$

where $\tilde{g} := 2g \quad f + c$ and $\tilde{f} := 2f - y - c$, with $c := g(0) + \|G(0)\|$.

Proof. Setting $y = 0$ in (6) we get

$$\|F(x) - G(x)\| - \|G(0)\| \le \|F(x) - G(x) - G(0)\| \le g(x) + g(0) - f(x)\,, \quad x, y \in S,$$

whence

$$\|F(x) - G(x)\| \le g(x) - f(x) + g(0) + \|G(0)\| = g(x) - f(x) + c\,, \quad x, y \in S.$$

Therefore, for all $x, y \in S$ one has

$$\begin{aligned}
&\|F(x+y) - F(x) - F(y)\| \\
&\quad \le \|F(x+y) - G(x) - G(y)\| + \|G(x) - F(x)\| + \|G(y) - F(y)\| \\
&\quad \le g(x) + g(y) - f(x+y) + (g(x) - f(x) + c) + (g(y) - f(y) + c) \\
&\quad \le (2g(x) - f(x) + c) + (2g(y) - f(y) + c) - f(x+y) \\
&\quad = \tilde{g}(x) + \tilde{g}(y) - f(x+y)\,,
\end{aligned}$$

which gives (7).

To prove (8), note that for all elements x, y, z from S we get

$$\|F(x+y+z) - G(x) - G(y+z)\| \le g(x) + g(y+z) - f(x+y+z)\,,$$

as well as

$$\|F(x+y+z) - G(x+y) - G(z)\| \le g(x+y) + g(z) - f(x+y+z)\,,$$

which easily implies that

$$\|G(x+y)+G(z)-G(x)-G(y+z)\| \le g(x)+g(y+z)+g(x+y)+g(z)-2f(x+y+z)\,.$$

Putting here $z = 0$ we conclude that

$$\begin{aligned}
&\|G(x+y) - G(x) - G(y)\| - \|G(0)\| \\
&\quad \le \|G(x+y) + G(0) - G(x) - G(y)\| \\
&\quad \le g(x) + g(y) + g(x+y) + g(0) - 2f(x+y)\,,
\end{aligned}$$

from which inequality (8) results immediately. Thus the proof has been completed.

3. Main results

Because of the tools used, in what follows we will need a group structure in the domain considered. In the reduced case the commutativity is weakened to amenability due to the fact that Theorem 1 is of no use. Recall that a semigroup $(S, +)$ is termed left (resp. right) amenable provided that there exists a real linear functional M on $\mathbb{B}(S, \mathbb{R})$ such that

$$\inf f(S) \le M(f) \le \sup f(S)\,, \qquad f \in \mathbb{B}(S, \mathbb{R})\,,$$

and M is left (resp. right) invariant in the sense that

$$M({}_af) = M(f) \qquad (\text{resp.}\, M(f_a) = M(f)\,)$$

for all $f \in \mathbb{B}(S, \mathbb{R})$ and all $a \in S$; here $({}_af)(x) := f(a+x)$ and $f_a(x) := f(x+a)$, $x, a \in S$.

It is well known that any commutative semigroup is amenable.

In what follows we will be using the following slightly modified version of Theorem 3 from [5]:

Theorem $(*)$. *Let $(S,+)$ be an amenable group and let $(X,\|\cdot\|)$ be a real normed linear space that is either reflexive or has the Hahn-Banach extension property. Suppose further that $F : S \longrightarrow X$ and $f : S \longrightarrow \mathbb{R}$ satisfy inequality*

$$\| F(x+y) - F(x) - F(y) \| \leq f(x) + f(y) - f(x+y)\,, \quad x, y \in S.$$

Then exists an additive mapping $A : S \longrightarrow X$ such that

$$\| F(x) - A(x) \| \leq 2f_e(x)\,, \quad x \in S\,,$$

where f_e stands for the even part of f.

Now, we are in a position to prove the following

Theorem 3. *Let $(S,+)$ be an amenable group and let $(X,\|\cdot\|)$ be a real normed linear space that is either reflexive or has the Hahn-Banach extension property. Given functions $f, g : S \longrightarrow \mathbb{R}$ and $F, G : S \longrightarrow X$ such that*

$$\|F(x+y) - G(x) - G(y)\| \leq g(x) + g(y) - f(x+y)$$

for all $x, y \in S$, and

$$c_0 := \sup\{g(x) - f(x) : x \in S\} < \infty\,,$$

there exist an additive map $A : S \longrightarrow X$ and real constants α, β such that

$$\|F(x) - A(x)\| \leq 2f_e(x) + \alpha\,, \quad x \in S,$$

and

$$\|G(x) - A(x)\| \leq 2g_e(x) + \beta\,, \quad x \in S\,.$$

Proof. By means of Theorem 2, for all $x, y \in S$ we have

$$\begin{aligned}
&\|G(x+y) - G(x) - G(y)\| \\
&\quad \leq g(x) + g(y) - \tilde{f}(x+y) \\
&\quad = g(x) + g(y) - 2f(x+y) + g(x+y) + c \\
&\quad = (g(x) + g(y) - g(x+y)) + 2(g(x+y) - f(x+y)) + c \\
&\quad \leq g(x) + g(y) - g(x+y) + 2c_0 + c \\
&\quad = \left(g + \frac{1}{2}\beta\right)(x) + \left(g + \frac{1}{2}\beta\right)(y) - \left(g + \frac{1}{2}\beta\right)(x+y)\,,
\end{aligned}$$

where $\beta := 4c_0 + 2c$. On account of Theorem $(*)$, there exists an additive map $A : S \longrightarrow X$ such that for every $x \in S$ one has

$$\|G(x) - A(x)\| \leq 2(g + \frac{1}{2}\beta)_e(x) = 2g_e(x) + \beta\,, \quad x \in S\,,$$

as claimed.

Moreover, by (6) applied for $y = 0$, for an arbitrary x from S we have also

$$\begin{aligned}\|F(x) - A(x)\| &\le \|F(x) - G(x)\| + \|G(x) - A(x)\| \\ &= g(x) - f(x) + c + 2g_e(x) + \beta \\ &\le c_0 + c + 2f_e(x) + 2c_0 + \beta \\ &= 2f_e(x) + \alpha\,,\end{aligned}$$

with $\alpha := 3c_0 + c + \beta$. This finishes the proof.

Theorem 4. Let $(S,+)$ be an Abelian group and let $(X, \|\cdot\|)$ be a real Banach space that is either reflexive or has the Hahn-Banach extension property. Given functions $f, g, h : S \longrightarrow \mathbb{R}$ and $F, G, H : S \longrightarrow X$ such that

$$\|F(x+y) - G(x) - H(y)\| \le g(x) + h(y) - f(x+y) \tag{5}$$

for all $x, y \in S$, and

$$c_0 := \sup\{g(x) + h(x) - 2f(x) : x \in S\} < \infty\,,$$

there exist an additive map $A : S \longrightarrow X$ and real constants $\alpha_0, \beta_0, \gamma_0$ such that

$$\|F(x) - A(x)\| \le 2f_e(x) + \alpha_0\,, \quad x \in S,$$

and

$$\begin{aligned}\|G(x) - A(x)\| &\le 2g_e(x) + \beta_0\,, \quad x \in S\,, \\ \|H(x) - A(x)\| &\le 2h_e(x) + \gamma_0\,, \quad x \in S\,,\end{aligned}$$

where f_e, g_e and h_e stand for the even parts of f, g and h, respectively.

Proof. By means of Theorem 1, for all $x, y \in S$ we have

$$\|F_1(x+y) - G_1(x) - G_1(y)\| \le g_1(x) + g_1(y) - f_1(x+y)\,, \quad x, y \in S,$$

where $F_1 := 2F$, $G_1 := G + H$, $f_1 := 2f$ and $g_1 := g + h$. Moreover, one has

$$\sup\{g_1(x) - f_1(x) : x \in S\} = \sup\{g(x) + h(x) - 2f(x) : x \in S\} = c_0 < \infty\,.$$

An appeal to Theorem 3 gives now the existence of an additive map $A_1 : S \longrightarrow X$ and real constants α and β such that

$$\|F_1(x) - A_1(x)\| \le 2(f_1)_e(x) + \alpha\,, \quad x \in S, \tag{9}$$

Obviously, the map $A := \frac{1}{2}A_1$ is additive as well and (9) says that

$$\|F(x) - A(x)\| \le 2f_e(x) + \frac{1}{2}\alpha = 2f_e(x) + \alpha_0\,, \quad x \in S\,,$$

where $\alpha_0 := \frac{1}{2}\alpha$. Now, with the aid of (5) applied for $y = 0$ we infer that the inequality

$$\begin{aligned}\|G(x) - A(x)\| &\le \|G(x) - F(x)\| + \|F(x) - A(x)\| \\ &\le g(x) - f(x) + h(0) + \|H(0)\| + 2f_e(x) + \alpha_0\,,\end{aligned}$$

holds true for every $x \in S$. On the other hand, (5) forces the difference $g(x) + h(y) - f(x+y)$ to be nonnegative for all x, y from S whence, in particular, $f(x) \leq h(x) + g(0)$, $x \in S$, and therefore, for every $x \in S$ one has

$$g(x) - f(x) \leq c_0 + f(x) - h(x) \leq c_0 + g(0)\,.$$

Consequently, because of the inequality $f(x) \leq g(x) + h(0)$ valid for every $x \in S$, , we get

$$\begin{aligned} \|G(x) - A(x)\| &\leq c_0 + g(0) + h(0) + \|H(0)\| + 2f_e(x) + \alpha_0 \\ &\leq 2g_e(x) + \beta_0\,, \quad x \in S, \end{aligned}$$

where $\beta_0 := c_0 + g(0) + 3h(0) + \|H(0)\| + \alpha_0\,.$

Along the same lines one may establish the inequality

$$\begin{aligned} \|H(x) - A(x)\| &\leq c_0 + g(0) + h(0) + \|G(0)\| + 2f_e(x) + \alpha_0 \\ &\leq 2h_e(x) + \gamma_0\,, \quad x \in S, \end{aligned}$$

where $\gamma_0 := c_0 + 3g(0) + h(0) + \|G(0)\| + \alpha_0\,.$ This completes the proof.

4. Concluding remarks

In [6] the stability of the so-called delta-convexity has been examined. These studies were focused on the functional inequality

$$\left\| F\left(\frac{x+y}{2}\right) - \frac{F(x) + F(y)}{2} \right\| \leq \frac{f(x) + f(y)}{2} - f\left(\frac{x+y}{2}\right)$$

with F and f defined on a nonempty open and convex subset D of a normed linear space $(E, \|\cdot\|)$. In the case where $D = E$ this inequality may equivalently be written in the form

$$\left\| F(x+y) - \frac{1}{2}F(2x) - \frac{1}{2}F(2y) \right\| \leq \frac{1}{2}f(2x) + \frac{1}{2}f(2y) - f(x+y) \qquad (10)$$

and no norm structure in the domain is needed any more. As an application of Theorem 3 we get the following result being, in a sense, complementary to those presented in [6].

Let $(S, +)$ be an amenable group and let $(X, \|\cdot\|)$ be a real Banach space that is either reflexive or has the Hahn-Banach extension property. Given functions $f : S \longrightarrow \mathbb{R}$ and $F : S \longrightarrow X$ such that inequality (10) *is satisfied for all $x, y \in S$, and*

$$\sup\{f(2x) - 2f(x) : x \in S\} < \infty\,,$$

there exist an additive map $A : S \longrightarrow X$ and a real constant α such that

$$\|F(x) - A(x)\| \leq 2f_e(x) + \alpha\,, \quad x \in S\,.$$

As an easy application of Theorem 4 we may consider the very simple case where $f = g = h =$ const getting a Hyers-Ulam stability result for the classical Pexider functional equation.

The assumption that the target space $(X, \|\cdot\|)$ is either reflexive or has the Hahn-Banach extension property, occurring here and in Theorems 3 and 4, may be replaced by the requirement that X admits a continuous projection of its second dual onto X (see F. Cabello Sánchez [2]). On the other hand, the basic tool we were using in the present paper (Theorem 3 from [5]) allows one to replace the assumption in question by another alternative one: $(X, \|\cdot\|)$ forms a boundedly complete Banach lattice with a strong unit. However, in this case, some additional coefficient will show up in the dominating function in the assertions of Theorems 3 and 4; for brevity of the statements, we have decided to omit that possible extension.

References

[1] T. Aoki, *On the stability of the linear transformation in Banach spaces*, J. Math. Soc. Japan **2** (1950), 64–66.

[2] F. Cabello Sánchez, *Stability of additive mappings on large subsets*, Proc. Amer. Math. Soc. **128** (1999), 1071–1077.

[3] Z. Gajda, *On the stability of additive mappings*, Internat. J. Math. & Math. Sci. **14** (1991), 431–434.

[4] R. Ger, *On functional inequalities stemming from stability questions*, General Inequalities 6, ISNM **103**, Birkhäuser Verlag, Basel, 1992, 227–240.

[5] R. Ger, *The singular case in the stability behaviour of linear mappings*, Grazer Math. Ber. **316** (1992), 59–70.

[6] R. Ger, *Stability aspects of delta-convexity*, in: "Stability of Hyers-Ulam type", A Volume dedicated to D.H. Hyers & S. Ulam, (ed. Th.M. Rassias & J. Tabor), Hadronic Press, Inc., Palm Harbor, 1994, 99–109.

[7] D.H. Hyers, *On the stability of the linear functional equation*, Proc. Nat. Acad. Sci. U.S.A. **27** (1941), 222–224.

[8] Kil-Woung Jun, Dong-Soo Shin & Byung-Do Kim, *On Hyers-Ulam-Rassias Stability of the Pexider Equation*, J. Math. Anal. Appl. **239** (1999), 20–29.

[9] Yang-Hi Lee & Kil-Woung Jun, *A Generalization of the Hyers-Ulam-Rassias Stability of the Pexider Equation*, J. Math. Anal. Appl. **246** (2000), 627–638.

[10] Th.M. Rassias, *On the stability of the linear mapping in Banach spaces*, Proc. Amer. Math. Soc. **72** (1978), 297–300.

[11] Th. M. Rassias and P. Šemrl, *On the behaviour of mappings which do not satisfy Hyers-Ulam stability*, Proc. Amer. Math. Soc. **114** (1992), 989–993.

Roman Ger
Institute of Mathematics
Silesian University
Bankowa 14
PL-40-007, Katowice, Poland
e-mail: `romanger@us.edu.pl`

International Series of Numerical Mathematics, Vol. 157, 291–298

A Characterization of the Exponential Distribution through Functional Equations

Gyula Maksa and Fruzsina Mészáros

Abstract. In this paper we give a characterization for the exponential distribution by using functional equations.

Mathematics Subject Classification (2000). 39B22.

Keywords. Exponential distribution, functional equation.

1. Introduction

Some years ago A.W. Marshall (University of British Columbia) and I. Olkin (Stanford University) raised the following problem (personal communication). Find all density functions f satisfying the following two properties

Property 1. *$f(u) = 0$ for almost all $u \in]-\infty, 0[$ (with respect to the Lebesgue measure)* and

Property 2. *There exist $0 \leq n \in \mathbb{Z}$ (the set of all integers) and $-1 < \beta \in \mathbb{R}$ (the set of all real numbers) such that the function p defined on $\mathbb{R}^2$ by $p(u,v) = 0$ if $u < 0$ or $v < 0$ and*

$$p(u,v) = \int_0^{+\infty} f(u)\left(F(u) - F(s+u)\right)^n f(s+u) f(s+u+v) F(s+u+v)^\beta ds \quad (1)$$

if $u, v \in [\,0, +\infty\,[$, where $F(u) = \int_u^{+\infty} f$, $u \geq 0$ is the survival function, is the joint density function of some two independent random variables.

It was only known that, if $f(u) = 0$ for $u < 0$ and $f(u) = e^{-u}$ for $u \geq 0$, then $p(u,v) = g(u)h(v)$ $(u \geq 0,\ v \geq 0)$ for some functions $g, h : [\,0, +\infty\,[\to \mathbb{R}$. We remark that the problem is similar to those ones which are extensively discussed in the book Azlarov-Volodin [2].

This research has been supported by the Hungarian Scientific Research Fund (OTKA) Grants NK 68040 and K 62316.

In this paper we give a solution of the problem by proving that all density functions f which are positive on $[\,0,+\infty\,[$ and have Properties 1–2 are exponential density functions. Furthermore we give a necessary and sufficient condition for the parameters n and β in (1) in order to the function p defined in Property 2, with some exponential density function f, be a density function itself, too.

We will need the following result of A. Járai (see [3] and [4]).

Theorem 1. *Let Z be a regular topological space, Z_i $(i = 1, 2, \ldots, n)$ be topological spaces and T be a first countable topological space. Let Y be an open subset of $\mathbb{R}^k$, X_i an open subset of $\mathbb{R}^{r_i}$, $r_i \in \mathbb{Z}$, $(i = 1, 2, \ldots, n)$ and D an open subset of $T \times Y$. Let furthermore $T' \subset T$ be a dense subset, $H\colon T' \to Z$, $g_i\colon D \to X_i$ and $h\colon D \times Z_1 \times \cdots \times Z_n \to Z$. Suppose that the function f_i is almost everywhere defined on X_i (with respect to the r_i-dimensional Lebesgue measure) with values in Z_i $(i = 1, 2, \ldots, n)$ and the following conditions are satisfied:*

(i) *for all $t \in T'$ and for almost all $y \in D_t = \{y \in Y : (t, y) \in D\}$*

$$H(t) = h(t, y, f_1(g_1(t, y)), \ldots, f_n(g_n(t, y))); \tag{2}$$

(ii) *for each fixed y in Y, the function h is continuous in the other variables;*

(iii) *f_i is Lebesgue measurable $(i = 1, 2, \ldots, n)$;*

(iv) *g_i and the partial derivative $\frac{\partial g_i}{\partial y}$ are continuous on D $(i = 1,\ 2,\ \ldots,\ n)$;*

(v) *for each $t \in T$ there exist a y such that $(t, y) \in D$ and the partial derivative $\frac{\partial g_i}{\partial y}$ has the rank r_i at $(t, y) \in D$ $(i = 1, 2, \ldots, n)$.*

Then there exists a unique continuous function $\widetilde{H}$ such that $H = \widetilde{H}$ almost everywhere on T, and if H is replaced by $\widetilde{H}$ then equation (2) is satisfied almost everywhere on D.

2. Results

We begin with our main result.

Theorem 2. *If the density function f has Property 1 and Property 2, and*

$$f\,(u) > 0 \quad \text{if} \quad u \geq 0 \tag{3}$$

then there exists $0 < \alpha \in \mathbb{R}$ such that

$$f(u) = \alpha e^{-\alpha u} \qquad \text{for almost all} \quad u \in [\,0, +\infty\,[\,, \tag{4}$$

that is, f is an exponential density function with expectation α^{-1}.

Proof. Since f is density function, the function F defined on $[\,0,+\infty\,[$ by

$$F(u) = \int_u^{+\infty} f$$

is absolutely continuous and $F' = -f$ a.e. on $[\,0,+\infty\,[$. On the other hand, with the substitution $t = s + u$ equation (1) can be written in the form

$$p(u,v) = \int_u^{+\infty} f(u)(F(u) - F(t))^n f(t) f(t+v) F(t+v)^\beta \, dt \tag{5}$$

for all $u, v \in [\,0,+\infty\,[$. Dividing both sides by $-f(u)$ and using the binomial theorem, (5) implies that

$$\frac{p(u,v)}{-f(u)} = \sum_{k=0}^{n} \binom{n}{k} (-1)^k F(u)^{n-k} R_k(u,v) \quad (u, v \in [\,0,+\infty\,[)$$

where

$$R_k(u,v) = -\int_u^{+\infty} F(t)^k f(t) f(t+v) F(t+v)^\beta \, dt$$

$(u, v \in [\,0,+\infty\,[\,,\ 0 \le k \in \mathbb{Z},\ k \le n)$. Therefore the function

$$u \longmapsto \frac{p(u,v)}{-f(u)} \quad (u \ge 0)$$

is differentiable a.e. on $[\,0,+\infty\,[$, too, and by the known identity $\sum_{k=0}^{n} \binom{n}{k} (-1)^k = 0$, after some calculation, we have that

$$\frac{\partial}{\partial u}\left(\frac{p(u,v)}{-f(u)}\right) = -f(u) \sum_{k=0}^{n} \binom{n}{k} (-1)^k (n-k) F(u)^{n-k-1} R_k(u,v) \tag{6}$$

a.e. on $[\,0,+\infty\,[\, \times [\,0,+\infty\,[$. Introducing the differential operator

$$Dq(u,v) = \frac{\partial}{\partial u}\left(\frac{q(u,v)}{-f(u)}\right),$$

(6) can be written as

$$Dp(u,v) = -f(u) \sum_{k=0}^{n} \binom{n}{k} (-1)^k (n-k) F(u)^{n-k-1} R_k(u,v).$$

This shows that the previous argument can be repeated for Dp instead of p, too, and, by the identity $\sum_{k=0}^{n} \binom{n}{k} (-1)^k (n-k) = 0$, we get that

$$D^2 p(u,v) = -f(u) \sum_{k=0}^{n} \binom{n}{k} (-1)^k (n-k)(n-k-1) F(u)^{n-k-2} R_k(u,v)$$

holds a.e. on $[\,0,+\infty\,[\, \times [\,0,+\infty\,[$. Finally, by the identity

$$\sum_{k=0}^{n} \binom{n}{k} (-1)^k (n-k) \dots (n-k-\ell) = 0 \quad (0 \le \ell < n),$$

we have that

$$D^n p(u,v) = -f(u)\sum_{k=0}^{n}\binom{n}{k}(-1)^k (n-k)\dots(-k+1)F(u)^{-k}R_k(u,v)$$
$$= -f(u)\,n!R_0(u,v),$$

whence

$$D^{n+1}p(u,v) = n!f(u)f(u+v)F(u+v)^{\beta} \tag{7}$$

follows for almost all $u,v \in [\,0,+\infty\,[$.

On the other hand, by Property 2, $p(u,v) = g(u)h(v)$ a.e. on $[\,0,+\infty\,[\,\times\,[\,0,+\infty\,[$ with some density functions g and h. Thus

$$\mathrm{D}^{n+1}\mathrm{p}(u,v) = g_n(u)h(v) \qquad \text{a.e. on } [\,0,+\infty\,[\,\times\,[\,0,+\infty\,[$$

with some function $g_n : [\,0,+\infty\,[\, \to \mathbb{R}$. Therefore, by (7), we obtain the exponential Pexider equation

$$f(u+v)F(u+v)^{\beta} = \frac{g_n(u)}{n!f(u)}h(v)$$

which holds for almost all $u,v \in [\,0,+\infty\,[$. This equation, with the notations

$$B(t) = f(t)F(t)^{\beta} \qquad \text{and} \qquad G_n(t) = \frac{g_n(t)}{n!f(t)},$$

yield the equation

$$B(u+v) = G_n(u)h(v), \tag{8}$$

which holds a.e. on the open set $]\,0,+\infty\,[\,\times\,]\,0,+\infty\,[$.

Since the measurable functions B, G_n, h satisfy equation (8) for almost all $(u,v) \in\,]\,0,+\infty\,[\,\times\,]\,0,+\infty\,[$, then, by Theorem 1, there exist unique continuous functions $\widetilde{B}, \widetilde{G}_n, \widetilde{h} :]\,0,+\infty\,[\, \to \mathbb{R}$, such that $\widetilde{B} = B$, $\widetilde{G}_n = G_n$, $\widetilde{h} = h$ almost everywhere on $]\,0,+\infty\,[$, and if B, G_n, h are replaced by $\widetilde{B}$, $\widetilde{G}_n$, $\widetilde{h}$, respectively, then equation (8) is satisfied everywhere on $]\,0,+\infty\,[\,\times\,]\,0,+\infty\,[$.

First we show that there exists a unique continuous function $\widetilde{B}$ which is equal to B a.e. on $]\,0,+\infty\,[$, and after replacing B by $\widetilde{B}$, equation (8) is satisfied almost everywhere. With the substitution

$$t = u+v, \qquad y = v$$

we get from (8) that

$$B(t) = G_n(t-y)h(y) \tag{9}$$

holds for almost all $(t,y) \in D$, where

$$D = \{(t,y)\,|\,t,y \in\,]\,0,+\infty\,[\,\}.$$

By Fubini's Theorem it follows that there exists $T' \subseteq\,]\,0,+\infty\,[$ of full measure such that, for all $t \in T'$ equation (9) is satisfied for almost every $y \in D_t$, where

$$D_t = \{y \in\,]\,0,+\infty\,[\,\,|(t,y) \in D\}.$$

Define the functions g_1, g_2, h by:

$$g_1(t,y) = t - y, \qquad g_2(t,y) = y,$$
$$h(t,y,z_1,z_2) = z_1 z_2,$$

and apply Theorem 1 to (9) with the following casting

$$\begin{aligned} B(t) &= H(t), & G_n(t) &= f_1(t), & h(t) &= f_2(t), \\ Z &= \mathbb{R}, & Z_i &= \mathbb{R}, & T &= \left]0,+\infty\right[, \\ Y &= \left]0,+\infty\right[, & X_i &= \left]0,+\infty\right[, & (i &= 1,2). \end{aligned}$$

Hence the first assumption in Theorem 1 with respect to (9) is valid. In the case of a fixed y, the function h is continuous in the other variables, so the second assumption holds, too. Since the functions in equation (9) are measurable, the third assumption is trivial. The functions g_i are continuous, the partial derivatives

$$\partial_2 g_1(t,y) = -1, \quad \partial_2 g_2(t,y) = 1,$$

are also continuous, so the fourth assumption holds, too.

For each $t \in \left]0,+\infty\right[$ there exist a $y \in \left]0,+\infty\right[$ such that $(t,y) \in D$ and the partial derivatives do not equal zero at (t,y), so they have the rank 1. Thus the last assumption is satisfied in Theorem 1. So we get that there exists a unique continuous function $\widetilde{B}$ which is almost everywhere equal to B on $\left]0,+\infty\right[$ and $\widetilde{B}, G_n, h$ satisfy equation (9) almost everywhere, which is equivalent to equation

$$\widetilde{B}(u+v) = G_n(u)\, h(v), \tag{10}$$

for almost all $(u,v) \in \left]0,+\infty\right[\times \left]0,+\infty\right[$.

There exist u_0 and v_0 so that, equation (10) implies that

$$h(v) = \frac{\widetilde{B}(u_0+v)}{G_n(u_0)}$$

holds for almost all $v \in \left]0,+\infty\right[$, and

$$G_n(u) = \frac{\widetilde{B}(u+v_0)}{h(v_0)}$$

holds for almost all $u \in \left]0,+\infty\right[$.

Due to the continuity of $\widetilde{B}$, we have proved before, there exist unique continuous functions $\widetilde{h} : \left]0,+\infty\right[\to \mathbb{R}$ and $\widetilde{G}_n : \left]0,+\infty\right[\to \mathbb{R}$, defined by the right-hand side of the last two equalities, which are almost everywhere equal to h and G_n on $\left]0,+\infty\right[$, respectively, and if we replace h and G_n by $\widetilde{h}$ and $\widetilde{G}_n$, respectively, the functional equation

$$\widetilde{B}(u+v) = \widetilde{G}_n(u)\, \widetilde{h}(v), \tag{11}$$

is fulfilled almost everywhere on $\left]0,+\infty\right[\times \left]0,+\infty\right[$. Both sides of (11) define continuous functions on $\left]0,+\infty\right[$, which are equal to each other on a dense subset of $\left]0,+\infty\right[$, therefore we obtain that (11) is satisfied everywhere on $\left]0,+\infty\right[\times \left]0,+\infty\right[$.

Therefore, by [1] (see pp. 28–31 and 42–46),

$$B(t) = ae^{bt}$$

a.e. on $]0, +\infty[$ with some $0 < a \in \mathbb{R}$, $b \in \mathbb{R}$. On the other hand $B(0) = G_n(0)\,h(0)$=a, and hence

$$f(t)\,F(t)^{\beta} = ae^{bt}$$

a.e. on $[0, +\infty[$ with some $0 < a \in \mathbb{R}$, $b \in \mathbb{R}$. Since f is density function and $F' = -f$ a.e. on $[0, +\infty[$, we get (4) with some $0 < \alpha \in \mathbb{R}$. □

Theorem 3. *Let f be an exponential density function with expectation α^{-1}. Then the function p defined in Property 2 is density function itself, too, if, and only if,*

$$(1+\beta)\cdots(n+3+\beta) = n!. \tag{12}$$

Furthermore, if (12) is satisfied then

$$p(u,v) = \alpha\,(n+3+\beta)\,e^{-\alpha(n+3+\beta)u}\alpha\,(1+\beta)\,e^{-\alpha(1+\beta)v}$$

(a.e. on $[0, +\infty[\times [0, +\infty[$), that is, p is the product of two exponential density functions with expectations $(\alpha\,(n+3+\beta))^{-1}$ and $(\alpha\,(1+\beta))^{-1}$, respectively.

Proof. We first prove, by induction on n, that

$$(2+\beta)\cdots(n+2+\beta)\sum_{k=0}^{n}\binom{n}{k}(-1)^k\frac{1}{k+2+\beta} = n! \tag{13}$$

holds for all $0 \le n \in \mathbb{Z}$ and $\beta \in \mathbb{R}\setminus\{-(n+2), \ldots, -2\}$. Equality (13) is obviously true for $n = 0$. Suppose that $n > 0$ and (13) holds, particularly with $\beta = -1$. Thus we have that

$$\sum_{k=0}^{n}\binom{n}{k}(-1)^k\frac{1}{k+1} = \frac{1}{n+1},$$

or equivalently (replacing here k by $n-k$)

$$\sum_{k=0}^{n}\binom{n}{k}(-1)^k\frac{1}{n+1-k} = \frac{(-1)^n}{n+1}. \tag{14}$$

Therefore

$$\begin{aligned}
&(2+\beta)\cdots(n+3+\beta)\sum_{k=0}^{n+1}\binom{n+1}{k}(-1)^k\frac{1}{k+2+\beta}\\
&= (2+\beta)\cdots(n+3+\beta)\left[\sum_{k=0}^{n}\binom{n+1}{k}(-1)^k\frac{1}{k+2+\beta} + (-1)^{n+1}\frac{1}{n+3+\beta}\right]\\
&= (-1)^{n+1}(2+\beta)\cdots(n+2+\beta) + (2+\beta)\cdots(n+3+\beta)\\
&\quad\times\sum_{k=0}^{n}\binom{n}{k}(-1)^k\frac{n+1}{(n+1-k)(k+2+\beta)}
\end{aligned}$$

$$= (-1)^{n+1}(2+\beta)\cdots(n+2+\beta) + (2+\beta)\cdots(n+3+\beta)$$
$$\times \sum_{k=0}^{n} \binom{n}{k} (-1)^k \frac{n+1}{n+3+\beta} \left(\frac{1}{n+1-k} + \frac{1}{k+2+\beta} \right)$$
$$= (-1)^{n+1}(2+\beta)\cdots(n+2+\beta) + (2+\beta)\cdots(n+2+\beta)(n+1)$$
$$\times \sum_{k=0}^{n} \binom{n}{k} (-1)^k \left(\frac{1}{n+1-k} + \frac{1}{k+2+\beta} \right)$$

whence, by (14) and the induction hypothesis,

$$(2+\beta)\cdots(n+3+\beta) \sum_{k=0}^{n+1} \binom{n+1}{k} (-1)^k \frac{1}{k+2+\beta}$$
$$= (-1)^{n+1}(2+\beta)\cdots(n+2+\beta) + (-1)^n(2+\beta)\cdots(n+2+\beta) + (n+1)(n!)$$
$$= (n+1)!$$

follows. Thus (13) is proved to hold.

Now the only thing we have to prove that $\int_{\mathbb{R}} p = 1$ if, and only if, (12) holds. Since $F(u) = e^{-\alpha u}$ for $u \geq 0$, after some calculation, (5) implies that

$$p(u,v) = \alpha^2 e^{-\alpha(u+v+v\beta)} \sum_{k=0}^{n} \binom{n}{k} (-1)^k \frac{1}{k+2+\beta} e^{-\alpha(n+2+\beta)u}$$

$(u \geq 0,\ v \geq 0)$. Therefore, by (13), we obtain that

$$p(u,v) = \alpha^2 \frac{n!}{(2+\beta)\cdots(n+2+\beta)} e^{-\alpha(n+3+\beta)u} e^{-\alpha(\beta+1)v}$$
$$= \frac{n!}{(1+\beta)\cdots(n+3+\beta)} \alpha(n+3+\beta) e^{-\alpha(n+3+\beta)u} \alpha(\beta+1) e^{-\alpha(\beta+1)v}$$

$(u \geq 0,\ v \geq 0)$. Now, it is easy to see that $\int_{\mathbb{R}^2} p = 1$ if, and only if, (12) holds and in this case

$$p(u,v) = \alpha(n+3+\beta) e^{-\alpha(n+3+\beta)u} \alpha(\beta+1) e^{-\alpha(\beta+1)v}$$

$(u \geq 0,\ v \geq 0)$ which was to be proved. □

Remark 1. *Since the function* φ *defined on* $]-1,+\infty[$ *by*

$$\varphi(t) = (1+t)\cdots(n+3+t) - n!$$

is continuous and strictly increasing, moreover $\lim_{t\to -1} \varphi(t) = -n! < 0$, $\varphi(1) = (n+4)! - n! > 0$ *therefore for all* $0 \leq n \in \mathbb{Z}$ *there exists exactly one* $-1 < \beta \in \mathbb{R}$ *such that* (12) *holds.*

References

[1] J. Aczél and J. Dhombres, *Functional Equations in Several Variables*, Cambridge University Press, Cambridge, 1989.

[2] T.A. Azlarov and N.A. Volodin, *Characterization Problems Associated with the Exponential Distribution*, Springer-Verlag, New York – Berlin – Heidelberg – Tokyo, 1986.

[3] A. Járai, *Measurable solutions of functional equations satisfied almost everywhere*, Mathematica Pannonica **10/1** (1999), 103–110.

[4] A. Járai, *Regularity Properties of Functional Equations in Several Variables*, Springer, Adv. Math. Dordrecht, **8** (2005).

Gyula Maksa and Fruzsina Mészáros
Institute of Mathematics
University of Debrecen
P.O. Box 12
H-4010 Debrecen, Hungary
e-mail: maksa@math.klte.hu
e-mail: mefru@math.klte.hu

International Series of Numerical Mathematics, Vol. 157, 299–304

Approximate Solutions of the Linear Equation

Dorian Popa

Abstract. In this paper we obtain a stability result for the general linear equation in Hyers-Ulam sense.

Mathematics Subject Classification (2000). Primary: 39B72, 39B05.

Keywords. Hyers-Ulam stability, general linear equation.

1. Introduction

The starting point of the stability theory of functional equations was the question of S.M. Ulam formulated at Wisconsin University in 1940 (see [10]):

Let $(X,\cdot)$ be a group and $(Y,\cdot,d)$ a metric group. Does for every $\varepsilon > 0$, there exist $\delta > 0$ such that if a function $f: X \to Y$ satisfies the inequality

$$d(f(xy), f(x)f(y)) \leq \delta \tag{1.1}$$

there exists a homomorphism $g: X \to Y$ such that

$$d(f(x), g(x)) \leq \varepsilon, \quad x \in X? \tag{1.2}$$

If the answer to this question is affirmative the equation $g(xy) = g(x)g(y)$ is called stable. A solution of the inequality (1.1) is called an **approximate solution** of the equation of homomorphism. In other words an equation is called stable if every approximate solution of the equation differs from a solution of the equation with a small error. A first answer to the problem of Ulam was given by D.H. Hyers in 1941 [4] in the case when X is a normed space and Y is a Banach space. Later the subject was strongly developed by many mathematicians, especially during the last 30 years.

In this paper we deal with the stability of the general linear equation

$$f(ax + by + k) = pf(x) + qf(y) + s \tag{1.3}$$

considered for the first time by J. Aczél [1] and studied later by Z. Daróczy and L. Losonczi (see [5]).

In some recent papers, Páles, Volkmann and Luce [7] and Páles [6] obtained very nice results on the stability of the Cauchy functional equation on square-symmetric groupoids, which leads as a particular case to the stability of the general linear equation. The role of square-symmetry was pointed out also by Rätz [9] and Borelli and Forti [2], [3]. A result on stability of the general linear equation, in Hyers-Ulam-Rassias sense, was obtained by the author in [8].

Let us recall this result.

Let X be a linear space over $\mathbb{R}$, Y a Banach space over $\mathbb{R}$, $\varphi : X \times X \to [0, \infty)$ be a given mapping, $a, b, p, q \in \mathbb{R} \setminus \{0\}$, $k \in X$, $s \in Y$, $a + b \neq 1$, $p + q \neq 1$ and $\psi(x) = \varphi(x + x_0, x + x_0)$, $x \in X$, $x_0 = \dfrac{k}{1 - a - b}$. Suppose that f satisfies

$$\|f(ax + by + k) - pf(x) - qf(y) - s\| \leq \varphi(x, y), \quad x, y \in X. \tag{1.4}$$

If

(i)

$$\lim_{n\to\infty} \frac{\varphi((a+b)^n x + x_0, (a+b)^n y + x_0)}{|p+q|^n} = 0, \quad x, y \in X, \tag{1.5}$$

(ii)

$$\sum_{n=1}^{\infty} \frac{\psi((a+b)^{n-1} x)}{|p+q|^n} = \mu(x), \quad x \in X \tag{1.6}$$

then there exist an additive mapping $g : X \to Y$ and a constant $c \in Y$ such that

$$\|f(x) - g(x) - c\| \leq \mu(x - x_0), \quad x \in X. \tag{1.7}$$

The goal of this paper is to obtain a result on the stability of the general linear equation, when the condition (1.4) is replaced by a condition of the form

$$f(ax + by + k) - pf(x) - qf(y) - s \in V \tag{1.8}$$

where V is some subset of a topological vector space Y.

2. Main result

In this section we denote by X a linear space over $\mathbb{R}$ and by Y a sequentially complete Hausdorff topological vector space over $\mathbb{R}$. Suppose that $a, b, p, q \in \mathbb{R} \setminus \{0\}$, $k \in X$ and $s \in Y$.

Remark 2.1. Let $p + q \neq 1$, $a + b \neq 1$, and suppose that $f : X \to Y$ satisfies the general linear equation (1.3). Then the mapping $h : X \to Y$ given by the relation

$$h(x) = f(x + x_0) - \frac{s}{1 - p - q}, \quad x \in X \tag{2.1}$$

where $x_0 = \dfrac{k}{1 - a - b}$, satisfies the equation

$$h(ax + by) = ph(x) + qh(y), \quad x, y \in X. \tag{2.2}$$

Proof. Suppose that f satisfies the general linear equation (1.3) and h is given by (2.1). Then

$$f(x) = h(x - x_0) + \frac{s}{1 - p - q}, \quad x \in X. \tag{2.3}$$

Replacing in (1.3) one gets

$$h(a(x - x_0) + b(y - x_0)) = ph(x - x_0) + qh(y - x_0), \quad x, y \in X, \tag{2.4}$$

which is equivalent to (2.2). □

Theorem 2.1. *Suppose that V is a nonempty bounded and convex subset of Y and $f : X \to Y$ satisfies*

$$f(ax + by + k) - pf(x) - qf(y) - s \in V, \quad x, y \in X. \tag{2.5}$$

If $a + b \notin \{0, 1\}$ and $p + q \geq 0$, $p + q \neq 1$, then there exists a unique mapping $g : X \to Y$ satisfying

$$g(ax + by + k) = pg(x) + qg(y) + s, \quad x, y \in X, \tag{2.6}$$

such that

$$g(x) - f(x) \in \frac{1}{p + q - 1} \cdot \overline{V}, \quad x \in X, \tag{2.7}$$

where $\overline{V}$ denotes the sequential closure of V.

Proof. Suppose that f satisfies (2.5), $a + b \notin \{0, 1\}$, $p + q \geq 0$, $p + q \neq 1$. Then in view of Lemma 2.1, the mapping $h : X \to Y$ given by (2.1) satisfies the relation

$$h(ax + by) - ph(x) - qh(y) \in V, \quad x, y \in X. \tag{2.8}$$

Taking in (2.8) $y = x \in X$, we get

$$h((a + b)x) - (p + q)h(x) \in V. \tag{2.9}$$

1) Suppose first that $p + q > 1$.

Replacing x by $(a+b)^n x$, $n \in \mathbb{N}$, and dividing by $(p+q)^{n+1}$ the relation (2.9) leads to

$$\frac{h((a + b)^{n+1}x)}{(p + q)^{n+1}} - \frac{h((a + b)^n x)}{(p + q)^n} \in \frac{1}{(p + q)^{n+1}} V. \tag{2.10}$$

Since V is convex, from (2.10) it follows that

$$\frac{h((a + b)^m x)}{(p + q)^m} - \frac{h((a + b)^n x)}{(p + q)^n} \in \left[\sum_{i=n+1}^{m} (p + q)^{-i}\right] V \tag{2.11}$$

for $m, n \in \mathbb{N}$, $m > n$, and $x \in X$.

We prove that $\left(\dfrac{h((a + b)^n x)}{(p + q)^n}\right)$ is a Cauchy sequence for every $x \in X$. Fix $x \in X$ and take U an arbitrary balanced neighbourhood of the origin of Y. Since

V is bounded, there exists $\alpha > 0$ such that $\alpha V \subseteq U$. As series $\sum_{i=0}^{\infty}(p+q)^{-i}$ is convergent, there exists $n_0 \in \mathbb{N}$ such that

$$\sum_{i=n+1}^{m} (p+q)^{-i} < \alpha \tag{2.12}$$

for every $n \geq n_0$ and every $m > n$. Then for $m > n \geq n_0$ we have

$$\frac{h((a+b)^m x)}{(p+q)^m} - \frac{h((a+b)^n x)}{(p+q)^n} \in \sum_{i=n+1}^{m} (p+q)^{-i} V \subseteq \alpha V \subseteq U. \tag{2.13}$$

Hence $\left(\dfrac{h((a+b)^n x)}{(p+q)^n}\right)$ is a Cauchy sequence, so it is convergent since Y is a sequentially complete topological vector space.

Let $w : X \to Y$ be given by

$$w(x) = \lim_{n\to\infty} \frac{h((a+b)^n x)}{(p+q)^n}, \quad x \in X. \tag{2.14}$$

Replacing x by $(a+b)^n x$ and y by $(a+b)^n y$ in (2.8) and dividing by $(p+q)^n$ we get

$$\frac{h((a+b)^n(ax+by))}{(p+q)^n} - p\frac{h((a+b)^n x)}{(p+q)^n} - q\frac{h((a+b)^n y)}{(p+q)^n} \in \frac{1}{(p+q)^n} V. \tag{2.15}$$

Taking the limit as $n \to \infty$ (2.15) gives

$$w(ax+by) = pw(x) + qw(y). \tag{2.16}$$

Putting $n = 0$ in (2.11) one gets

$$\frac{h((a+b)^m x)}{(p+q)^m} - h(x) \in \sum_{i=1}^{m} (p+q)^{-i} V \tag{2.17}$$

and so letting $m \to \infty$ one obtains

$$w(x) - h(x) \in \frac{1}{p+q-1}\overline{V}, \quad x \in X. \tag{2.18}$$

Now taking $g(x) = w(x - x_0) + \dfrac{s}{1-p-q}$, $x \in X$, by (2.18) follows

$$g(x) - f(x) \in \frac{1}{p+q-1}\overline{V}, \quad x \in X. \tag{2.19}$$

In this way we have proved the existence of g. To prove the uniqueness of g it is sufficient to show that the function w from (2.18) is unique. So suppose that there exist two mappings $w_1, w_2 : X \to Y$ satisfying (2.16) with the property

$$w_k(x) - h(x) \in \frac{1}{p+q-1}\overline{V}, \quad k \in \{1,2\}. \tag{2.20}$$

By (2.18) and (2.20) we get

$$w_1(x) - w_2(x) \in \frac{1}{p+q-1}(\overline{V} - \overline{V}). \tag{2.21}$$

Since V is bounded, so is $\overline{V}$. Therefore the set $\frac{1}{p+q-1}(\overline{V} - \overline{V})$ is bounded. Note that by (2.16) the functions w_1, w_2 satisfy the relations

$$w_k((a+b)^n x) = (p+q)^n w_k(x), \quad k \in \{1,2\} \tag{2.22}$$

for every $x \in X$ and every $n \in \mathbb{N}$. Now by (2.21) and (2.22) follows

$$w_1(x) - w_2(x) \in \frac{1}{(p+q)^n} \cdot \frac{1}{p+q-1}(\overline{V} - \overline{V}), \quad x \in X,\ n \in \mathbb{N}. \tag{2.23}$$

Fixing x in (2.23) and letting $n \to \infty$ one gets $w_1(x) = w_2(x)$, thus the uniqueness is proved.

2) Now suppose that $0 \leq p+q < 1$.

Replacing in (2.9) x by $\frac{x}{(a+b)^{n+1}}$ and multiplying (2.9) by $(p+q)^n$ one gets

$$(p+q)^n h\left(\frac{x}{(a+b)^n}\right) - (p+q)^{n+1} h\left(\frac{x}{(a+b)^{n+1}}\right) \in (p+q)^n V,\ x \in X,\ n \in \mathbb{N}. \tag{2.24}$$

Now, arguing as in the first case, we obtain

$$(p+q)^n h\left(\frac{x}{(a+b)^n}\right) - (p+q)^m h\left(\frac{x}{(a+b)^m}\right) \in \sum_{i=n}^{m-1} (p+q)^i V \tag{2.25}$$

for $x \in X$, $m, n \in \mathbb{N}$, $n > m$. Hence $\left((p+q)^n h\left(\frac{x}{(a+b)^n}\right)\right)_{n \geq 0}$ is a Cauchy sequence and since Y is sequentially complete topological vector space it is convergent. Define

$$w(x) = \lim_{n\to\infty} (p+q)^n h\left(\frac{x}{(a+b)^n}\right), \quad x \in X. \tag{2.26}$$

Taking in (2.25) $n = 0$ and letting $n \to \infty$ we get

$$h(x) - w(x) \in \frac{1}{1-p-q}\overline{V}, \quad x \in X. \tag{2.27}$$

Replacing in (2.8) x by $\frac{x}{(a+b)^n}$, y by $\frac{y}{(a+b)^n}$, multiplying (2.8) by $(p+q)^n$ and letting $n \to \infty$ follows

$$w(ax+by) = pw(x) + qw(y), \quad x, y \in X. \tag{2.28}$$

Now taking $g(x) = w(x - x_0) + \frac{s}{1-p-q}$ the existence is proved. The uniqueness follows as in the first part of the proof. $\square$

Remark 2.2. Let Y be a Banach space over $\mathbb{R}$, ε is a positive number, $a+b \notin \{0,1\}$, $p+q \geq 0$, $p+q \neq 1$. Suppose that $f : X \to Y$ satisfies

$$\|f(ax+by+k) - pf(x) - qf(y) - s\| \leq \varepsilon, \quad x, y \in X. \tag{2.29}$$

Then there exists a unique mapping $g : X \to Y$ satisfying (2.6) such that

$$\|g(x) - f(x)\| \leq \frac{\varepsilon}{|p+q-1|}, \quad x \in X. \tag{2.30}$$

Proof. The result follows from Theorem 2.1 taking $V = B(0,\varepsilon)$, where $B(0,\varepsilon)$ is the closed ball of center 0 and radius ε in Y.

The result from Remark 2.2 is analogous to Corollary 3 from [6], where this corollary is a consequence of a more general result obtained by Zs. Páles for the stability of Cauchy functional equation on square-symmetric groupoids. □

References

[1] J. Aczél, *Über eine Klasse von Functionalgleichungen*, Comment. Math. Helv. **21** (1948), 247–252.

[2] C. Borelli-Forti and G.L. Forti, *On a general Hyers-Ulam stability result*, Internat. J. Math. Sci. **18** (1995), 229–236.

[3] G.L. Forti, *Hyers-Ulam stability of functional equations in several variables*, Aequationes Math. **50** (1995), 143–190.

[4] D.H. Hyers, *On the stability of the linear functional equation*, Proc. Math. Acad. Sci., U.S.A. **27** (1941), 222–224.

[5] M. Kuczma, *An introduction to the theory of functional equations and inequalities*, PWN – Uniwersytet Ślaski, Warszawa – Kraków – Katowice, 1985.

[6] Zs. Páles, *Hyers-Ulam stability of the Cauchy functional equation on square-symmetric groupoids*, Publ. Math. Debrecen **4** (2001), 651–666.

[7] Zs. Páles, P. Volkmann and R.D. Luce, *Hyers-Ulam stability of functional equations with a square-symmetric operation*, Proc. Natl. Acad. Sci. U.S.A. **95** (1998), 12772–12775 (electronic).

[8] D. Popa, *Hyers-Ulam-Rassias stability of the general linear equation*, Nonlinear Funct. Anal. & Appl. **4** (2002), 581–588.

[9] J. Rätz, *On approximately additive mappings*, General Inequalities 2 (E.F. Beckenbach, ed.), International Series in Numerical Mathematics **47** Birkhäuser Verlag, Basel-Boston, 1980, 233–251.

[10] S.M. Ulam, *Problems in Modern Mathematics*, Wiley, New York, 1964.

Dorian Popa
Department of Mathematics
Technical University of Cluj-Napoca
Str. C. Daicoviciu 15
400020 Cluj-Napoca, Romania

International Series of Numerical Mathematics, Vol. 157, 305–315

On a Functional Equation Containing Weighted Arithmetic Means

Adrienn Varga and Csaba Vincze

Abstract. In this paper we solve the functional equation

$$\sum_{i=1}^{n} a_i f(\alpha_i x + (1-\alpha_i)y) = 0$$

which holds for all $x, y \in I$, where $I \subset \mathbb{R}$ is a non-void open interval, $f\colon I \to \mathbb{R}$ is an unknown function and the weights $\alpha_i \in (0,1)$ are arbitrarily fixed $(i = 1, \ldots, n)$. It will be proved that all solutions are generalized polynomials of degree at most $n-2$. Furthermore we give a sufficient condition for the existence of nontrivial solutions.

Mathematics Subject Classification (2000). 39B22.

Keywords. Functional equation, p-Wright and Jensen affine functions.

1. Introduction and preliminaries

Consider the functional equation

$$\sum_{i=1}^{n} a_i f(\alpha_i x + (1-\alpha_i)y) = 0 \tag{1.1}$$

which holds for all $x, y \in I$, where $I \subset \mathbb{R}$ is a non-void open interval, $f\colon I \to \mathbb{R}$ is an unknown function and the parameters $\alpha_i \in [0,1]$ are arbitrarily fixed $(i = 1, \ldots, n)$. The particular case $n = 4$, $a_1 = a_2 = 1$, $a_3 = a_4 = -1$ and $\alpha_3 = 1$, $\alpha_4 = 0$ has been investigated in Daróczy-Maksa-Páles [3], Daróczy-Lajkó-Lovas-Maksa-Páles [8], and also in Maksa [9] in connection with the equivalence of certain functional equations involving means. The result have been extended for the case of arbitrary α_3, $\alpha_4 \in (0,1)$ in the paper [10]. The purpose of this paper is to extend these results for equation (1.1) and to give nonzero additive solutions of (1.1) by generalizing a result of Daróczy [1]. We remark that equation, with possibly different functions

This research has been supported by the Hungarian Scientific Research Fund (OKTA) Grants F049212, NK-68040.

but similar to (1.1), has been investigated in [2], however it is supposed that some of the weights are equal to 1 therefore some of the unknown functions at the point x can explicitly be expressed from the equation. In our case it is impossible.

The paper is organized as follows. First of all we make some simple remarks, which indicate that the most important is to investigate the case $a_i \neq 0$ $(i = 1, \ldots, n)$ such that $\sum_{i=1}^n a_i = 0$ and $\alpha_i \neq \alpha_j$ if $i \neq j$ $(i, j = 1, \ldots, n)$. We mention that the method in [10] could be applied to this more general problem. As we shall see, any solution of (1.1) is a generalized polynomial of degree at most $n-2$, that is the sum of the diagonalizations of symmetric k-additive functions, where $k = 0, \ldots, n-2$. Here we introduce some basic notions and results we need in the following. Throughout the paper I denotes a non-void open interval.

Definition 1.1. Let k be a positive integer and $A_k \colon \mathbb{R}^k \to \mathbb{R}$ be a symmetric k-additive function, i.e., A_k is additive in each variable.

(i) The function $D(A_k)$ defined by

$$D(A_k)(x) := A_k(\underbrace{x, \ldots, x}_{k \text{ times}}) \quad (x \in \mathbb{R})$$

is said to be the diagonal of A_k at x.

(ii) Let

$$A_{k,l}(x, y) := A_k(\underbrace{x, \ldots, x}_{l \text{ times}}, \underbrace{y, \ldots, y}_{k-l \text{ times}}) \quad (x, y \in \mathbb{R}).$$

We use the phrase "0-additive function" for constant functions.

The following result is a particular case of Lemma 1.3 in Székelyhidi [7].

Lemma 1.2. *If $k \geq 0$ is an integer and $A_k \colon \mathbb{R}^k \to \mathbb{R}$ is k-additive and symmetric then we have that*

$$D(A_k)(x + y) = \sum_{l=0}^{k} \binom{k}{l} A_{k,l}(x, y).$$

Remark 1.3. It is well known that every k-additive symmetric function is rational homogeneous in each variable.

The following theorem is very important for us. It says that the function $f \colon I \to \mathbb{R}$ is a locally generalized polynomial of degree at most n on I if and only if it is a globally generalized polynomial of degree at most n on I.

Theorem 1.4. *Let $f \colon I \to \mathbb{R}$ be a function. Suppose that for any point $\xi \in I$, there is an $\varepsilon > 0$ such that f has the form*

$$f(x) = \sum_{k=0}^{n} D(A_k^{\xi})(x) \qquad x \in J_\xi := (\xi - \varepsilon, \xi + \varepsilon) \subset I,$$

where $A_k^{\xi} \colon \mathbb{R}^k \to \mathbb{R}$ is k-additive and symmetric $(k = 0, \ldots, n)$. Then

$$f(x) = \sum_{k=0}^{n} D(A_k)(x) \qquad x \in I,$$

where $A_k\colon \mathbb{R}^k \to \mathbb{R}$ *is a uniquely determined* k*-additive and symmetric function for any* $k = 0, \dots, n$.

Proof. Let $f\colon I \to \mathbb{R}$ be a function and suppose that $\xi_1, \xi_2 \in I$ such that $(\xi_1 - \varepsilon_1, \xi_1 + \varepsilon_1) \subset I$ and $(\xi_2 - \varepsilon_2, \xi_2 + \varepsilon_2) \subset I$ for some $\varepsilon_1, \varepsilon_2 > 0$. For the simplicity assume that

$$f(x) = \sum_{k=0}^{n} D(A_k)(x) \quad x \in (\xi_1 - \varepsilon_1, \xi_1 + \varepsilon_1)$$

$$\text{and} \qquad f(x) = \sum_{k=0}^{n} D(B_k)(x) \quad x \in (\xi_2 - \varepsilon_2, \xi_2 + \varepsilon_2)$$

where A_k and $B_k\colon \mathbb{R}^k \to \mathbb{R}$ $(k = 0, \dots, n)$ are symmetric k-additive functions. The proof can be divided into two parts:

I. $(\xi_1 - \varepsilon_1, \xi_1 + \varepsilon_1) \cap (\xi_2 - \varepsilon_2, \xi_2 + \varepsilon_2) \neq \emptyset$

II. $(\xi_1 - \varepsilon_1, \xi_1 + \varepsilon_1) \cap (\xi_2 - \varepsilon_2, \xi_2 + \varepsilon_2) = \emptyset$.

I. Denote the intersection of $(\xi_1 - \varepsilon_1, \xi_1 + \varepsilon_1)$ and $(\xi_2 - \varepsilon_2, \xi_2 + \varepsilon_2)$ by M. Then M is an open interval and

$$\sum_{k=0}^{n} D(A_k)(x) = \sum_{k=0}^{n} D(B_k)(x) \qquad (x \in M). \tag{1.2}$$

In the first step we show that (1.2) implies that

$$D(A_k)(x) = D(B_k)(x) \qquad (x \in M), \quad (k = 0, \dots, n). \tag{1.3}$$

In the second step it is proved that if (1.3) holds then

$$D(A_k)(x) = D(B_k)(x) \qquad (x \in \mathbb{R}), \quad (k = 0, \dots, n). \tag{1.4}$$

Therefore

$$f(x) = \sum_{k=0}^{n} D(A_k)(x) = \sum_{k=0}^{n} D(B_k)(x) \quad (x \in I).$$

First step. So, assume that (1.2) holds. Let $x \in M$ be fixed. Then there exists $\varepsilon > 0$ such that for all $r \in (1 - \varepsilon, 1 + \varepsilon) \cap \mathbb{Q}$ we have that $rx \in M$. Replacing x by rx in equation (1.2) and using remark 1.3 we get that

$$\sum_{k=0}^{n} r^k D(A_k)(x) = \sum_{k=0}^{n} r^k D(B_k)(x). \tag{1.5}$$

Since every real number $z \in (1 - \varepsilon, 1 + \varepsilon)$ can be approximated by rational sequences, we can replace $r \in \mathbb{Q}$ by z and thus (1.3) holds.

Second step. Let $y \in \mathbb{R}$, $y \neq 0$ be arbitrarily fixed. Then there exist $r \in \mathbb{Q}$ such that $ry \in M$. Therefore, by (1.3) we get that

$$r^k D(A_k)(y) = r^k D(B_k)(y) \qquad (k = 0, \dots, n).$$

Thus

$$D(A_k)(y) = D(B_k)(y) \qquad (k = 0, \dots, n)$$

which, of course, holds also for $y = 0$.

II. Suppose that $\xi_1 < \xi_2$. Then there exist

$$x_1, \dots, x_m \in [\xi_1, \xi_2],\ x_1 < \dots < x_m \ \text{ such that } \ [\xi_1, \xi_2] \subset \cup_{i=1}^{m} J_{x_i},$$

where f has the special (polynomial) form on each of the intervals $J_{x_i} := (x_i - \delta_i, x_i + \delta_i)$ for some $\delta_i > 0$ $(i = 1, \dots, m)$ and

$$(\xi_1 - \varepsilon_1, \xi_1 + \varepsilon_1) \cap (x_1 - \delta_1, x_1 + \delta_1) \neq \emptyset$$
$$(x_i - \delta_i, x_i + \delta_i) \cap (x_{i+1} - \delta_{i+1}, x_{i+1} + \delta_{i+1}) \neq \emptyset \quad (i = 1, \dots, m-1)$$
$$(x_m - \delta_m, x_m + \delta_m) \cap (\xi_2 - \varepsilon_2, \xi_2 + \varepsilon_2) \neq \emptyset$$

hold. Then the first part of the proof implies the statement. □

In order to construct non-trivial solutions of (1.1) we will use particular field isomorphisms. The existence of such a field isomorphism will also be discussed. We need the following notions.

Definition 1.5. Let m be a positive integer, μ_i and ν_i be real numbers $(i = 1, \dots, m)$.

(i) The ideal

$$\mathcal{I} := \{\, p \in \mathbb{Q}[x_1, \dots, x_m] \mid p(\mu_1, \dots, \mu_m) = 0 \,\}$$

of the polynomial ring $\mathbb{Q}[x_1, \dots, x_m]$ is called the *defining ideal* of $\mu_1, \dots, \mu_m$.

(ii) If the defining ideals of $\mu_1, \dots, \mu_m$ and $\nu_1, \dots, \nu_m$ are the same then we say that they are *algebraic conjugate* of each other.

Remark 1.6. In the particular case $m = 1$ the ideal $\mathcal{I}$ can be generated by the minimal polynomial and Definition 1.5 (ii) gives back the following notion: μ_1 and ν_1 are algebraic conjugate if both of them are transcendent or they are algebraic and their defining polynomials are the same. For the details see [5].

2. The solutions of equation (1.1)

Replacing y by x in (1.1) we get that

$$\left(\sum_{i=1}^{n} a_i\right) f(x) = 0 \quad (x \in I),$$

therefore if f is a not identically zero solution of (1.1), then $\sum_{i=1}^{n} a_i = 0$.

On the other hand if $\alpha_i = \alpha_j$ for some indices i and j then we can reduce equation (1.1).

Therefore without loss of generality we may assume that

$$a_i \neq 0 \ (i = 1, \dots, n); \ \sum_{i=1}^{n} a_i = 0 \ \text{ and } \ \alpha_1 < \alpha_2 < \dots < \alpha_n, \tag{2.1}$$

where $2 \leq n$.

At first we prove the following lemma.

Lemma 2.1. *Suppose that condition* (2.1) *holds. Let*

$$\beta_i := \frac{\alpha_n - \alpha_i}{\alpha_n - \alpha_1} \qquad (i = 2, \dots, n).$$

If $f\colon I \to \mathbb{R}$ *satisfies functional equation* (1.1) *then for any point* $\xi \in I$ *there is an* $\varepsilon > 0$ *such that*

$$a_1 f(u) + \sum_{i=2}^{n} a_i f\big(\beta_i u + (1-\beta_i)v\big) = 0 \tag{2.2}$$

holds for all $u, v \in J_\xi := (\xi - \varepsilon, \xi + \varepsilon) \subset I$.

Proof. Using the transformation

$$u = \alpha_1 x + (1-\alpha_1)y, \qquad v = \alpha_n x + (1-\alpha_n)y \qquad (x,y) \in I^2 \tag{2.3}$$

functional equation (1.1) goes over into the form

$$a_1 f(u) + \sum_{i=2}^{n} a_i f\big(\beta_i u + (1-\beta_i)v\big) = 0 \quad (u,v) \in P(I^2) \tag{2.4}$$

where $P(I^2)$ is the image of I^2 under transformation (2.3). Transformation (2.3) has the matrix

$$P := \begin{pmatrix} \alpha_1 & 1-\alpha_1 \\ \alpha_n & 1-\alpha_n \end{pmatrix}$$

and $\det P = \alpha_1 - \alpha_n$. As $\alpha_1, \alpha_n \in (0,1)$ are different real numbers (because of condition (2.1)) our linear transformation is regular. Since every regular linear transformation is an open mapping and

$$P\begin{pmatrix} x \\ x \end{pmatrix} = \begin{pmatrix} x \\ x \end{pmatrix} \qquad (x \in I),$$

every point of

$$\operatorname{diag} I^2 := \{(\xi,\xi) \mid \xi \in I\}$$

is an interior point of the set $P(I^2)$ (the image of I^2 under P). Thus for any point $\xi \in I$ there is an $\varepsilon > 0$ such that

$$(\xi - \varepsilon, \xi + \varepsilon)^2 \subset P(I^2).$$

On the other hand $P(I^2) \subset I^2$ because both u and v are between x and y. Thus equation (2.4) holds for all $u, v \in J_\xi := (\xi - \varepsilon, \xi + \varepsilon) \subset I$. □

Lemma 2.2. *Suppose that condition* (2.1) *holds. Let* $\xi \in I$ *be arbitrarily fixed and assume that* f *satisfies functional equation* (2.2) *for all* $u, v \in J_\xi$. *Then there exists a unique extension* $\tilde{f}\colon \mathbb{R} \to \mathbb{R}$ *such that* $\tilde{f}$ *satisfies* (2.2) *for all* $u, v \in \mathbb{R}$ *and* $\tilde{f}|_{J_\xi} = f$.

Proof. The lemma is a simple consequence of Theorem 5 in Páles [4] in the following settings:

$$F = X = \mathbb{R}, \quad K = J_\xi, \quad Y = (\mathbb{R}, +),$$

$$\varphi_0 = 0, \quad \varphi_1 = 0, \quad \varphi_i \colon \mathbb{R} \to \mathbb{R}, \quad \varphi_i(t) = -\frac{a_i}{a_1} t \ \ (i = 2, \ldots, n). \qquad \square$$

Before we can state our theorem in its final form we need the following lemma.

Lemma 2.3. *Let* $\varphi_i, \psi_i \colon \mathbb{R} \to \mathbb{R}$ *be additive mappings of* $\mathbb{R}$ *onto itself such that*

$$\mathrm{Rg}\ (\psi_j \circ \psi_i^{-1} - \varphi_j \circ \varphi_i^{-1}) = \mathbb{R} \ \ \textit{for} \ \ i \neq j \quad (i, j = 1, 2 \ldots, n-1), \tag{2.5}$$

where Rg *denotes the range of the transformation.*

If the functions $f_i \colon \mathbb{R} \to \mathbb{R}$ $(i = 0, 1, \ldots, n-1)$ *satisfy the functional equation*

$$f_0(x) + \sum_{i=1}^{n-1} f_i\big(\varphi_i(x) + \psi_i(y)\big) = 0 \qquad (x, y \in \mathbb{R})$$

then there exist $A_k^i \colon \mathbb{R}^k \to \mathbb{R}$ $(k = 0, 1, \ldots, n-2;\ i = 0, 1, \ldots, n-1)$ *k-additive symmetric functions such that*

$$f_i(x) = \sum_{k=0}^{n-2} D(A_k^i)(x) \quad (i = 0, 1, \ldots, n-1) \quad (x \in \mathbb{R}).$$

Proof. The lemma is an easy consequence of Theorem 3.9 in Székelyhidi [6]. $\square$

About the solutions of (1.1) we can state the following

Theorem 2.4. *Suppose that condition* (2.1) *holds. The function* $f \colon I \to \mathbb{R}$ *satisfies functional equation* (1.1) *for all* $x, y \in I$ *if and only if there exist uniquely determined symmetric k-additive functions*

$$A_k \colon \mathbb{R}^k \to \mathbb{R} \quad (k = 0, 1, \ldots, n-2)$$

such that

$$f(x) = \sum_{k=0}^{n-2} D(A_k)(x) \tag{2.6}$$

and, with the notation $p_i := \frac{\alpha_i - \alpha_1}{\alpha_n - \alpha_1}$ $(i = 2, \ldots, n)$, *the equations*

$$\sum_{i=2}^{n} a_i A_{k,k-l}(s, tp_i) = 0 \quad (s,\ t \in \mathbb{R}) \tag{2.7}$$

hold for any $k = 1, \ldots, n-2$ *and* $l = 1, \ldots, k$, *provided that* $n \geq 3$. *If* $n = 2$ *then, by* (2.6), f *is constant.*

Proof. Let $f \colon I \to \mathbb{R}$ be a solution of equation (1.1) and $\xi \in I$ be arbitrarily fixed. According to Lemma 2.1 and Lemma 2.2 there exists $\tilde{f} \colon \mathbb{R} \to \mathbb{R}$ such that

$$a_1 \tilde{f}(u) + \sum_{i=2}^{n} a_i \tilde{f}\big(\beta_i u + (1 - \beta_i) v\big) = 0 \tag{2.8}$$

holds for all $u, v \in \mathbb{R}$, where $\beta_i := \frac{\alpha_n - \alpha_i}{\alpha_n - \alpha_1}$, $(i = 2, \dots, n)$; moreover $\tilde{f}|_{J_\xi} = f$ where $J_\xi := (\xi - \varepsilon, \xi + \varepsilon) \subset I$ for some $\varepsilon > 0$. Using the substitutions

$$u = s \quad \text{and} \quad v = s - t \qquad (u, v \in \mathbb{R})$$

in (2.8), we obtain

$$a_1 \tilde{f}(s) + \sum_{i=2}^{n} a_i \tilde{f}\big(s - t(1 - \beta_i)\big) = 0 \quad (s, t \in \mathbb{R}). \tag{2.9}$$

Because of condition (2.1) equation (2.9) is equivalent to the equation

$$\tilde{f}(s) + \sum_{i=2}^{n} \frac{a_i}{a_1} \tilde{f}\big(s - t(1 - \beta_i)\big) = 0 \quad (s, t \in \mathbb{R}). \tag{2.10}$$

Now we can apply Lemma 2.3 for equation (2.10) in the following settings:

$$f_0 = \tilde{f}, \quad f_i = \frac{a_{i+1}}{a_1} \tilde{f} \quad (i = 1, \dots, n-1)$$

$$\psi_i(x) = -(1 - \beta_{i+1})x \quad \text{and} \quad \varphi_i(x) = x \quad (i = 1, \dots, n-1).$$

It is easy to see that conditions (2.5) hold because of (2.1). Thus we get that there exist symmetric k-additive functions

$$A_k^\xi \colon \mathbb{R}^k \to \mathbb{R} \quad (k = 0, 1, \dots, n-2)$$

such that

$$\tilde{f}(x) = \sum_{k=0}^{n-2} D(A_k^\xi)(x) \qquad (x \in \mathbb{R}). \tag{2.11}$$

Recall that ξ was arbitrarily fixed. At this point of the proof we need Theorem 1.4 which implies that

$$f(x) = \sum_{k=0}^{n-2} D(A_k)(x) \qquad (x \in I),$$

where $A_k \colon \mathbb{R}^k \to \mathbb{R}$ $(k = 0, 1, \dots, n-2)$ are symmetric k-additive functions and $\tilde{f} \colon \mathbb{R} \to \mathbb{R}$ is the unique extension of f. Therefore substituting expression (2.11) of $\tilde{f}$ into (2.10) we get that

$$\sum_{i=1}^{n} \frac{a_i}{a_1} \left(\sum_{k=0}^{n-2} D(A_k)\big(s - t(1 - \beta_i)\big) \right) = 0 \qquad (s, t \in \mathbb{R}),$$

where $\beta_1 := 1$. Because of Lemma 1.2 and condition (2.1) we get that

$$\sum_{i=2}^{n} a_i \left(\sum_{k=1}^{n-2} \Big(\sum_{l=1}^{k} \binom{k}{l} (-1)^l A_{k,k-l}(s, tp_i) \Big) \right) = 0 \quad (s, t \in \mathbb{R}), \tag{2.12}$$

where $p_i := 1 - \beta_i \ \ (i = 2, \dots, n)$. Replacing s by xs and t by yt, where $x,\ y \in \mathbb{Q}$, the rational homogeneity implies that

$$\sum_{i=2}^{n} a_i \Big(\sum_{k=1}^{n-2} \big(\sum_{l=1}^{k} \binom{k}{l} (-1)^l x^{k-l} y^l A_{k,k-l}(s, tp_i) \big) \Big) = 0 \quad (s,\ t \in \mathbb{R}),$$

where $p_i := 1 - \beta_i \ \ (i = 2, \dots, n)$. Since every $z,\ w \in \mathbb{R}$ can be approximated by rational sequences, the expression on the left-hand side can be considered as a polynomial of the variables z and w. Therefore

$$\sum_{i=2}^{n} a_i A_{k,k-l}(s, tp_i) = 0 \quad (s,\ t \in \mathbb{R})$$

holds for any $k = 1, \dots, n-2$ and $l = 1, \dots, k$ provided that $n \geq 3$. The converse is trivial. □

3. Sufficient conditions for the existence of non-trivial solutions

According to Theorem 2.4 if we give a nonzero additive $A\colon \mathbb{R} \to \mathbb{R}$ solution of the equation

$$\sum_{i=2}^{n} a_i A(tp_i) = 0 \quad (t \in \mathbb{R}), \quad \text{where} \ \ n \geq 3,$$

then we have a solution of (1.1) with nonzero additive part. This means that we are looking for nonzero additive functions satisfying (2.7) for $k = 1$. We remark that the case $n = 3$, i.e., when there are two members in the sum, the problem is solved in Daróczy [1]. Here we generalize his result.

Lemma 3.1. *Let $3 \leq n \in \mathbb{N}$, γ_i and $\delta_i \in \mathbb{R}$ $(i = 2, \dots, n-1)$ be arbitrarily fixed. There exists a field isomorphism*

$$\delta\colon \mathbb{Q}(\gamma_2, \dots, \gamma_{n-1}) \to \mathbb{Q}(\delta_2, \dots, \delta_{n-1})$$

such that

$$\delta(\gamma_i) = \delta_i \ \ \textit{for all} \ \ i = 2, \dots, n-1$$

if and only if $\gamma_2, \dots, \gamma_{n-1}$ and $\delta_2, \dots, \delta_{n-1}$ are algebraic conjugate (see Definition 1.5).

Proof. For the simplicity denote the defining ideals of $\gamma_2, \dots, \gamma_{n-1}$ and $\delta_2, \dots, \delta_{n-1}$ by $\mathcal{I}_\gamma$ and $\mathcal{I}_\delta$, respectively. It is well known that for any field isomorphism δ we have that $\delta(q) = q$ for all $q \in \mathbb{Q}$. Therefore we have to see that the mapping

$$z = \frac{w(\gamma_2, \dots, \gamma_{n-1})}{k(\gamma_2, \dots, \gamma_{n-1})} \mapsto \delta(z) := \frac{w(\delta_2, \dots, \delta_{n-1})}{k(\delta_2, \dots, \delta_{n-1})}, \tag{3.1}$$

where $w,\ k \in \mathbb{Q}[x_2, \dots, x_{n-1}]$, $k \notin \mathcal{I}_\gamma$ and $k \notin \mathcal{I}_\delta$ is well defined if and only if $\mathcal{I}_\gamma = \mathcal{I}_\delta$.

First we prove that if $\mathcal{I}_\gamma = \mathcal{I}_\delta$ then (3.1) is well defined. To see this let $z \in \mathbb{Q}(\gamma_2, \ldots, \gamma_{n-1})$ such that

$$z = \frac{p(\gamma_2, \ldots, \gamma_{n-1})}{q(\gamma_2, \ldots, \gamma_{n-1})} = \frac{w(\gamma_2, \ldots, \gamma_{n-1})}{k(\gamma_2, \ldots, \gamma_{n-1})} \tag{3.2}$$

for some $p,\ q,\ w,\ k \in \mathbb{Q}[x_2, \ldots, x_{n-1}]$, where $q,\ k \notin \mathcal{I}_\gamma$. We have to show that if $\mathcal{I}_\gamma = \mathcal{I}_\delta$ then

$$\frac{p(\delta_2, \ldots, \delta_{n-1})}{q(\delta_2, \ldots, \delta_{n-1})} = \frac{w(\delta_2, \ldots, \delta_{n-1})}{k(\delta_2, \ldots, \delta_{n-1})}. \tag{3.3}$$

(3.2) means that $(pk - wq)(\gamma_2, \ldots, \gamma_{n-1}) = 0$ and thus $pk - wq \in \mathcal{I}_\gamma$. Using the condition $\mathcal{I}_\gamma = \mathcal{I}_\delta$ we get that $(pk - wq)(\delta_2, \ldots, \delta_{n-1}) = 0$ also holds. According to the conditions $q,\ k \notin \mathcal{I}_\gamma$ and $\mathcal{I}_\gamma = \mathcal{I}_\delta$ we get that (3.3) holds. It is easy to see that (3.1) is a field isomorphism with the required property.

Conversely, assume that (3.1) is well defined and $p \in \mathcal{I}_\gamma$. Then

$$z = p(\gamma_2, \ldots, \gamma_{n-1}) \ \mapsto \ \delta(z) = p(\delta_2, \ldots, \delta_{n-1}).$$

The additivity of δ and the condition $p \in \mathcal{I}_\gamma$ imply that $p \in \mathcal{I}_\delta$. So we have just seen that $\mathcal{I}_\gamma \subset \mathcal{I}_\delta$. Using the inverse of δ we get that $\mathcal{I}_\delta \subset \mathcal{I}_\gamma$. Thus the proof is completed. □

Theorem 3.2. *Let $3 \le n \in \mathbb{N}$ be arbitrarily fixed and $a_i,\ p_i \in \mathbb{R}$ be nonzero real numbers, $i = 2, \ldots, n$. If there exists a field isomorphism*

$$\delta\colon \mathbb{Q}\left(\frac{p_2}{p_n}, \ldots, \frac{p_{n-1}}{p_n}\right) \to \mathbb{Q}\left(\frac{a_2}{a_n}, \ldots, \frac{a_{n-1}}{a_n}\right)$$

such that

$$\frac{a_2}{a_n}\delta\left(\frac{p_2}{p_n}\right) + \cdots + \frac{a_{n-1}}{a_n}\delta\left(\frac{p_{n-1}}{p_n}\right) = -1,$$

then there exists a not identically zero additive function $A\colon \mathbb{R} \to \mathbb{R}$ such that

$$\sum_{i=2}^{n} a_i\ A(tp_i) = 0 \quad (t \in \mathbb{R}).$$

Proof. Consider $\mathbb{R}$ as the vector space over $\mathbb{Q}\left(\frac{p_2}{p_n}, \ldots, \frac{p_{n-1}}{p_n}\right)$ with the basis $\mathcal{H}$. Define $A\colon \mathbb{R} \to \mathbb{R}$ as follows: on the elements of $\mathcal{H}$ we define it arbitrarily and for $t = \sum_j c_j h_j$, where $c_j \in \mathbb{Q}\left(\frac{p_2}{p_n}, \ldots, \frac{p_{n-1}}{p_n}\right)$ and $h_j \in \mathcal{H}$ let

$$A(t) := \sum_j \delta(c_j) A(h_j) \quad (t \in \mathbb{R});$$

it is easy to see that for any $i = 2, \ldots, n-1$

$$A\left(\frac{p_i}{p_n}t\right) = \delta\left(\frac{p_i}{p_n}\right)A(t) \quad (t \in \mathbb{R}).$$

Indeed,

$$A\Big(\frac{p_i}{p_n}t\Big) = A\Big(\sum\nolimits_j \frac{p_i}{p_n}c_jh_j\Big) = \sum\nolimits_j \delta\Big(\frac{p_i}{p_n}c_j\Big)A(h_j)$$
$$= \sum\nolimits_j \delta\Big(\frac{p_i}{p_n}\Big)\delta(c_j)A(h_j) = \delta\Big(\frac{p_i}{p_n}\Big)\sum\nolimits_j \delta(c_j)A(h_j) = \delta\Big(\frac{p_i}{p_n}\Big)A(t)$$

holds for all $t \in \mathbb{R}$, where $i = 2, \dots, n-1$. Therefore

$$A(t) + \frac{a_2}{a_n}A\Big(\frac{p_2}{p_n}t\Big) + \dots + \frac{a_{n-1}}{a_n}A\Big(\frac{p_{n-1}}{p_n}t\Big)$$
$$= A(t)\Big(1 + \frac{a_2}{a_n}\delta\Big(\frac{p_2}{p_n}\Big) + \dots + \frac{a_{n-1}}{a_n}\delta\Big(\frac{p_{n-1}}{p_n}\Big)\Big) = 0$$

for all $t \in \mathbb{R}$, i.e.,

$$\sum_{i=2}^{n} \frac{a_i}{a_n}A\Big(\frac{p_i}{p_n}t\Big) = 0 \quad (t \in \mathbb{R}).$$

Multiplying by a_n and substituting t by p_nt this equation is equivalent to

$$\sum_{i=2}^{n} a_i\ A(tp_i) = 0 \quad (t \in \mathbb{R})$$

which was to be stated. □

Lemma 3.3. *Let $3 \le n \in \mathbb{N}$, γ_i and $\delta_i \in \mathbb{R}$ $(i = 2, \dots, n-1)$ be arbitrarily fixed. Suppose that*

$$\delta\colon \mathbb{Q}(\gamma_2, \dots, \gamma_{n-1}) \to \mathbb{Q}(\delta_2, \dots, \delta_{n-1})$$

is a field isomorphism such that $\delta(\gamma_i) = \delta_i$ $(i = 2, \dots, n-1)$. Then, with the notations

$$\frac{p_i}{p_n} = \gamma_i \quad \text{and} \quad -\frac{a_ia_n}{\sum_{j=2}^{n-1} a_j^2} = \delta_i \quad (i = 2, \dots, n-1),$$

where a_j, $p_j \in \mathbb{R}$ $(j = 2, \dots, n)$ are non-zero real numbers,

$$\frac{a_2}{a_n}\delta\Big(\frac{p_2}{p_n}\Big) + \dots + \frac{a_{n-1}}{a_n}\delta\Big(\frac{p_{n-1}}{p_n}\Big) = -1.$$

Proof. The proof is trivial. □

Theorem 3.4. *Suppose that condition (2.1) holds and $3 \le n \in \mathbb{N}$. If the elements*

$$\gamma_i := \frac{\alpha_i - \alpha_1}{\alpha_n - \alpha_1} \quad \text{and} \quad \delta_i := -\frac{a_ia_n}{\sum_{j=2}^{n-1} a_j^2}, \quad (i = 2, \dots, n-1)$$

are algebraic conjugate then there exists a not identically zero additive function $A\colon \mathbb{R} \to \mathbb{R}$ such that

$$f(x) = A(x) + c \quad (x \in I)$$

satisfies equation (1.1) for any constant $c \in \mathbb{R}$.

Proof. Recall that $p_n = 1$; see Theorem 2.4. The result is a simple consequence of Lemma 3.1, Theorem 3.2 and Lemma 3.3. □

Acknowledgements

The authors would like to thank Professor Gyula Maksa for his valuable help to prepare this paper.

References

[1] Z. Daróczy, *Notwendige und hinreichende Bedingungen für die Existenz von nichtkonstanten Lösungen linearer Funktionalgleichungen*, Acta Sci. Math. Szeged **22** (1961), 31–41.

[2] Z. Daróczy and Gy. Maksa, *Functional equations on convex sets*, Acta Math. Hungar. **68** (1995), 187–195.

[3] Z. Daróczy, Gy. Maksa and Zs. Páles, *Functional equations involving means and their Gauss composition*, Proc. Amer. Math. Soc. **134** (2006), 521–530.

[4] Zs. Páles, *Extension theorems for functional equations with bisymmetric operations*, Aequationes Math. **63** (2002), 266–291.

[5] M. Kuczma, *An Introduction to the Theory of Functional Equations and Inequalities*, Prace Naukowe Universitetu Śląskiego w Katowicach **CDLXXXIX**, Państwowe Wydawnictwo Naukowe – Universitet Śląski, Warszawa – Kraków – Katowice, 1985.

[6] L. Székelyhidi, *On a class of linear functional equations*, Publ. Math. **29** (1982), 19–28.

[7] L. Székelyhidi, *Convolution type functional equations on topological Abelian groups*, World Scientific Publishing Co. Inc., Teaneck, NJ, 1991.

[8] Z. Daróczy, K. Lajkó, R.L. Lovas, Gy. Maksa and Zs. Páles, *Functional equations involving means*, Acta Math. Hungar. **116** (2007), 79–87.

[9] Gy. Maksa, *Functional equations involving means*, (in Hungarian), Talks at the Academy, Hungarian Academy of Sciences, Budapest, 2006, to appear.

[10] A. Varga, *On a functional equations containing four weighted arithmetic means*, Banach J. Math. Appl. **2**(1) (2008), 21–32, `www.math-analysis.org`.

Adrienn Varga and Csaba Vincze
Institute of Mathematics
University of Debrecen
P. O. Box 12
Debrecen, Hungary
e-mail: `vargaa@math.klte.hu`

International Series of Numerical Mathematics (ISNM)

Edited by

Karl-Heinz Hoffmann, TU München, Germany

Hans D. Mittelmann, Arizona State University, Tempe, USA

BIRKHÄUSER

ISNM is open to all aspects of numerical mathematics. Some of the topics of particular interest include free boundary value problems for differential equations, phase transitions, problems of optimal control and optimization, other nonlinear phenomena in analysis, nonlinear partial differential equations, efficient solution methods, bifurcation problems and approximation theory. When possible, the topic of each volume is discussed from three different angles, namely those of mathematical modeling, mathematical analysis, and numerical case studies.

ISNM 156: Barbu, L. / Moroşanu, B., Singularly Perturbed Boundary-Value Problems (2007). ISBN 978-3-7643-8330-5

ISNM 155: Kunisch, K. / Leugering, G. / Sprekels, J. / Tröltsch, F. (Eds.), Control of Coupled Partial Differential Equations (2007). ISBN 978-3-7643-7720-5

ISNM 154: Figueiredo, I.N. / Rodrigues, J.F. / Santos, L. (Eds.), Free Boundary Problems. Theory and Applications (2006). ISBN 978-3-7643-7718-2

ISNM 153: Roubícek, T., Nonlinear Partial Differential Equations with Applications (2005). ISBN 978-3-7643-7293-4

ISNM 152: Nowak, I., Relaxation and Decomposition Methods for Mixed Integer Nonlinear Programming (2005). ISBN 978-3-7643-7238-5

ISNM 151: Mache, D.H. / Szabados, J. / de Bruin, M.G. (Eds.), Trends and Applications in Constructive Approximation (2005). ISBN 978-3-7643-7124-1

ISNM 150: Rohwer, C., Nonlinear Smoothing and Multiresolution Analysis (2005). ISBN 978-3-7643-7229-3

ISNM 149: Voigt, A. (Ed.), Multiscale Modeling in Epitaxial Growth (2005). ISBN 978-3-7643-7208-8

ISNM 148: Lagnese, J.E. / Leugering, G., Domain Decomposition in Optimal Control of Partial Differential Equations (2004). ISBN 978-3-7643-2194-9

ISNM 147: Colli, P. / Verdi, C. / Visintin, A. (Eds.), Free Boundary Problems. Theory and Applications (2003). ISBN 978-3-7643-2193-2

ISNM 146: Antreich, K. / Bulirsch, R. / Gilg, A. / Rentrop, P. (Eds.), Modeling, Simulation and Optimization of Integrated Circuits (2003). ISBN 978-3-7643-2192-5

ISNM 145: Haussmann, W. / Jetter, K. / Reimer, M. / Stöckler, J. (Eds.), Modern Developments in Multivariate Approximation (2003). ISBN 978-3-7643-2195-6

ISNM 144: Reimer, M., Multivariate Polynomial Approximation (2003). ISBN 978-3-7643-1638-9

ISNM 143: Desch, W. / Kappel, F. / Kunisch, K. (Eds.), Control and Estimation of Distributed Parameter Systems (2002). ISBN 978-3-7643-7004-6

ISNM 142: Buhmann, M.D. / Mache, D. (Eds.), Advanced Problems in Constructive Approximation (2003). ISBN 978-3-7643-6648-3

ISNM 141: Freistühler, H. / Warnecke, G. (Eds.), Hyperbolic Problems: Theory, Numerics, Applications. Volume II (2001). ISBN 978-3-7643-6710-7

ISNM 140: Freistühler, H. / Warnecke, G. (Eds.), Hyperbolic Problems: Theory, Numerics, Applications. Volume I (2001). ISBN 978-3-7643-6709-1

ISNM 140/141: (Set) ISBN 978-3-7643-6711-4

ISNM 139: Hoffmann, K.-H. / Lasiecka, I. / Leugering, G. / Sprekels, J. (Eds.), Optimal Control of Complex Structures (2001). ISBN 978-3-7643-6682-7